STUDENT RESOURCE MANUAL

APPLIED
COLLEGE ALGEBRA

A Graphing Approach

GARETH WILLIAMS

Gareth Williams

Stetson University

Carolyn Hamilton

Utah Valley State College

Saunders College Publishing

A Division of Harcourt College Publishers

Fort Worth Philadelphia San Diego New York Orlando Austin
San Antonio Toronto Montreal London Sydney Tokyo

Printed in the United States of America

ISBN 0-03-026037-x

012 202 7654321

Table of Contents

Part One: Solutions to all odd-numbered section Exercises, all Review Exercises, and all Chapter Tests

Part Two: *Math in Practice: An Applied Video Companion CD-ROM*

Preface

Our goal in creating this manual was to offer students more than just the standard collection of solutions. The study of college algebra is not easy and requires each student to work hard to ensure a firm grasp of the material. As has often been said, *mathematics is not a spectator sport*. You cannot expect to learn mathematics without doing mathematics, any more that you can learn to swim without getting wet. Like swimming or dancing or reading or any other skill, mathematics takes practice. We believe that the parts of this manual will give you the tools needed to excel in your study of college algebra.

On the next several pages you will find two correlation charts. The first chart (pages v-xii) deals with the *Electronic Companion to College Algebra*™ CD-ROM that is inserted in the back of the main text. The CD-ROM consists of 12 modules and this chart shows where specific modules match up with the chapters in the text. The second chart (pages xiii-xiv) deals with the videotape series that was created in conjunction with the text. These videotapes should be available at your Media Center, within the Mathematics Department or in your library. This chart shows you which examples and exercises from the main text are developed on the videotapes. Selected examples and exercises are also available on the Core Concepts videotape (see description below).

Following these charts, detailed solutions to all the odd-numbered section exercises, all Review Exercises, and all Chapter Tests are given. Gareth Williams has tried to make these solutions as complete as possible. For example, suggestions for appropriate graphing calculator windows have been included for many exercises. Discussions of projects and suggestions for the Group Discussions have been included where appropriate.

The second part of the manual begins on page 449 and contains the printed versions of the problems that are found on the *Math in Practice: An Applied Video Companion* CD-ROM. This CD-ROM is inserted on the inside back cover of this manual. The pages are formatted to allow room for you to complete the problems and turn them in to your instructor. The CD-ROM was designed by Utah Valley State College's Lori Palmer and Carolyn Hamilton and is designed to show you that college algebra concepts arise in real life. Lori has conducted over 20 engaging interviews with professionals in such fields as aviation, food services, banking, and environmental sciences. Each video clip is accompanied by two problems written by Carolyn. Answers to the problems are presented at the back of the manual starting on page A-97.

Acknowledgments

Special thanks to Mike Montano (Riverside Community College) and Larry Friesen (Butler County Community College) for their accuracy checks of the actual answers. Thanks also to Donna Williams for her accuracy check of answers, proofreading, and support and encouragement, as well as to Henry Smith (Southeastern Louisiana University) for checking the accuracy of the problems on the CD.

Additional Resources

There are several other supplements available for purchase that you may want to consider as you complete this course. They are as follows:

Johanna Halsey of Dutchess Community College has written the **TI-83 Graphing Calculator Manual**. This manual explains how to use your TI-83 graphing calculator through explanations and numerous worked examples. Key menus are given on the inside covers of the manual.

In the back of your main text, there is a business reply card for the **ESATUTOR 2000 CD-ROM**. This CD-ROM contains hundreds of problems and answers that correspond with every section of the text. To learn more about this exciting product and how to order it, turn to the back of your text.

As mentioned above, a videotape series was created in conjunction with this text. Additionally, a **Core Concepts Video** is available for you to purchase. This single videotape contains the most important topics covered in the full video series. This take-home tutorial can be used as a preview of what is to be covered in class, as an aid to completing homework assignments, or as a tool to review for a test.

If you are interested in purchasing the TI-83 Graphing Calculator Manual or the Core Concepts videotape, please call our Customer Service department at 1-800-782-4479 or visit our website at http://www.saunderscollege.com/math/williams

Correlation Chart for *An Electronic Companion to College Algebra CD-ROM*

Applied College Algebra		*Elec. Companion CD-ROM*		
Section	**Topic**	**Review Module**	**Topic**	**Subtopic**
Chapter R Review of Basic Algebra				
R.1 Real Numbers and Their Properties	Real Numbers	Algebra Review	Real Numbers	
R.2 Rational Exponents, Scientific Notation, and Doses of Medicine	Rational Exponents	Algebra Review	Exponents Radical Expressions	
R.3 Polynomials	Polynomials	Algebra Review	Polynomials	
R.4 Factoring	Factoring	Algebra Review	Factoring Polynomials	
R.5 Rational Expressions	Rational Expressions	Algebra Review	Rational Expressions	
R.6 Equations, Variation, Scuba Diving, and Fermat's Last Theorem	Equations	Preliminaries	Solving Equations	Linear Rational
R.7 Functions, Tables, and Skeleton Sizes	Definition, Domain, Range	Functions	Definitions	
Chapter 1 Algebra and Geometry				
1.1 Coordinate Systems and Graphs	Coordinate Systems and Graphs	Preliminaries	Rectangular Coordinates	
		Functions	Graphing	Tables Intercepts
1.2 Lines, Least-Squares Fit, and the Year 2000 Problem	Slope Equations	Linear Functions	Graphing	
	Intercepts	Linear Functions	Equation of a Line	

Correlation Chart for *An Electronic Companion to College Algebra CD-ROM*

Section	Topic	Review Module	Topic	Subtopic
1.3 Linear Functions, Reservoir Monitoring and Cost-Volume Analysis	Linear Functions and Applications	Linear Functions	Applications	
1.4 Parabolas and Quadratic Equations	Parabolas and Quadratic Equations Vertex, Intercepts	Quadratic Functions	Graphing	
1.6 Distances, Circles and Typesetting	Distances Circles	Conic Sections	The Distance Formula Circles	
Chapter 2 Solving Equations and Inequalities				
2.1 Quadratic Equations	Quadratic Equations	Preliminaries	Solving Equations	Quadratic
2.2 Applications of Quadratic Equations Including Motion and Architecture	Applications of Quadratic Equations Motion under Gravity	Quadratic Functions	Applications	
2.3 Further Types of Equations	Further Types of Equations	Preliminaries	Solving Equations	Radical
2.4 Inequalities, Temperature Range of a Computer, and Spread of a Disease	Inequalities	Preliminaries	Solving Inequalities	Linear Compound Quadratic
2.5 Equations and Inequalities with Absolute Values	Equations with Absolute Values	Preliminaries	Solving Equations	Absolute Value
	Inequalities with Absolute Values		Solving Inequalities	Absolute Value
Chapter 3 Further Development of Functions				
3.1 Polynomial Functions	Polynomial Functions	Linear Functions	Definitions	
		Quadratic Functions	Definitions	
		Polynomial and Rational Functions	Definitions Graphing Polynomials	

Correlation Chart for *An Electronic Companion to College Algebra CD-ROM*

Section	Topic	Review Module	Topic	Subtopic
3.2 Construction of Functions, Optimization, and Drug Administration	Functions Increasing/Decreasing	Functions	Graphing	Increase/Decrease
	Optimization	Polynomial and Rational Functions	Applications	Volume
3.3 Special Functions, Symmetry, and Consumer Awareness	Symmetry	Functions	Graphing	Symmetry Reflections
3.4 Shifting and Stretching Graphs, Supply and Demand for Wheat	Shifting and Stretching Graphs	Functions	Graphing	Translations
3.5 Rational Functions	Rational Functions Asymptotes	Polynomial and Rational Functions	Graphing Rational Functions	
			Applications	Weight Loss
3.6 Operations on Functions and an Introduction to the Field of Chaos	Operations on Functions	Functions	Combining Functions	
Chapter 4 Further Theory of Polynomials				
4.1 Synthetic Division, Zeros, and Factors	Division Zeros and Factors	Polynomial and Rational Functions	Zeros of a Polynomial	Division/Zeros/Factors Real Zeros
4.2 Complex Numbers and the Mandelbrot Set	Complex Numbers	Preliminaries	Complex Numbers	
4.3 Complex Zeros, Complex Factors, and the Fundamental Theorem of Algebra	Complex Zeros Factors	Polynomial and Rational Functions	Zeros of a Polynomial	Complex Zeros

Correlation Chart for *An Electronic Companion to College Algebra CD-ROM*

Section	Topic	Review Module	Topic	Subtopic
Chapter 5 Exponential and Logarithmic Functions				
5.1 Exponential Functions and the Growth of *Escheriscia coli*	Exponential Functions	Exponential/Log Functions	Definitions	Domain/Range
		Exponential/Log Functions	Graphing Exponentials	
5.2 The Natural Exponential Function and Solving Exponential Equations	The Natural Exponential Function	Exponential/Log Functions	Definitions	The Number e
	Compound Interest			
		Exponential/Log Functions	Applications	Compound Interest
5.3 Logarithmic Functions and Seismography	Logarithmic Functions	Exponential/Log Functions	Logarithmic Functions	
	Seismography			
		Exponential/Log Functions	Applications	Earthquakes
5.4 Properties of Logarithms and Solving Logarithmic Equations	Properties of Logarithms	Exponential/Log Functions	Properties of Logs	
		Exponential/Log Functions	Solving Equations	
	Population Growth	Exponential/Log Functions	Applications	Growth Models
	Carbon Dating	Exponential/Log Functions	Applications	Decay Models
Chapter 6 Systems of Equations and Inequalities				
6.1 Systems of Linear Equations in Two Variables	Linear Equations in Two Variables	Systems of Equations	Definitions	
			Solving Systems Algebraically	

Correlation Chart for *An Electronic Companion to College Algebra CD-ROM*

Section	Topic	Review Module	Topic	Subtopic
6.2 Systems of Linear Equations in Three Variables	General Systems of Equations	Systems of Equations	Solving Systems Algebraically	Elimination by Subtraction (3x3)
		Matrices	Definitions	Augmented Matrix Row Operations
		Systems of Equations	Matrices (page 1)	
6.3 Gauss-Jordan Elimination and Electrical Networks	Electrical Networks	Systems of Equations	Applications	Electric Circuits
6.5 Systems of Linear Inequalities	Linear Inequalities	Systems of Equations	Linear Inequalities	
6.6 Linear Programming and Optimal Use of Resources	Linear Programming	Systems of Equations	Linear Programming	
Chapter 7 Matrices				
7.1 Matrices and Group Relationships in Sociology	Matrices	Matrices	Definitions	
		Matrices	Matrix Operations	Matrix Addition Scalar Multiplication
7.2 Multiplication of Matrices and Population Movements	Multiplication of Matrices	Matrices	Matrix Operations	Matrix Multiplication
	Population Movement	Matrices	Applications	Markov Chains
7.3 The Inverse of a Matrix and the Interdependence of Industries	The Inverse of a Matrix	Matrices	Inverse of a Matrix	
		Matrices	Applications	Cryptanalysis
		Systems of Equations	Matrices (page 2)	

Correlation Chart for *An Electronic Companion to College Algebra CD-ROM*

Section	Topic	Review Module	Topic	Subtopic
Chapter 8 Sequences and Series				
8.1 Sequences, Series, and the Fibonacci Sequence	Sequences and Series	Sequences and Series	Sequences	
8.2 Arithmetic Sequences	Arithmetic Sequences	Sequences and Series	Series	Finite Arithmetic Series
8.3 Geometric Sequences, Count of Bacteria, and the Flaw in the Pentium Chip	Geometric Sequences	Sequences and Series	Series	Finite Geom.Series Infinite Geom.Series
	Group Discussion: Ancestor Analysis	Sequences and Series	Applications	The Number of One's Ancestors
8.5 The Binomial Theorem	The Binomial Theorem	Sequences and Series	The Binomial Theorem	
Chapter 9 Permutations, Combinations, and Probability				
9.1 Permutation and Combinations	Permutation and Combinations	Counting and Probability	Principle of Counting Permutations Combinations	
9.2 Probability and Blood Groups	Probability Probability of Outcome of Interest	Counting and Probability	Probability	Empirical Probability
	Sample Space and Events	Counting and Probability	Probability	Theoretical Probability
	Probability of AND	Counting and Probability	Probability	Multiplication Rule
	Probability of OR	Counting and Probability	Probability	Addition Rule of Probability

x

Correlation Chart for *An Electronic Companion to College Algebra CD-ROM*

Section	Topic	Review Module	Topic	Subtopic
Appendix A Conic Sections				
Parabola	Parabola	Conic Sections	Parabolas	
Ellipse	Ellipse	Conic Sections	Ellipses	
Hyperbola	Hyperbola	Conic Sections	Hyperbolas	

Correlation Chart for Videotape Series

By selecting *Applied College Algebra: A Graphing Approach* by Gareth Williams, your school received a series of videotapes. They should be available in your Media Center, library, or the Mathematics department. Ask your instructor if you are unable to locate the video series. The following chart shows you which examples and exercises from the text are developed on the videotapes. A selection of these examples and exercises is also available on the *Core Concepts Videotape* ; you can purchase it through your bookstore or visit our website: http://www.harcourtcollege.com/math/williams/

Note: EX = example, EC = exercise, EXP = exploration, SC = self-check

Chapter R Review of Basic Algebra								
R.1 Real Numbers and Their Properties	EX 1	EC 6	EX 2	EX 3	EX 4	SC 3a	EC 59	
R.2 Rational Exponents, Scientific Notation, and Doses of Medicine	EX 1	EX 2	EX 3, EX 4	EX5	EX6	EX 8	EX 9	
R.3 Polynomials	EX 1	EX 2	EX 3	EX 4	EX 6	EX 8		
R.4 Factoring	EX 1	EX 3	EX 5	EX 6	EX 7	EC 63	EX 10	
R.5 Rational Expressions	EX 1	EX 2	EX 3	EX 5	EX 6	EX 8		
R.6 Equations, Variation, Scuba Diving, and Fermat's Last Theorem	EX 2	EX 3	EX 5	EX 8	EX 11			
R.7 Functions, Tables, and Skeleton Sizes	EX 1	EX 3	EX 4	EX 7	EC 65			
Chapter 1 Algebra and Geometry								
1.1 Coordinate Systems and Graphs	EX 1	EX 2	EX 3	EX 6	EC 39			
1.2 Lines, Least-Squares Fit, and the Year 2000 Problem	EX 2	EX 3a, EC 8	EX 3b	EX 3c	EC 27	EX 5	EX 9, EC 61	EX 11
1.3 Linear Functions, Reservoir Monitoring and Cost-Volume Analysis	EX 1	EX 2						
1.4 Parabolas and Quadratic Equations	EX 1	EX 4						
1.5 Quadratic Functions, Learning German, and Fluid Flow	EX 2	EX 4						
1.6 Distances, Circles and Typesetting	EX 6	EX 7	EX 8					
Chapter 2 Solving Equations and Inequalities								
2.1 Quadratic Equations	EX 1	EX 3	EX 5	EX 6	EX 7	EX 8		
2.2 Applications of Quadratic Equations Including Motion and Architecture	EX 1	EX 2, EC 9	EX 5					
2.3 Further Types of Equations	EX 1	EX 3	EX 5	EX 9	EX 10			
2.4 Inequalities, Temperature Range of a Computer, and Spread of a Disease	EX 2	EX 3	EX 5	EX 6				
2.5 Equations and Inequalities with Absolute Values	EX 1	EX 4	EX 5					
Chapter 3 Further Development of Functions								
3.1 Polynomial Functions	EX 1	EX 3	EX 4	EX 5				
3.2 Construction of Functions, Optimization, and Drug Administration	EX 1	EX 2	EX 6	EC 32				
3.3 Special Functions, Symmetry, and Consumer Awareness	EX 1	EX 3	EX 4	EX 5				
3.4 Shifting and Stretching Graphs, Supply and Demand for Wheat	EX 1	EX 2	EX 3	EX 4	EX 5			
3.5 Rational Functions	EX 2	EX 3	EX 4	EX 6b	EX 8			
3.6 Operations on Functions and an Introduction to the Field of Chaos	EX 1	EX 2	EX 3	EX 4				
3.7 Inverse Functions, One-to-One, and Cryptography	EX 1	EX 3	EX 6	EX 7				

Solutions to Odd-numbered Section Exercises, all Review Exercises, and all Tests

Chapter R

Section R.1

<u>Group Discussion</u> $4\boxed{\wedge}3\boxed{\wedge}2$ is $(4^3)^2$. Both are equal to 4096, see screen below. Natural to select $(4^3)^2$ since $4\boxed{\wedge}3\boxed{\wedge}2 = (4\boxed{\wedge}3)\boxed{\wedge}2$, reading left to right.

   In general, a^{b^c} is $a\boxed{\wedge}b\boxed{\wedge}c$ and is calculated as $(a\boxed{\wedge}b)\boxed{\wedge}c$.

```
4^3^2
            4096
(4^3)^2
            4096
4^(3^2)
          262144
```

1.(a) $\{2,9\}$ (b) $\{-9,-2,0,2,9\}$ (c) $\{-9,-\frac{7}{3},-2,0,2,\frac{8}{3},9\}$

3.(a) $\{4/2,5,9\}$ (b)$\{-60,-21,-\frac{15}{3},0,\frac{4}{2},5,9\}$ (c)$\{-60,-21,-\frac{15}{3},0,\frac{17}{6}\}$

5.

$[2,9)$

7.

$(-5,\infty)$

9.

$(-\infty,8)$

11. $23.45687 = 23.4569$, to four decimal places.

13. $29.5555555 = 29.555556$, to six decimal places.

15. $539.649 = 539.6$, to one decimal place.

17. $12.64 \div 9.23 = 1.3694$

19. $53.16 \div -4.37 = -12.1648$

21. $3\cdot4 = 4\cdot3$, commutative property of multiplication.

23. $2\cdot(2+9) = (2\cdot2) + (2\cdot9)$, distributive property.

25. $x+(y+z) = (x+y)+z$, associative property of addition.

27. $(a+b)+4 = a+(b+4)$, associative property of addition.

29. $2.3\text{X}(4.15+7.93) = 27.784$

31. $1.23\text{X}(-4.56+7.91-2.3576) = 1.220652$

33. $\sqrt{3} = 1.7321$, to 4 decimal places.

35. $(\sqrt{37} + 3.2) \div 1.67 = 5.56$, to 2 decimal places.

37. $\dfrac{1}{2.35} = 0.43$, to 2 decimal places.

39. $|2.67 - 9.86| = 7.19$

41. $2^6 = 2 \cdot 2 \cdot 2 \cdot 2 \cdot 2 \cdot 2 = 64$ 43. $-4^3 = -(4 \cdot 4 \cdot 4) = -64$

45. $(-3)^4 = (-3)(-3)(-3)(-3) = 81$ 47. $8 + (9/3) - 1 = 10$

49. $5 + ((9/3)x\ 4) = 17$ 51. $2 + 7 - (3x5) = -6$

53. $7 + 2^3 = 15$ $55. 1 + (4 \times 5^2) = 101$

57. $(2x7) - 2^5 + 1 = -17$

59. $\dfrac{1}{\sqrt{7}} - |-3 - 5| = -7.6220$ 61. $\sqrt{|-5.8-3.2|}+(2.4)(-8.3)=-16.92$

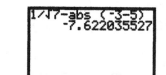   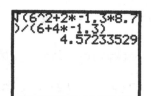

63. $a = \sqrt{b^2 + c^2} = \sqrt{3.2^2 + 4.8^2} = 5.7689$

65. $a = \sqrt{b^2 + c^2} = \sqrt{4.365^2 + 9.871^2} = 10.7930$

67. $\dfrac{\sqrt{4^2 + 2(1.12)(3.6)}}{4 + 4(1.12)} = 0.5785$ 69. $\dfrac{\sqrt{6^2 + 2(-1.3)(8.7)}}{6 + 4(-1.3)} = 4.5723$

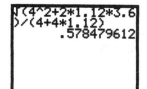   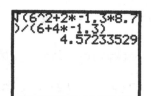

71. (a) $\pi = 3.141592654$, to 9 decimal places on TI-82/83.

    (b) $\pi = 3.14159$, to 5 decimal places.

    (c) $\pi = 3.1415927$, to 7 decimal places.

    (d) $(\sqrt{2} - 9\pi)^2 \div 3 = 240.4887$, to 4 decimal places.

73. (a) $\sqrt{3}$ = 1.7321 to 4 dec places. See screen below.

(b) error message since square root of negative number is not a real number.

(c) $\sqrt{|-3|}$ = 1.7321 = $\sqrt{3}$, since $|-3| = |3|$.

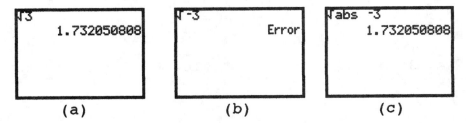

      (a)   (b)   (c)

75. Each term is the square of the previous term.

6th term will be 4294967296. See screen below.

77. (a) $2^{2^n}+1$ for n = 1,2,3,4,5 is 5,17,257,65537,4294967297.
(b) 4294967297/641=6700417. Thus $2^{2^5}+1$ is not prime.

79. π = 3.141592654 on a TI-83 calculator. The calculator displays eleven characters.
$\pi - 3$ = .1415926536."Getting rid" of the known digit 3 enables us to find π to one more decimal place.
$10(\pi - 3.1)$ = .4159265359. The $(\pi - 3.1)$ gets rid of the known digits 3.1 in π. We need to multiply by 10 to make sure that the calculator does not use up valuable spaces for zeros. Thus π = 3.14159265359 at the moment. Continuing,
$100(\pi - 3.14)$= .159265359. No new additional places revealed.
$1000(\pi - 3.141)$= .5926535898. Two new places revealed.
$10000(\pi - 3.1415)$= .926535898. No new places revealed.
$100000(\pi - 3.14159)$= .26535898. No further places revealed after this stage. TI-83 has π = 3.1415926535898 in memory.

    TI-85 does not behave like this. π gives 3.14159265359. $\pi -3$ gives no new info. $10(\pi -3.1)$ gives .415926535898. No more info after this. TI-85 has π = 3.1415926535898 in memory.

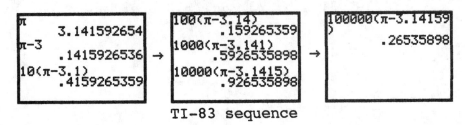

TI-83 sequence

Section R.2

1. $2^{-3} = \dfrac{1}{2^3} = \dfrac{1}{8}$

3. $3^{-2} \cdot 3^0 \cdot 3^5 = 3^{(-2+0+5)} = 3^3 = 27$

5. $5^{-3} \cdot 5 \cdot 25^2 = 5^{-3} \cdot 5 \cdot (5^2)^2 = 5^{-3} \cdot 5 \cdot 5^4 = 5^{(-3+1+4)} = 5^2 = 25$

7. $\sqrt{25} = \sqrt{5 \cdot 5} = 5$

9. $\sqrt[3]{-1000} = -10$

11. $(\sqrt[3]{27})^5 = (3)^5 = 243$

13. $\sqrt[4]{-16}$. Not a real number.

15. $\sqrt[5]{(3)^5} = 3$

17. $16^{1/2} = \sqrt{16} = 4$

19. $(-27)^{1/3} = \sqrt[3]{-3^3} = -3$

21. $81^{1/4} = \sqrt[4]{3^4} = 3$

23. $16^{3/2} = (16^{1/2})^3 = 4^3 = 64$

25. $27^{-2/3} = (27^{1/3})^{-2} = 3^{-2} = \dfrac{1}{3^2} = \dfrac{1}{9}$ (=.1111... for calculator check)

27. $-9^{5/2} = -(9^{1/2})^5 = -3^5 = -243$

29. $(-16)^{-5/4} = ((-16)^{1/4})^{-5}$. DNE. $(-16)^{1/4}$ is not a real number.

31. $3^2 \cdot 3^{-3} = 3^{(2-3)} = 3^{-1} = \dfrac{1}{3}$ (= .333...)

33. $4^3 \cdot (\dfrac{3}{4})^2 = 4^3 \cdot \dfrac{3^2}{4^2} = 4^3 \cdot 4^{-2} \cdot 3^2 = 4^{(3-2)} \cdot 3^2 = 4 \cdot 9 = 36$

35. $(7 \cdot 2)^3 (\dfrac{2}{7})^4 = 7^3 \cdot 2^3 \cdot \dfrac{2^4}{7^4} = 7^3 \cdot 2^3 \cdot 2^4 \cdot 7^{-4} = 7^{-1} \cdot 2^7 = \dfrac{128}{7}$ (=18.28571429)

37. $3^{1/2} \cdot 3^{3/2} = 3^{(1/2+3/2)} = 3^{4/2} = 3^2 = 9$

39. $6^{1/4} \cdot 6^{1/2} \cdot 6^{-3/4} = 6^{(1/4+2/4-3/4)} = 6^0 = 1$

41. $\dfrac{27^{4/3}}{27^{2/3}} = 27^{(4/3-2/3)} = 27^{2/3} (27^{1/3})^2 = 3^2 = 9$

43. $x^3 x^{-1} = x^{(3-1)} = x^2$

45. $x^2 y^3 x^{-7} = x^{(2-7)} y^3 = x^{-5} y^3 = \dfrac{y^3}{x^5}$

47. $\dfrac{x^4 x^{-2}}{x^{-3}} = x^4 x^{-2} x^3 = x^{(4-2+3)} = x^5$

4

49. $x^{1/3}x^{5/6} = x^{(2/6+5/6)} = x^{7/6}$

51. $x^2x^{-5/6} = x^{(2-5/6)} = x^{(12/6-5/6)} = x^{7/6}$

53. $\dfrac{(x^2y^{1/3})^2}{y^{2/3}} = \dfrac{x^4y^{2/3}}{y^{2/3}} = x^4$

55. $\sqrt[3]{8x^5} = \sqrt[3]{2^3x^3x^2} = 2x(\sqrt[3]{x^2})$

57. $\sqrt{x^4y^7} = \sqrt{x^4y^6y} = \sqrt{(x^2)^2(y^3)^2y} = x^2|y^3|\sqrt{y} = x^2y^3(\sqrt{y})$, since y cannot be negative if $\sqrt{y}$ exists.

59. $\sqrt[4]{2x^3y^5z^{13}} = \sqrt[4]{2x^3y^4y(z^3)^4z} = |y||z^3|(\sqrt[4]{2x^3yz})$

61. $\dfrac{1}{x^{-n}} = (x^{-n})^{-1} = x^{(-n\cdot-1)} = x^n$

63. $\dfrac{1}{(\frac{1}{x})} = \dfrac{1}{x^{-1}} = (x^{-1})^{-1} = x$

65. $4.1^5 = 1158.5620$

67. $5.18^{4.7} = 2276.9310$

69. $4.93^{2/5} = 1.8929$

71. $3.5426^{-5/7} = 0.4052$

73. $\sqrt{5.62} = 2.3707$

75. $\sqrt[4]{16.4532} = (16.4532)^{1/4} = 2.0140$

77. $\sqrt[5]{(-19.81^3)} = (-19.81)^{3/5} = -5.9997$

79. $25,000,000 = 2.5 \times 10^7$

81. $0.00043 = 4.3 \times 10^{-4}$

83. $63,400,000,000 = 6.34 \times 10^{10}$

85. $2 \times 10^4 = 20,000$

87. $-4.7 \times 10^3 = -4700$

89. $9.3471 \times 10^{-4} = 0.00093471$

91. $(4 \times 10^2) \times (3 \times 10^3) = 12 \times 10^5 = 1.2 \times 10^6$

93. $(5 \times 36 \times 10^4) \times (2.81 \times 10^{-5}) = 5 \times 36 \times 2.81 \times 10^4 \times 10^{-5}$

    $= 505.8 \times 10^{-1} = 50.58 = 5.058 \times 10^1$

95. (a) miles in one light year $= 186,262 \times 60 \times 60 \times 24 \times 365$
       $= 5.873958432 \times 10^{12}$
    (b) distance $= 410 \times 5.873958432 \times 10^{12} = 2.408322957 \times 10^{15}$ miles

5

97. total mass = $(5.26 \times 10^{21}) \times (1.673 \times 10^{-24}) = 8.7998 \times 10^{-3}$ grams.

99. (a) Amount in account after 1 period = $P(1 + r/n)$.
Amount in account after 2 periods = $(P(1 + r/n))(1 + r/n)$
$$= P(1 + r/n)^2.$$
There will be nt such periods in t years.
Thus amount in account after t years = $P(1 + r/n)^{nt}$.

   (b) $8750(1 + 0.08/4)^{(4 \times 5)} = \$13,002.04$
   (c) $8750(1 + 0.08/12)^{(12 \times 5)} = \$13,036.15$

101. volume of cylinder = $p(5)^2 18 = 1413.716694$
$$= 1413.72 \text{ cubic inches, to 2 decimal places.}$$

103. N=2500, p=15 and q=225. $\sqrt{\dfrac{2qN}{p}} = \sqrt{\dfrac{2(225)(2500)}{15}} = 273.86$.
Rounded to 274. The economic ordering quantity is 274. The store should plan to order 274 televisions at a time, when inventory is getting low.

Section R.3

1. $2x^2 - 3x + 1$. Degree 2, first term $2x^2$, constant term 1.

3. $7x^4 - 2x + 5$. Degree 4, first term $7x^4$, constant term 5.

5. $4x^8 + 9x^4$. Degree 8, first term $4x^8$, no constant term.

7. Polynomial

9. Not a polynomial. It is a quotient of polynomials.

11. Polynomial

13. $(2x^2 + 3x + 1) + (5x^2 + x + 3)$
$= (2x^2 + 5x^2) + (3x + x) + (1 + 3) = 7x^2 + 4x + 4$

15. $(3x^2 + 5x + 7) - (2x^2 - 2x + 4)$
$= (3x^2 - 2x^2) + (5x + 2x) + (7 - 4) = x^2 + 7x + 3$

17. $(3y^3 + 2) + (5y^2 + y + 3)$
$= 3y^3 + 5y^2 + y + (2 + 3) = 3y^3 + 5y^2 + y + 5$

19. $(t^2 + t + 2) + (3t^2 - 4t + 1) + (5t + 3)$
$= (t^2 + 3t^2) + (t - 4t + 5t) + (2 + 1 + 3) = 4t^2 + 2t + 6$

21. $(4x + 3)(x + 2) = 4x \cdot x + 4x \cdot 2 + 3 \cdot x + 3 \cdot 2$

Answers to *Math in Practice: An Applied Video Companion CD-ROM*

#1
Math Application by Topic: Math on Stage: Acoustics
Math Application by Concept: Basics
Answer: 1.16 seconds

#2
Math Application by Topic: Math at Work: Aeronautics
Math Application by Concept: Basics
Answer: Approximately 24 seconds

#3
Math Application by Topic: Math on Stage: Piano Pitch
Math Application by Concept: Basics
Answers: a. $k = 3728$
                b. 1.88 inches

#4
Math Application by Topic: Math in the Environment: Beetle Infestation
Math Application by Concept: Basics
Answers: a. 12.5 million trees per year
                b. 5 million trees
                c. $10 per tree

#5
Math Application by Topic: Math in the Environment: Air Pollution
Math Application by Concept: Basics
Answers: a. 2.45 cars per family
                b. 0.914 pounds per car
                c. 26,666,667 miles

#6
Math Application by Topic: Math at Work: Fighting Fires
Math Application by Concept: Basics
Answers: a. Approximately 169 gallons per minute
                b. About 4,225 gallons

#7
Math Application by Topic: Math of Matter and Motion: Line of Sight
Math Application by Concept: Roots and Radicals
Answer: 216 feet

#8
Math Application by Topic: Math at Work: Fighting Fires
Math Application by Concept: Roots and Radicals

Answers: a. 16.96 psi
             b. 248 gallons per minute
             c. $G = 2.2D^2v$

#9
Math Application by Topic: Math of Matter and Motion: Falling Objects
Math Application by Concept: Roots and Radicals
Answers: a. 54.2 m/s
             b. 91.8 kg

#10
Math Application by Topic: Math of Matter and Motion: Pendulums
Math Application by Concept: Roots and Radicals
Answer: 0.99 meters. Thus, one meter is the length of a pendulum that has a perio
T of two seconds, thus 'beating' one second on every swing.

#11
Math Application by Topic: Math Over Time: Dinosaurs
Math Application by Concept: Roots and Radicals
Answer: 2.1 meters per second

#12
Math Application by Topic: Math Over Time: Dinosaurs
Math Application by Concept: Roots and Radicals
Answer: 0.268 feet per second = 3.2 inches per second

#13
Math Application by Topic: Math on Stage: Piano Pitch
Math Application by Concept: Roots and Radicals
Answers: a. 220 Hz
             b. Increase the tension by 145.76 N;
                Decrease the linear mass density by 0.0015 kg/m

---

#14
Math Application by Topic: Math on Stage: Lighting a Stage
Math Application by Concept: Polynomial and Rational Functions
Answers: a. $k = 25,000$
             b. 1.5 foot-candles
             c. 15.8 feet

#15
Math Application by Topic: Math of Matter and Motion: Optical Lab
Math Application by Concept: Polynomial and Rational Functions
Answers: a. 66.67 cm
             b. 1.5 diopters

#16
Math Application by Topic: Math of Matter and Motion: Electronics
Math Application by Concept: Polynomial and Rational Functions
Answers: a. 15 ohms, 1.67 amps
                b. 3.33 ohms, 7.5 amps
                c. 12 ohms

#17
Math Application by Topic: Math in the Environment: Oceanographer
Math Application by Concept: Polynomial and Rational Functions
Answers: a. $D = 1$
                b. 0.295 square miles

#18
Math Application by Topic: Math Over Time: Map Evolution
Math Application by Concept: Polynomial and Rational Functions
Answers: a. $2.4 million
                b. During the year 2005 ($x = 5.594$)

#19
Math Application by Topic: Math of Matter and Motion: Falling Objects
Math Application by Concept: Solving Quadratic Equations
Answers: a. 7.6 seconds
                b. 253 ft/sec, or 173 mph

#20
Math Application by Topic: Math Over Time: Atlatl
Math Application by Concept: Solving Quadratic Equations
Answers: a. 12.25 meters per second
                b. Energy increases by a factor of 225

#21
Math Application by Topic: Math Over Time: World War II Bombers
Math Application by Concept: Solving Quadratic Equations
Answer: 1.9 seconds

#22
Math Application by Topic: Math Over Time: Atlatl
Math Application by Concept: Solving Quadratic Equations
Answers: a. 3.24 seconds
                b. 86.9 feet per second

#23
Math Application by Topic: Math at Work: Aeronautics
Math Application by Concept: Solving Quadratic Equations
Answers: a. $8,000 per week

b. $R = -h^2 + 5h + 50$

c. $h = 2.5$ resulting in an increase of \$1.25 per ticket

#24

Math by Application Topic: Math at Work: Designing Heating Ducts

Math by Application Concept: Solving Quadratic Equations

Answer: 11 ¼ inches

#25

Math Application by Topic: Math of Matter and Motion: Line of Sight

Math Application by Concept: Solving Quadratic Equations

Answers: a. 12.25 miles

b. 600 feet

#26

Math Application by Topic: Math of Matter and Motion: Optical Lab

Math Application by Concept: Polynomial and Rational Functions

Answer: 6 hours

#27

Math Application by Topic: Math in the Environment: Beetle Infestation

Math Application by Concept: Exponential / Logarithmic Functions

Answer: $x = 4.15$, so five applications will be needed

#28

Math Application by Topic: Math in the Environment: Population Growth

Math Application by Concept: Exponential / Logarithmic Functions

Answers: a. 1960: 70.2 years, 1970: 72.4 years

b. 2028 years

#29

Math Application by Topic: Math in the Environment: Air Pollution

Math Application by Concept: Exponential / Logarithmic Functions

Answers: a. 0.08 ppm

b. 10:20 a.m.

c. Yes, at approximately 11:30 a.m.

#30

Math Application by Topic: Math of Matter and Motion: Metal Alloys

Math Application by Concept: Exponential / Logarithmic Functions

Answers: a. 2200° F

b. 614° F

c. 9.8 hours

#31

Math Application by Topic: Math in the Environment: Population Growth

Math Application by Concept: Exponential / Logarithmic Functions
Answers: a. $k \approx 0.0214$
            b. The year 2005

#32
Math Application by Topic: Math in the Environment: Earthquakes
Math Application by Concept: Exponential / Logarithmic Functions
Answers: a. Richter measure $R = 5.4$
            b. $3,162,277,660 i_0$ or 3.16 billion times that of i_0.

#33
Math Application by Topic: Math in the Environment: Earthquakes
Math Application by Concept: Exponential / Logarithmic Functions
Answers: a. 650 kilometers per hour
            b. Approximately 5 times the intensity

#34
Math Application by Topic: Math at Work: Investing and Debt
Math Application by Concept: Systems of Equations / Matrices
Answer: Invest $5,818.18 in the stock market and $6,181.82 in the bond

---

#35
Math Application by Topic: Math in the Environment: Living Soil Crust
Math Application by Concept: Systems of Equations / Matrices
Answer: Approximately 19 ¾ gallons of the 12% mixture and 30 ¼ gallons of the
            50% mixture

#36
Math Application by Topic: Math of Matter and Motion: Metal Alloys
Math Application by Concept: Systems of Equations / Matrices
Answer: 7.16 parts nickel, 17.91 parts cobalt, 14.93 parts chromium

---

#37
Math Application by Topic: Math at Work: Flying to Hawaii
Math Application by Concept: Systems of Equations / Matrices
Answer: 6.11 hours

#38
Math Application by Topic: Math of Matter and Motion: Electronics
Math Application by Concept: Systems of Equations / Matrices
Answers: $I_1 = 6.5$, $I_2 = 5.2$, $I_3 = 1.3$

#39
Math Application by Topic: Math Over Time: Premature Infant Formula
Math Application by Concept: Systems of Equations / Matrices

Answer: 5 units of Formula A, 2 units of Formula B and 3 units of Formula C

#40
Math Application by Topic: Math Over Time: Premature Infant Formula
Math Application by Concept: Systems of Equations / Matrices

Answers: a. $A^{-1} = \begin{bmatrix} 0.6 & -0.2 & 0 \\ -1.6 & 0.2 & 1 \\ 1.4 & 0.2 & -1 \end{bmatrix}$

b. 0.8 ounces of Formula I, 7.2 ounces of Formula II and 3.2 ounces of Formula III

#41
Math Application by Topic: Math at Work: Investing and Debt
Math Application by Concept: Series & Sequences / Conics

Answers: a. $A = \dfrac{Bi}{(1+i)^n - 1}$

b. $21.36 per month
c. $54.62 per month
d. $516.76 per month

#42
Math Application by Topic: Math on Stage: Sound Dish
Math Application by Concept: Series & Sequences / Conics
Answer: 9 inches

#43
Math Application by Topic: Math on Stage: Sound Dish
Math Application by Concept: Series and Sequences / Conics
Answers: a. $a = 0.0224$
             b. Approximately 11.2 meters

#44
Math Application by Topic: Math on Stage: Acoustics
Math Application by Concept: Series & Sequences / Conics
Answer: 200 feet

---

*. Math Application by Topic: Math of Matter and Motion: Pendulums
Math Application by Concept: Trigonometry
Answers: a. 38.47 hours
             b. 42.69° latitude

---

*. Math Application by Topic: Math at Work: Designing Heating Ducts
Math Application by Concept: Trigonometry
Answers: a. $x = 12\sin\theta$, $h = 12\cos\theta$

b. $A = 6\cos\theta(42 + 24\sin\theta)$

c. $\theta = 23.2°$

---

*. Math Application by Topic: Math at Work: Flying to Hawaii
Math Application by Concept: Trigonometry
Answers: a. 185.5 mph
            b. 6 hours 26 minutes
            c. 5 hours 41 minutes

---

*. Math Application by Topic: Math Over Time: World War II Bombers
Math Application by Concept: Trigonometry
Answer: 66.17°

MULTIMEDIA CD-ROM SINGLE USER LICENSE AGREEMENT

1. **NOTICE.** WE ARE WILLING TO LICENSE THE MULTIMEDIA CD-ROM PRODUCT TO YOU ONLY ON THE CONDITION THAT YOU ACCEPT ALL OF THE TERMS CONTAINED IN THIS LICENSE AGREEMENT. PLEASE READ THIS LICENSE AGREEMENT CAREFULLY BEFORE OPENING THE SEALED DISC PACKAGE. BY OPENING THAT PACKAGE YOU AGREE TO BE BOUND BY THE TERMS OF THIS AGREEMENT. IF YOU DO NOT AGREE TO THESE TERMS WE ARE UNWILLING TO LICENSE THE MULTIMEDIA PRODUCT TO YOU, AND YOU SHOULD NOT OPEN THE DISC PACKAGE. IN SUCH CASE, PROMPTLY RETURN THE UNOPENED DISC PACKAGE AND ALL OTHER MATERIAL IN THIS PACKAGE ALONG WITH PROOF OF PAYMENT, TO THE AUTHORIZED-DEALER FROM WHOM YOU OBTAINED IT FOR A FULL REFUND OF THE PRICE YOU PAID.

2. **Ownership and License.** This is a license agreement and NOT an agreement for sale. It permits you to use one copy of the Multimedia Product on a single computer. The Multimedia Product and its contents are owned by us or our licensors, and are protected by U.S. and international copyright laws. Your rights to use the Multimedia Product are specified in this Agreement, and we retain all rights not expressly granted to you in this Agreement.

- You may use one copy of the Multimedia Product for your own internal informational use.
- You may not copy any portion of the Multimedia Product to your computer hard disk, network, or any other media other than printing out or downloading insubstantial portions of the text and images in the Multimedia Product for your own internal informational use.
- You may not copy any of the documentation or other printed materials accompanying the Multimedia Product.

Neither concurrent use on two or more computers nor use in a local area network or other network is **permitted** *without* separate authorization and **the payment** of additional license fees. **You may** obtain **such authorization by contacting Customer** Service at 1-800-447-9457 and paying the applicable **network license** fee.

3. **Transfer and Other Restrictions.** You may not rent, lend, or lease this Multimedia Product. You may not and you may not permit others to (a) disassemble, decompile, or otherwise derive source code from the software included in the Multimedia Product (the "Software"), (b) reverse engineer the Software, (c) modify or prepare derivative works of the Multimedia Product, (d) use the Software in an on-line system or (e) use the Multimedia Product in any manner that infringes the intellectual property or other rights of another party.

However, you may transfer this license to use the Multimedia Product to another party on a permanent basis by transferring this copy of the License Agreement, the Multimedia Product, and all documentation. Such transfer of possession terminates your license from us. Such other party shall be licensed under the terms of this Agreement upon its acceptance of this Agreement by its initial use of the Multimedia Product. If you transfer the Multimedia Product, you must remove the installation files from your hard disk and you may not retain any copies of those files for your own use.

4. **Limited Warranty and Limitation of** Liability. For a period of sixty (60) days from the date you acquired the Multimedia Product from us or our authorized dealer, we warrant that the media containing the Multimedia Product will be free from defects that prevent you from installing the Multimedia Product on your computer. If the disc fails to conform to this warranty, you may, as your sole and exclusive remedy, obtain a replacement free of charge if you return the defective disc to us with a dated proof of purchase. Otherwise the Multimedia Product is licensed to you on an "AS IS" basis without any warranty of any nature.

WE DO NOT WARRANT THAT THE MULTIMEDIA PRODUCT WILL MEET YOUR REQUIREMENTS OR THAT ITS OPERATION WILL BE UNINTERRUPTED OR ERROR-FREE. WE EXCLUDE AND EXPRESSLY DISCLAIM ALL EXPRESS AND IMPLIED WARRANTIES NOT STATED HEREIN, INCLUDING THE IMPLIED WARRANTIES OF MERCHANTABILITY AND FITNESS FOR A PARTICULAR PURPOSE.

WE SHALL NOT BE LIABLE FOR ANY DAMAGE OR LOSS OF ANY KIND ARISING OUT OF OR RESULTING FROM YOUR POSSESSION OR USE OF THE MULTIMEDIA PRODUCT (INCLUDING DATA LOSS OR CORRUPTION), REGARDLESS OF WHETHER SUCH LIABILITY IS BASED IN TORT, CONTRACT OR OTHERWISE AND INCLUDING, BUT NOT LIMITED TO ACTUAL, SPECIAL, INDIRECT, INCIDENTAL OR CONSEQUENTIAL DAMAGES. IF THE FOREGOING LIMITATION IS HELD TO BE UNENFORCEABLE, OUR MAXIMUM LIABILITY TO YOU SHALL NOT EXCEED THE AMOUNT OF THE LICENSE FEE PAID BY YOU FOR THE MULTIMEDIA PRODUCT. THE REMEDIES AVAILABLE TO YOU AGAINST US AND THE LICENSORS OF MATERIALS INCLUDED IN THE MULTIMEDIA PRODUCT ARE EXCLUSIVE.

Some states do not allow the limitation or exclusion of implied warranties or liability for incidental or consequential damages, so the above limitations or exclusions may not apply to you.

5. **United States Government Restricted Rights.** The Multimedia Product and documentation are provided with Restricted Rights. Use, duplication or disclosure by the U.S. Government or any agency or instrumentality thereof is subject to restrictions as set forth in subdivision (c)(1)(ii) of the Rights in Technical Data and Computer Multimedia Product clause at 48 C. F.R. 252.227-7013, or in subdivision (c)(1) and (2) of the Commercial Computer Multimedia Product '-Restricted Rights Clause at 48 C.F.R. 52.227-19, as applicable. Manufacturer is Harcourt, Inc., 301 Commerce Street, Fort Worth, TX 76102.

6. **Termination.** This license and your right to use this Multimedia Product automatically terminate if you fail to comply with any provisions of this Agreement, destroy the copy of the Multimedia Product in your possession, or voluntarily return the Multimedia Product to us. Upon termination you will destroy all copies of the Multimedia Product and documentation.

7. **Miscellaneous Provisions.** This Agreement will be governed by and construed in accordance with the substantive laws of the Commonwealth of Pennsylvania. This is the entire agreement between us relating to the Multimedia Product, and supersedes any prior purchase order, communications, advertising or representations concerning the contents of this package. No change or modification of this Agreement will be valid unless it is in writing, and is signed by us.

SYSTEM REQUIREMENTS

For Windows:

Requires IBM or Compatible Pentium Base
200 MHz or higher
Windows 95, 98, or NT Operating System
16 MG of RAM
16 bit sound card
256 color or greater Video Card (High Color or
        greater recommended)
6X CD-ROM Drive
10 MG free Hard Drive space

INSTALLATION INSTRUCTIONS

This CD-ROM requires QuickTime 4.0 for the video playback. If you do not have have it already installed, it is provided on the CD-ROM. If you are not sure if QuickTime 4.0 has already been installed on your computer, it is recommended that you install it. It will not cause any problems if QuickTime 4.0 was previously installed.

WINDOWS 95 or higher
To Install the QuickTime Player (Skip if QuickTime is already installed.)
- Insert the CD-ROM.
- After the CD-ROM auto-starts, click Exit button.
- Return to Windows Desktop.
- Double click on the My Computer icon.
- Double click on the CD-ROM drive icon.
- Double click on the PCFULLQT folder.
- Double click on the QuickTimeInstaller.exe icon.
- Follow the screen prompts.

To run Math In Practice: An Applied Video Companion CD-ROM.
1. Insert the CD-ROM
2. CD-ROM will auto-start.

TECHNICAL SUPPORT

For support call 1-800-447-9457 from 7am to 6pm Central Time Monday through Friday or visit our website at http://www.hbtechsupport.com. Questions and concerns can be faxed to 800-352-1680 or 800-354-1774. You may also email your concerns to tschbc@hbtechsupport.com

$= 4x^2 + 8x + 3x + 6 = 4x^2 + 11x + 6$

23. $(7x + 2)(x + 3) = 7x \cdot x + 7x \cdot 3 + 2 \cdot x + 2 \cdot 3$
$= 7x^2 + 21x + 2x + 6 = 7x^2 + 23x + 6$

25. $(x^2 + 2)(x^2 - 4) = x^2 \cdot x^2 + x^2 \cdot (-4) + 2 \cdot x^2 + 2 \cdot (-4)$
$= x^4 - 4x^2 + 2x^2 - 8 = x^4 - 2x^2 - 8$

27. $2x(x^2 - 3x + 4) = 2x \cdot x^2 + 2x \cdot (-3x) + 2x \cdot 4 = 2x^3 - 6x^2 + 8x$

29. $(4x^2 + 2)(x + 3) = 4x^2 \cdot x + 4x^2 \cdot 3 + 2 \cdot x + 2 \cdot 3$
$= 4x^3 + 12x^2 + 2x + 6 = 4x^3 + 12x^2 + 2x + 6$

31. $(x + 3)(x^2 - x + 2) = x(x^2 - x + 2) + 3(x^2 - x + 2)$
$= x^3 - x^2 + 2x + 3x^2 - 3x + 6 = x^3 + 2x^2 - x + 6$

33. $(2x^2 - 3x + 2)(2x^2 - x + 4)$
$= 2x^2(2x^2 - x + 4) - 3x(2x^2 - x + 4) + 2(2x^2 - x + 4)$
$= 4x^4 - 2x^3 + 8x^2 - 6x^3 + 3x^2 - 12x + 4x^2 - 2x + 8$
$= 4x^4 - 8x^3 + 15x^2 - 14x + 8$

35. $2y(y + 3) + 4(y + 2) = 2y^2 + 6y + 4y + 8$
$= 2y^2 + 10y + 8$

37. $(a + b)(a - b) = a \cdot a - a \cdot b + b \cdot a - b \cdot b$
$= a^2 - a \cdot b + a \cdot b - b^2 = a^2 - b^2$

39. $(a + b)^3 = (a + b)(a + b)^2 = (a + b)(a^2 + 2ab + b^2)$
$= a(a^2 + 2ab + b^2) + b(a^2 + 2ab + b^2)$
$= a^3 + 2a^2b + ab^2 + ba^2 + 2ab^2 + b^3$
$= a^3 + 2a^2b + ab^2 + a^2b + 2ab^2 + b^3$
$= a^3 + 3a^2b + 3ab^2 + b^3$

41. $(4x + 1)^2 = (4x)^2 + 2(4x)(1) + 1^2 = 16x^2 + 8x + 1$

43. $(x + 4)^3 = x^3 + 3(x^2)(4) + 3(x)(4)^2 + 4^3$
$= x^3 + 12x^2 + 48x + 64$

45. $(4x + 5y)^2 = (4x)^2 + 2(4x)(5y) + (5y)^2 = 16x^2 + 40xy + 25y^2$

47. $(x - 1)^3 = (x)^3 - 3(x)^2(1) + 3(x)(1)^2 - (1)^3$
$= x^3 - 3x^2 + 3x - 1$

49. $(x - y)(x + y)^3 = (x - y)(x^3 + 3x^2y + 3xy^2 + y^3)$
$= x(x^3 + 3x^2y + 3xy^2 + y^3) - y(x^3 + 3x^2y + 3xy^2 + y^3)$
$= x^4 + 3x^3y + 3x^2y^2 + xy^3 - x^3y - 3x^2y^2 - 3xy^3 - y^4$
$= x^4 + 2x^3y - 2xy^3 - y^4$

51.

$$\begin{array}{r} x + 2 \\ x+3 \overline{\smash{\big)}\ x^2 + 5x + 8} \\ \underline{x^2 + 3x} \\ 2x + 8 \\ \underline{2x + 6} \\ 2 \end{array}$$

$$\frac{x^2+5x+8}{x+3} = x+2 + \frac{2}{x+3}$$

53.

$$\begin{array}{r} x^2 - 2x - 2 \\ x-3 \overline{\smash{\big)}\ x^3 - 5x^2 + 4x - 7} \\ \underline{x^3 - 3x^2} \\ -2x^2 + 4x \\ \underline{-2x^2 + 6x} \\ -2x - 7 \\ \underline{-2x + 6} \\ -13 \end{array}$$

$$\frac{x^3-5x^2+4x-7}{x-3} = x^2-2x-2 - \frac{13}{x-3}$$

55.

$$\begin{array}{r} 3x^2 - 1 \\ 2x^2+1 \overline{\smash{\big)}\ 6x^4 + x^2 + 1} \\ \underline{6x^4 + 3x^2} \\ -2x^2 + 1 \\ \underline{-2x^2 - 1} \\ 2 \end{array}$$

$$\frac{6x^4+x^2+1}{2x^2+1} = 3x^2-1+ \frac{2}{2x^2+1}$$

57.

$$\begin{array}{r} 2x^3 + 2x^2+ 2x + 5 \\ x-1 \overline{\smash{\big)}\ 2x^4 + 0x^3 + 0x^2 + 3x + 1} \\ \underline{2x^4 - 2x^3} \\ 2x^3 + 0x^2 \\ \underline{2x^3 - 2x^2} \\ 2x^2 + 3x \\ \underline{2x^2 - 2x} \\ 5x + 1 \\ \underline{5x - 5} \\ 6 \end{array}$$

$$\frac{2x^4+3x+1}{x-1} = 2x^3+2x^2+2x+5+ \frac{6}{x-1} + \frac{6}{x-1}$$

59. False: $(x^3 + x + 1) + (-x^3 + x + 2) = 2x + 3$,
a polynomial of degree 1.

61. True: $(ax + b)(cx + d)$ with $a \neq 0$, $c \neq 0$,
$\qquad = acx^2 + (bc + ad)x + bd$, and $ac \neq 0$.

63. (a) False. Let $P = 2x+3$, $Q = 5x$. Then $P+Q = 7x+3$.
(b) True. The constant term of PQ is obtained by multiplying constant terms of P and Q.

Project (a) $a=p^2-q^2$, $b=2pq$, $c=p^2+q^2$. Need $p>q$ if a is to be positive. Thus $a^2+b^2=(p^2-q^2)^2+(2pq)^2=(p^2)^2-2(p^2)(q^2)+(-q^2)^2+4p^2q^2=p^4-2p^2q^2+q^4+4p^2q^2=p^4+2p^2q^2+q^4=(p^2+q^2)^2 = c^2$. {a,b,c} is a set of Pythagorean Numbers.
(b) Let $p=3$, $q=2$. Then $a=3^2-2^2=5$, $b=2(3)(2)=12$, $c=3^2+2^2=13$. Let $p=7$, $q=4$. Then $a=7^2-4^2=33$, $b=2(7)(4)=56$, $c=7^2+4^2=65$.

Let p=8, q=3. Then a=8^2-3^2=55, b=2(8)(3)=48, c=8^2+3^2=73.
Check: $5^2+56^2=13^2$=169, $33^2+56^2=65^2$=4225, $55^2+48^2=73^2$=5329.
(c) Let {a,b,c} be a set of Pythagorean Numbers. Then
$(na)^2+(nb)^2=n^2a^2+n^2b^2=n^2(a^2+b^2)=n^2c^2=(nc)^2$. Thus (na,nb,nc} is
a set of Pythagorean Numbers.
Start with the set {3,4,5} of Pythagorean Numbers. Then
multiply by 2:{6,8,10}. multiply by 3:{9,12,15}. multiply by
7: {21,28,35}.
Calculator: $6^2+8^2=10^2$=100, $9^2+12^2=15^2$=4225, $21^2+28^2=35^2$=5329.

Section R.4

1. $4x^3 + 2x^2 = 2x^2(2x+1)$

3. $3x^4 + 6x^2 + 9 = 3(x^4 + 2x^2 + 3)$

5. $24x^6 + 12x^5 + 18x^4 + 30x^3 = 6x^3(4x^3 + 2x^2 + 3x + 5)$

7. $5y^2(y+4) + 10y^3(y+4)^2 = 5y^2(y+4)(1 + 2y(y+4))$
   $= 5y^2(y+4)(2y^2+8y+1)$

9. $x^2 + x - 2 = (x+2)(x-1)$ 11. $x^2 + 5x + 6 = (x+3)(x+2)$

13. $x^2 + 7x + 12 = (x+4)(x+3)$ 15. $3x^2 - 5x - 2 =(3x+1)(x-2)$

17. $10y^2 + 13y - 3 = (5y-1)(2y+3)$ 19. $x^2 - xy - 2y^2=(x-2y)(x+y)$

21. $2x^2 + 5xy - 3y^2 = (2x-y)(x+3y)$

23. $8x^2 + 2xy - y^2 = (4x-y)(2x+y)$

25. $x^2 - xy + 4x - 4y = x(x-y) +4(x-y) = (x-y)(x+4)$

27. $3x^2 + xy - 12x - 4y = x(3x+y) -4(3x+y) = (3x+y)(x-4)$

29. $5x + 10y - yx - 2y^2 = 5(x+2y) - y(x+2y) = (x+2y)(5-y)$

31. $xy^2 + 2y^2 - 3x - 6 = y^2(x+2) - 3(x+2) = (x+2)(y^2-3)$

33. $x^3 + x^2 - 2x = x(x^2 + x - 2) = x(x+2)(x-1)$

35. $x^3 - 4x^2 + 3x = x(x^2 - 4x + 3) = x(x-3)(x-1)$

37. $6x^2 + 27x + 12 = 3(2x^2 + 9x + 4) = 3(2x+1)(x+4)$

9

39. $3x^2 + 3xy^3 - 9x + xy + y^4 - 3y = 3x(x+y^3-3) + y(x+y^3-3)$
    $= (x+y^3-3)(3x+y)$

41. $u^2v+uv-6v+2u^2+2u-12 = v(u^2+u-6) + 2(u^2+u-6)$
    $= (u^2+u-6)(v+2) = (u+3)(u-2)(v+2)$

43. $(a+b)(a-b) = a^2 + ba - ab - b^2 = a^2 - b^2$

45. $(a-b)^2 = (a-b)(a-b) = a^2 -ba -ab + b^2 = a^2 - 2ab + b^2$

47. $(a - b)(a^2 + ab + b^2) = a(a^2 + ab + b^2) - b(a^2 + ab + b^2)$
    $= a^3 + a^2b + ab^2 - ba^2 - ab^2 - b^3 = a^3 - b^3$

49. $9x^2 - y^2 = (3x+y)(3x-y)$

51. $(3x+2y)^2 - 4z^2 = ((3x+2y)+2z)((3x+2y)-2z)$
    $= (3x+2y+2z)(3x+2y-2z)$

53. $16u^6 - 9v^4 = (4u^3)^2 - (3v^2)^2 = (4u^3+3v^2)(4u^3-3v^2)$

55. $x^2 + 2xy + y^2 = (x+y)^2$

57. $x^2 - 4xy + 4y^2 = (x-2y)^2$

59. $(x+2y)^2 - 6z(x+2y) + 9z^2 = (x+2y-3z)^2$

61. $8x^3 + y^3 = (2x)^3 + y^3 = (2x+y)((2x)^2 - (2x)(y) + y^2)$
    $= (2x+y)(4x^2 - 2xy + y^2)$

63. $125a^3 - 8b^3 = (5a)^3 - (2b)^3$
    $= (5a-2b)((5a)^2 + (5a)(2b) + (2b)^2)$
    $= (5a-2b)(25a^2 + 10ab + 4b^2)$

65. $(x+y)^3 - z^3 = (x+y)^3 - z^3$
    $= ((x+y)-z)((x+y)^2 + (x+y)z + z^2)$
    $= (x+y-z)(x^2+ 2xy + y^2 + xz + yz + z^2)$

67. $x^2 + 2x - 3 = (x+3)(x-1)$

69. $x^2 + 6xy + 9y^2 = (x+3y)(x+3y) = (x+3y)^2$

71. $4t^2 - 4t - 8 = 4(t^2 - t - 2) = 4(t-2)(t+1)$

73. $u^3 - 27v^3 = u^3 - (3v)^3 = (u-3v)(u^2 + (u)(3v) + (3v)^2)$
    $= (u-3v)(u^2 + 3uv + 9v^2)$

75. $9x^2 + 12x - 21 = 3(3x^2 + 4x - 7) = 3(3x+7)(x-1)$
77. $9x^2 - (3y+4z)^4 = (3x)^2 - ((3y+4z)^2)^2$

$$= (3x+(3y+4z)^2)(3x-(3y+4z)^2)$$

79. $x^2 + 4xy + 4y^2 - 9 = (x+2y)^2 - 3^2 = (x+2y+3)(x+2y-3)$

81. $u^2 + u - 2v - 4v^2 = u^2 - 4v^2 + u - 2v$
$$= (u+2v)(u-2v) + u - 2v = (u-2v)(u+2v+1)$$

83. $x^2 + 4x$. Half the coefficient of x is 2, thus add $(2)^2$.
$x^2 + 4x + (2)^2 = (x+2)^2$.

85. $x^2 + 2x$. Add $(1)^2$. Get $x^2 + 2x + (1)^2 = (x+1)^2$

87. $x^2 - 10x$. Add $(-5)^2$. Get $x^2 - 10x + (-5)^2 = (x-5)^2$

Section R.5

1. $\dfrac{3x^2}{2x + 5x^2} = \dfrac{3x^2}{x(2 + 5x)} = \dfrac{3x}{2 + 5x}$

3. $\dfrac{2x + 2}{4x^2 + 6x + 2} = \dfrac{2(x + 1)}{2(2x^2 + 3x + 1)} = \dfrac{2(x + 1)}{2(2x+1)(x+1)} = \dfrac{1}{2x + 1}$

5. $\dfrac{x + 1}{x^2 + 5x + 4} = \dfrac{x + 1}{(x+1)(x+4)} = \dfrac{1}{x + 4}$

7. $\dfrac{x^2 - 1}{x^2 + 4x + 3} = \dfrac{(x-1)(x+1)}{(x+1)(x+3)} = \dfrac{x - 1}{x + 3}$

9. $\dfrac{z^2 + 2z - 3}{z^2 + z - 6} = \dfrac{(z-1)(z+3)}{(z-2)(z+3)} = \dfrac{z - 1}{z - 2}$

11. $\dfrac{(x + 2)^2}{2} \cdot \dfrac{4}{x + 2} = 2(x + 2)$

13. $\dfrac{2x - 1}{4x + 1} \div \dfrac{4(2x - 1)}{x} = \dfrac{2x - 1}{4x + 1} \cdot \dfrac{x}{4(2x - 1)} = \dfrac{x}{4(4x + 1)}$

15. $\dfrac{2u^2 + 3u - 2}{u^2 - 2u - 3} \cdot \dfrac{4u^2 + 3u - 1}{2u^2 + 9u - 5} = \dfrac{(2u-1)(u+2)}{(u+1)(u-3)} \cdot \dfrac{(4u-1)(u+1)}{(2u-1)(u+5)}$

$$= \dfrac{(u+2)(4u-1)}{(u-3)(u+5)}$$

17. $\dfrac{1}{x} + \dfrac{3}{x - 1} = \dfrac{(x-1) + 3x}{x(x - 1)} = \dfrac{4x - 1}{x(x - 1)}$

19. $\dfrac{1}{x - 1} + \dfrac{4x}{x + 1} = \dfrac{(x+1) + 4x(x-1)}{(x-1)(x+1)} = \dfrac{4x^2 - 3x + 1}{(x-1)(x+1)}$

21. $\dfrac{x - 2}{x + 4} - \dfrac{x - 3}{2x - 3} = \dfrac{2x^2 - 7x + 6 - x^2 - x + 12}{(x+4)(2x-3)} = \dfrac{x^2 - 8x + 18}{(x+4)(2x-3)}$

23. $\dfrac{y^2 + y - 3}{y} - \dfrac{3y - 1}{3} = \dfrac{3(y^2+y-3) - y(3y-1)}{3y}$

$= \dfrac{3y^2 + 3y - 9 - 3y^2 + y}{3y} = \dfrac{4y - 9}{3y}$

25. $\dfrac{x}{(x+1)(x+2)} + \dfrac{2x+1}{(x+1)(x-1)} = \dfrac{x(x-1) + (2x+1)(x+2)}{(x+1)(x+2)(x-1)}$

$= \dfrac{(x^2-x) + (2x^2+5x+2)}{(x+1)(x+2)(x-1)} = \dfrac{3x^2 + 4x + 2}{(x+1)(x+2)(x-1)}$

27. $\dfrac{4x-1}{(x+2)(x-1)} + \dfrac{3x+2}{(x-1)(x+3)} = \dfrac{(4x-1)(x+3) + (3x+2)(x+2)}{(x+2)(x-1)(x+3)}$

$= \dfrac{(4x^2+11x-3) + (3x^2+8x+4)}{(x+2)(x-1)(x+3)} = \dfrac{7x^2 + 19x + 1}{(x+2)(x-1)(x+3)}$

29. $\dfrac{2z}{z^2 - 2z - 3} + \dfrac{3}{z^2 + 3z + 2} = \dfrac{2z}{(z+1)(z-3)} + \dfrac{3}{(z+1)(z+2)}$

$= \dfrac{2z(z+2) + 3(z-3)}{(z+1)(z-3)(z+2)} = \dfrac{(2z^2+4z) + (3z-9)}{(z+1)(z-3)(z+2)} = \dfrac{2z^2 + 7z - 9}{(z+1)(z-3)(z+2)}$

31. $\dfrac{x^2+x-2}{x^2-x-12} \cdot \dfrac{x^2+3x}{x^2-4} = \dfrac{(x-1)(x+2)}{(x-4)(x+3)} \cdot \dfrac{x(x+3)}{(x-2)(x+2)} = \dfrac{x(x-1)}{(x-4)(x-2)}$

33. $\dfrac{6x^2-11x-2}{x^2-3x} \div \dfrac{6x^2-5x-1}{x^2-x} = \dfrac{6x^2-11x-2}{x^2-3x} \cdot \dfrac{x^2-x}{6x^2-5x-1}$

$= \dfrac{(6x+1)(x-2)}{x(x-3)} \cdot \dfrac{x(x-1)}{(6x+1)(x-1)} = \dfrac{(x-2)}{(x-3)}$

35. $\dfrac{6v+1}{v-2} + \dfrac{2}{v} + 5v-4 = \dfrac{v(6v+1) + 2(v-2) + v(v-2)(5v-4)}{v(v-2)}$

$= \dfrac{(6v^2+v) + (2v-4) + v(5v^2-14v+8)}{v(v-2)} = \dfrac{6v^2+v+2v-4+5v^3-14v^2+8v}{v(v-2)}$

$= \dfrac{5v^3-8v^2+11v-4}{v(v-2)}$

37. $\dfrac{x^2 + x}{\left(\dfrac{1}{x}\right)} = (x^2 + x)(x) = x^3 + x^2$

39. $\dfrac{(x - 1)}{(1 - \frac{1}{x})} = \dfrac{(x - 1)}{(\frac{x - 1}{x})} = \dfrac{(x - 1)}{(x - 1)} (x) = x$

41. $\dfrac{(1 - \frac{1}{x})}{(1 + \frac{1}{x})} = \dfrac{(\frac{x-1}{x})}{(\frac{x+1}{x})} = \dfrac{x-1}{x} \cdot \dfrac{x}{x+1} = \dfrac{x-1}{x+1}$

43. $\dfrac{(\frac{1}{x} - \frac{3}{x^2})}{(\frac{2}{x})} = \dfrac{(\frac{x-3}{x^2})}{(\frac{2}{x})} = \dfrac{x-3}{x^2} \cdot \dfrac{x}{2} = \dfrac{x-3}{2x}$

45. $\dfrac{(1 + \frac{1}{s})}{(\frac{s + 1}{s - 1})} = \dfrac{(\frac{s + 1}{s})}{(\frac{s + 1}{s - 1})} = (\frac{s + 1}{s})(\frac{s - 1}{s + 1}) = \dfrac{s - 1}{s}$

47. $\dfrac{1}{x - 1} + \dfrac{4}{2x + 3} = \dfrac{2x + 3 + 4(x - 1)}{(x - 1)(2x + 3)} = \dfrac{6x - 1}{(x - 1)(2x + 3)}$

49. $\dfrac{3x}{(x - 1)(x + 2)} + \dfrac{x + 1}{x(x - 1)} = \dfrac{3x^2 + (x+1)(x+2)}{x(x - 1)(x + 2)}$

$= \dfrac{3x^2+x^2+3x+2}{x(x - 1)(x + 2)} = \dfrac{4x^2+3x+2}{x(x - 1)(x + 2)}$

51. $\dfrac{x}{(x - 1)(x + 2)} - \dfrac{x + 1}{(x - 1)(x - 2)} = \dfrac{x(x-2) - (x+1)(x+2)}{(x - 1)(x + 2)(x - 2)}$

$= \dfrac{x^2-2x - (x^2+3x+2)}{(x - 1)(x + 2)(x - 2)} = \dfrac{-5x - 2}{(x - 1)(x + 2)(x - 2)}$

53. $\dfrac{x + 1}{x^2 - 2x - 3} - \dfrac{x - 3}{x^2 + 3x + 2} = \dfrac{x + 1}{(x-3)(x+1)} - \dfrac{x - 3}{(x+2)(x+1)}$

$= \dfrac{(x+1)(x+2) - (x-3)(x-3)}{(x-3)(x+2)(x+1)} = \dfrac{x^2+3x+2 - (x^2-6x+9)}{(x-3)(x+2)(x+1)}$

$= \dfrac{9x - 7}{(x-3)(x+2)(x+1)}$

55. $\dfrac{(x + 1)}{(1 + \frac{1}{x})} - x + 2 = \dfrac{(x + 1)}{(\frac{x + 1}{x})} - x + 2 = (x + 1)(\dfrac{x}{x + 1}) - x + 2 = 2$

Section R.6
Group discussion: Quick rule for conversion from Celsius to
　　　　　　　Fahrenheit. Two suggestions-

(a) F=2C+30.
(b) Remember base degrees. 10C=50F, 20C=68F,30C=86F,40C=104F. Add 2 for ever degree C over these. e.g. 15C≈50+10≈60F, 22C≈68+4≈72F, 33C≈86+6≈92F.

1. $x = 1$, $2x + 4 = 6$. Left side $= 2(1) + 4 = 6$.
   Thus $x = 1$ is a solution.

3. $x = 1$, $x^2 - 2x - 1 = 0$. Left side $= 1^2 - 2(1) - 1 = -2 \neq 0$.
   Thus $x = 1$ is not a solution.

5. $x = -3$, $x^2 + x - 6 = 0$.
   Left side $= (-3)^2 - 3 - 6 = 0$. Thus $x = -3$ is a solution.
   $x = 2$, $x^2 + x - 6 = 0$.
   Left side $= 2^2 + 2 - 6 = 0$. Thus $x = 2$ is a solution.

7. $x = 2$, $\dfrac{x - 3}{x + 1} + 4 = \dfrac{x + 9}{x + 1}$.
   Left side $= \dfrac{2 - 3}{2 + 1} + 4 = \dfrac{-1}{3} + 4 = \dfrac{-1}{3} + \dfrac{12}{3} = \dfrac{11}{3}$.
   Right side $= \dfrac{2 + 9}{2 + 1} = \dfrac{11}{3}$. Thus $x = 2$ is a solution.

9. $2x - 3 = 5$, $2x - 3 + 3 = 5 + 3$, $2x = 8$, $x = 4$.

11. $-2x + 9 = -11$, $-2x + 9 - 9 = -11 - 9$, $-2x = -20$, $x = 10$.

13. $7x - 1 = 12x - 3$, $7x - 12x = -3 + 1$, $-5x = -2$, $x = 2/5$.

15. $3(x + 2) = 2x - 1$, $3x + 6 = 2x - 1$, $3x - 2x = -1 - 6$, $x = -7$.

17. $5x + 7 = 0$. Linear equation in standard form.

19. $2x - 3 = 8$. Linear equation. Standard form is $2x - 11 = 0$.

21. $2(x - 3) + 4 = x + 6$. Simplify and rewrite with all the terms on the left. $2x - 6 + 4 = x + 6$, $2x - 2 = x + 6$, $x - 8 = 0$. Linear equation with standard form $x - 8 = 0$.

23. $z + 2 = z^2 - 3z + 1$. Rewrite with all the terms on the left. $-z^2 + 4z + 1 = 0$. Non-linear because of the $-z^2$ term.

25. $5(x - 2) = 3(x + 4) - 6$, $5x - 10 = 3x + 12 - 6$,
    $5x - 3x = 12 - 6 + 10$, $2x = 16$, $x = 8$.

27. $8x - 1 = 2x - 3 + 9(x - 1)$, $8x - 1 = 2x - 3 + 9x - 9$,
    $8x - 1 = 11x - 12$, $-3x = -11$, $x = \dfrac{11}{3}$.

29. $3x^2 - 2x + 4 = (x^2 - 4) + (2x^2 - x + 5)$,
    $3x^2 - 2x + 4 = 3x^2 - x + 1$, $-2x + 4 = -x + 1$, $-x = -3$, $x = 3$.

14

31. $\dfrac{3x}{4} + \dfrac{1}{2} = \dfrac{x}{2} - 1$, $\dfrac{3x}{4} - \dfrac{x}{2} = -1 - \dfrac{1}{2}$, $\dfrac{x}{4} = -\dfrac{3}{2}$, $x = -6$.

33. $\dfrac{2x - 1}{x + 3} + \dfrac{3}{x + 3} = 4$, $\dfrac{(2x - 1) + 3}{x + 3} = 4$, $(2x-1)+3 = 4(x+3)$,
    $2x + 2 = 4x + 12$, $-2x = 10$, $x = -5$. This satisfies the
    original equation. Solution is $x = -5$.

35. $\dfrac{x(x - 1)}{x^2 + x - 2} + 3 = \dfrac{14}{x + 2}$, $\dfrac{x(x - 1)}{(x - 1)(x + 2)} + 3 = \dfrac{14}{x + 2}$,
    $\dfrac{x}{x + 2} + 3 = \dfrac{14}{x + 2}$, $x + 3(x + 2) = 14$, $x + 3x + 6 = 14$
    $4x = 8$, $x = 2$. This satisfies the original equation.
    Solution is $x = 2$.

37. $\dfrac{4}{x + 2} - \dfrac{2}{x - 1} = \dfrac{3}{x + 2}$. Multiply both sides by $(x+2)((x-1)$.
    $4(x-1) - 2(x+2) = 3(x-1)$, $4x - 4 - 2x - 4 = 3x - 3$,
    $4x - 2x - 3x = -3 + 4 + 4$, $-x = 5$, $x = -5$.
    This satisfies the original equation. Solution is $x = -5$.

39. $C = 5n + 1500$. $C - 1500 = 5n$. $n = C/5 - 300$.
    When $C = 5000$, $n = 000/5 - 300 = 1000 - 300 = 700$.
    Number of items = 700.

41. $e = .75r + 1.2$. e is estimated time, r is real time.
    When $e = 50$, $50 = .75r + 1.2$, $.75r = 48.8$, $r = 65.066666667$.
    The class will last 65 minutes.

43. Golfer has rounds of 78, 74, and 68 on the first three rounds.
    Par is 72. Suppose needs x on 4th round to get par average.
    Average for 4 rounds = $\dfrac{78 + 74 + 68 + x}{4}$.
    Thus $\dfrac{78 + 74 + 68 + x}{4} = 72$, $78+74+68+x = 4(72)$, $220+x = 288$,
    $x = 68$. Golfer needs a score of 68 on the final round.

45. Let the overall percentage growth be g. Then actual growth in
    $62,000 = sum of the growths in all the individual stocks.
    Thus $62000g =(10000X38.1+8000X28.2+12000X19.9+7000X75.2+$
    $10000X47.3+6500X4.2+8500X7.2)$. Solve for g.
    $g=(10000X38.1+...+8500X7.2)/62000 = 31.18225806$.
    Overall growth is 31.2%.

47. Let E be extension when weight is W. Then $E = kW$.
    When $W=20$, $E=4$. Thus $4 = 20k$. $k = 1/5$. $E = W/5$.
    When $W = 35$, $E = 35/5 = 7$. Extension = 7 inches.

49. $E = kts^2$. When $t=1$ and $s=50$ then $E=2$. Thus $2 = k(1)(50)^2$.
    $k = \dfrac{1}{1250}$. $E = \dfrac{ts^2}{1250}$. When $s=55$ and $t=3$, $E = \dfrac{3(55)^2}{1250} = 7.26$

mistakes per page on average.

51. $T = k\sqrt{d^3}$. $T = kd^{3/2}$. Earth: When d=93 then T=365.
Thus $365 = (93)^{3/2}k$. $k = \dfrac{365}{(93)^{3/2}}$. $T = \dfrac{365}{(93)^{3/2}} d^{3/2}$.

$T = 365(\dfrac{d}{93})^{3/2}$. Mercury: When d=36, $T = 365(\dfrac{36}{93})^{3/2} = 87.91$

days. Pluto: When d=3675, $T = 365(\dfrac{3675}{93})^{3/2} = 90668.09$ days.

53. $V=kr^3$. $V_1=k(r_1)^3$. For $3r_1$, $V_2 = k(3r_1)^3 = 27k(r_1)^3 = 27V_1$.
Volume is multiplied by 27.

55. $P = kAv^2$. When A=10 and v=20, $P = k(10)(20)^2 = 4000k$.
$v = \sqrt{\dfrac{P}{kA}}$. When 2P and A=80, $v = \sqrt{\dfrac{8000k}{k80}} = \sqrt{100} = 10$.
Wind velocity would be 10 miles per hour.

Section R.7

1. $f(x)=3x-3$. (a) $f(1)=3(1)-3=0$. (b) $f(4)=3(4)-3=9$.
(c) $f(0)=3(0)-3=-3$. (d) $f(-2)=3(-2)-3=-9$.

3. $f(x)=x^2+2x-3$. (a) $f(2)=(2)^2+2(2)-3=5$. (b) $f(5)=(5)^2+2(5)-3=32$.
(c) $f(0)=(0)^2+2(0)-3=-3$. (d) $f(-3)=(-3)^2+2(-3)-3=0$.

5. $g(x)=x^3+2x^2-4x+3$.
(a) $g(-4.1)=(-4.1)^3+2(-4.1)^2-4(-4.1)+3=-15.901$.
(b) $g(-2.8)=(-2.8)^3+2(-2.8)^2-4(-2.8)+3=7.928$.
(c) $g(0)=(0)^3+2(0)^2-4(0)+3=3$.
(d) $g(2.3)=(2.3)^3+2(2.3)^2-4(2.3)+3=16.547$.

7. $h(x) = \sqrt{3x - 2}$. (a) $h(2) = \sqrt{3(2) - 2} = \sqrt{4} = 2$.
(b) $h(9) = \sqrt{3(9) - 2} = \sqrt{25} = 5$.
(c) $h(-1) = \sqrt{3(-1) - 2} = \sqrt{-5}$, not defined.

9. $f(x) = \dfrac{2x + 4}{x}$. (a) $f(1) = \dfrac{2(1) + 4}{1} = 6$.
(b) $f(-2) = \dfrac{2(-2) + 4}{-2} = 0$. (c) $f(0) = \dfrac{2(0) + 4}{0}$, not defined.

11. $f(x) = \dfrac{2x}{x^2 - 3x}$. (a) $f(-2) = \dfrac{2(-2)}{(-2)^2 - 3(-2)} = -\dfrac{4}{10} = -0.4$.
(b) $f(2) = \dfrac{2(2)}{(2)^2 - 3(2)} = -\dfrac{4}{2} = -2$.
(c) $f(3) = \dfrac{2(3)}{(3)^2 - 3(3)} = \dfrac{6}{0}$, not defined.

13. $f(x) = 4x+1$. $f(a) = 4(a)+1 = 4a+1$.

15. $g(x) = 2x^2-x+3$. $g(-2a) = 2(-2a)^2-(-2a)+3 = 8a^2+2a+3$.

17. $g(t) = 3t-4$. $g(3s+2) = 3(3s+2)-4 = 9s+2$.

19. $f(x)=3x-1$. Domain: Set of real numbers, $(-\infty,\infty)$.

21. $f(x)=3x^2+4x-1$. Domain: Set of real numbers, $(-\infty,\infty)$.

23. $f(x)=\sqrt{x - 3}$. Need $x-3 \geq 0$, $x \geq 3$. Domain $[3,\infty)$.

25. $f(x)= \dfrac{2x}{x - 6}$. Need $x-6\neq0$. All reals but 6. Set $(-\infty,6)\cup(6,\infty)$.

27. $f(x)= \dfrac{1}{\sqrt{x - 4}}$. Need $x>4$. Domain $(4,\infty)$.

29. $f(x) = 3x-2$. Domain: Set of real numbers, $(-\infty,\infty)$.
    Range: Set of real numbers, $(-\infty,\infty)$.

31. $g(x) = 7$. Domain: Set of real numbers, $(-\infty,\infty)$.
    Range: Set is 7.

33. $g(x) = x^2+8$. Domain: Set of real numbers, $(-\infty,\infty)$.
    Range: $x^2+8 \geq 8$. Set $[8,\infty)$.

35. $f(x) = \sqrt{x - 3}$. Domain: $x-3\geq0$. $x\geq3$. Set $[3,\infty)$.
    Range: $\sqrt{x - 3} \geq 0$. Set $[0,\infty)$.

37. $f(x) = |x|$. Domain: Set of real numbers, $(-\infty,\infty)$.
    Range: $|x|\geq0$. Set $[0,\infty)$.

39. $g(x) = -|2x-3|$. Domain: Set of real numbers, $(-\infty,\infty)$.
    Range: $|2x-3|\geq0$. $-|2x-3|\leq0$. Set $(-\infty,0]$.

41. $f(x) = x^4-2$. Domain: Set of real numbers, $(-\infty,\infty)$.
    Range: Have $x^4\geq0$. $x^4-2 \geq -2$. Set $[-2,\infty)$.

43. $f(x) = \dfrac{1}{x}$. Domain: $x\neq0$. Set $(-\infty,0)\cup(0,\infty)$.
    Range: f can never be zero. Set $(-\infty,0)\cup(0,\infty)$.

45. $f(x) = 3.2x^2 + 4.1$. (a) $f(2.5)=24.1000$.
    (b) $f(3.92)=53.2725$. (c) $f(-7.81)=199.2875$.

47. $f(x) = \sqrt{x - 2}$. (a) $f(3.5) = 1.2247$. (b) $f(8.75) = 2.5981$.
    (c) $f(0)$ gives an error message. Check:When $x=0$, $\sqrt{x - 2} = \sqrt{-2}$,
    not a real number. Thus 0 is not in the domain of f.

49. $f(x) = \sqrt{x^2 - 3x - 4}$. (a) $f(5)=2.4495$. (b) $f(4)=0$.
    (c) $f(2)$ gives an error message.
    Check: When $x=2$, $\sqrt{x^2-3x-4} = \sqrt{2^2-3(2)-4} = \sqrt{-6}$, not a real

number. Thus 2 not in the domain of f.

51. $f(x) = x^2+3$. $\dfrac{f(x+h) - f(x)}{h} = \dfrac{((x+h)^2+3) - (x^2+3)}{h}$

$= \dfrac{x^2+2xh+h^2+3-x^2-3}{h} = \dfrac{2xh+h^2}{h} = 2x+h$.

53. $f(x) = 2x^2-x+1$. $\dfrac{f(x+h) - f(x)}{h} = \dfrac{(2(x+h)^2-(x+h)+1) - (2x^2-x+1)}{h}$

$= \dfrac{2x^2+4xh+2h^2-x-h+1-2x^2+x-1}{h} = \dfrac{4xh+2h^2-h}{h} = 4x+2h-1$.

55. $f(x) = \dfrac{1}{25}x^2$. (a) $f(10) = \dfrac{1}{25}\ 10^2 = 4$. $f(15) = \dfrac{1}{25}\ 15^2 = 9$. $f(40) = \dfrac{1}{25}\ 40^2 = 64$.

$f(72) = \dfrac{1}{25}\ 72^2 = 207.36$. (b) Mathematical domain of f is $(-\infty, \infty)$.
We are told that the function is valid for speeds of 10mph to
85mph. The meaningful domain in this application is [10,85].

57. $s(t) = -16t^2 + 160t$. (a) Table below. (b) s=0 when t=0 and
t=10. t=0 corresponds to launch time, t=10 is the time when it
returns to earth. (c) Meaningful domain of the function is
[0, 10]. We get negative values of s outside this interval.
These numbers have no physical meaning.

time	0	1	2	3	4	5	6	7	8	9	10
height	0	144	256	336	384	400	384	336	256	144	0

X	Y1
0	0
1	144
2	256
3	336
4	384
5	400
6	384

X	Y1
6	384
7	336
8	256
9	144
10	0
11	-176
12	-384

t=11 and t=12 give
negative values of s.
These times are not in
the meaningful domain
of s.

59. $N(t) = 48t^2 - t^3$.
(a)

t	1	2	3	4	5	6	7
N(t)	47	184	405	704	1075	1512	2009

(b) We see that the number of cases is still increasing after
seven months. Further investigation gives

t	30	31	32	33	34
N(t)	16299	16337	16384	16335	16184

The disease peaks after 32 months (to nearest month).
(c)

t	42	43	44	45	46	47	48
N(t)	10584	9245	7744	6075	4232	2209	0

There will be no cases of the disease after 48 months.

61. (a) $N(t) = 6t^2 + 72t + 9000$. t is time measured from 1986.

	1989	1990	1991	1992	1993	1994	1995
t	3	4	5	6	7	8	9
N(t)	9270	9384	9510	9648	9798	9960	10134

18

Total number of cars = 9270+9384+9510+9648+9798+9960+10134
= 67704 hundreds.

(b) Further investigation gives,

	2000	2001	2002	2003	2004	2005	2006
t	14	15	16	17	18	19	20
N(t)	11184	11430	11688	11958	12240	12534	12840

The number of new cars produced will likely exceed 1,200,000 in the year 2004.

63. $P(t) = 0.0057t^3 - 0.302t^2 + 6.087t + 22.6$. (a) In 1995, $P(15)=65.1925$. Approx 65.2% of U.S. households with cable TV in 1995. (b) Predictions for 2000 and 2010, $P(20)=69.14\%$, $P(30)=87.31\%$. (c) $P(35)=110.0825$. This would imply that 110% of households would have cable in 2015. Cannot of course have more than 100%. The function has domain [0,15] - it was known to closely describe the percentages of households with cable from 1980 to 1995. It can be expected to become increasingly unreliable outside the interval [0, 15].

65. (a) Let height be y. Surface area = 4(area of side)+(area of base) $= 4xy + x^2$. Thus $4xy + x^2 = 64$. $y = \dfrac{64-x^2}{4x}$.

Volume, $V(x) = x^2y = \dfrac{x(64-x^2)}{4}$.

(b)

x	0	1	2	3	4	5	6	7	8
V(x)	0	15.75	30	41.25	48	48.75	42	26.25	0

(c) Domain of V is [0,8] since cannot have negative volume.
(d) Volume largest when x=5. Volume is then 48.75 cubic inches.

Chapter R Project

(a) $t_E=\dfrac{d}{v}$, $t_S=\dfrac{d}{v}\sqrt{1-v^2}$. Distance of Sirius is 8 light-years. Total distance traveled there and back is 16 light years, d=16. Velocity is 0.9 speed of light, v=0.9. Thus $t_E=\dfrac{16}{0.9} = 17.78$ (to 2 dec places). $t_S=\dfrac{16}{0.9}\sqrt{1-0.9^2} = 7.75$. Duration of voyage for person on Earth is 17.78 years. For person on the space ship the voyage lasts 7.75 years.

(b) Capella is 45 light-years from Earth. Total distance traveled there and back is 90 light years, d=90. Velocity is 0.99 speed of light, v=0.99. Thus $t_E=\dfrac{45}{0.99} = 45.45$ (to 2 dec places). $t_S=\dfrac{45}{0.99}\sqrt{1-0.99^2} = 6.41$. Duration of voyage for person on Earth is 45.45 years. For person on the space ship the voyage lasts 6.41 years.

(c) e.g. The star cluster Pleiades in the constellation Taurus is 410 light-years from Earth. Suppose space ship travels at 0.999 the speed of light. Then d=820, v=0.999.

$t_E = \dfrac{820}{0.999} = 828.28$ (to 2 dec places). $t_S = \dfrac{820}{0.999} \sqrt{1-0.999^2} =$
36.70. Duration of voyage for person on Earth is 828.28 years.
For person on the space ship the voyage lasts 36.70 years.

The star cluster Praesepe in the constellation Cancer is 515
light-years from Earth. Suppose space ship travels at 0.99999
the speed of light. Then d=1030, v=0.99999.
$t_E = \dfrac{1030}{0.99999} = 1030.01$ (to 2 dec places). $t_S = \dfrac{1030}{0.99999} \sqrt{1-0.99999^2}$
=4.61. Duration of voyage for person on Earth is 1030.01 years.
For person on the space ship the voyage lasts 4.61 years.
The farther one goes, and the greater the velocity, the larger
will be the the difference between earth and space times.

Chapter R Review Exercises

1. a) {2,12}   b) {-7,0,2,12}   c) {-7,0}
   d) $\{-7,\frac{3}{4},0,2,\frac{5}{4},12\}$ e) $\{\sqrt{2}\}$

2. a) number line b) number line c) number line
      $(-3,8]$ $[-2,\infty)$ $(-\infty,4)$

3. a) 4 + 7 = 7 + 4, commutative property of addition
   b) 4·(3 + 7) = 4·3 + 4·7, distributive property
   c) 3 + (4 + 7) = (3 + 4) + 7, associative property of addition
   d) (xy)z = x(yz), associative property of multiplication
   e) (x + y)z = xz + yz, distributive property

4. a) 5.793 x (2.58+3.46) = 34.98972

   b) $\sqrt{37} \div (2.3+1.78)$ = 1.49087 to five decimal places.

   c) $(\sqrt{7.9} + 1.3^{-1})$ x 4.65 = 16.65 to two decimal places.

5. a) 8 $\boxplus$ 9 $\boxdiv$ 3 = 8 + $(\frac{9}{3})$ = 8 + 3 = 11

   b) 7 $\boxtimes$ 3 $\boxplus$ 12 $\boxdiv$ 6 = (7x3) + $(\frac{12}{6})$ = 21 + 2 = 23

   c) 4 $\boxtimes$ 3 $\boxed{\wedge}$ 2 $\boxplus$ 4 $\boxdiv$ 2 = (4×3^2) + $(\frac{4}{2})$ = 36 + 2 = 38

6. a) $\sqrt{121}$ = 11   b) $\sqrt[3]{27}$ = 3

   c) $\sqrt[3]{\dfrac{-125}{8}} = -\dfrac{5}{2}$ d) $(\sqrt[4]{\dfrac{16}{81}})^3 = (\dfrac{2}{3})^3 = \dfrac{8}{27}$

7. a) $25^{1/2}$ = 5   b) $(-8)^{1/3}$ = -2

c) $243^{1/5} = 3$ $\qquad\qquad$ d) $(-128)^{1/7} = -2$

8. a) $27^{2/3} = (27^{1/3})^2 = 3^2 = 9$

   b) $25^{-3/2} = (25^{1/2})^{-3} = 5^{-3} = \dfrac{1}{5^3} = \dfrac{1}{125}$

   c) $(-125)^{4/3} = ((-125)^{1/3})^4 = -5^4 = 625$

   d) $(-128)^{-5/4} = ((-128)^{1/4})^{-5}$. DNE. $(-128)^{1/4}$, not a real number.

9. a) $\dfrac{4^3}{5} \cdot 5^7 = 4^3 \cdot 5^6 = 1{,}000{,}000$

   b) $(2^3) \cdot (4)^{-1} = (2^3) \cdot \dfrac{1}{4} = 8 \cdot \dfrac{1}{4} = 2$

   c) $(3 \cdot 4 \cdot 5)^{-2} \cdot 4^5 = \dfrac{1}{(3 \cdot 4 \cdot 5)^2} \cdot 4^5 = \dfrac{1}{3^2 \cdot 4^2 \cdot 5^2} \cdot 3^2 \cdot 5 = \dfrac{1}{4^2 \cdot 5} = \dfrac{1}{80}$

10. a) $2^{1/2} \cdot 2^{5/2} = 2^{(1/2+5/2)} = 2^{6/2} = 2^3 = 8$

    b) $5^{11/4} \cdot 5^{-3/4} = 5^{(11/4-3/4)} = 5^{8/4} = 5^2 = 25$

    c) $4^{1/3} \cdot 2^{4/3} = (2^2)^{1/3} \cdot 2^{4/3} = 2^{2/3} \cdot 2^{4/3} = 2^{(2/3+4/3)} = 2^2 = 4$

11. a) $x^2 x^{-3} = x^{(2-3)} = x^{-1} = \dfrac{1}{x}$

    b) $(x^3)^{-2} = x^{(3 \cdot -2)} = x^{-6} = \dfrac{1}{x^6}$

    c) $x^{2/3} x^{5/6} x^2 = x^{(2/3+5/6+2)} = x^{(4/6+5/6+12/6)} = x^{(21/6)} = x^{7/2}$

    d) $\dfrac{(x^4 y^{-3})^2}{(x^{-1} y^{-2})^{-3}} = \dfrac{x^{(4 \cdot 2)} y^{(-3 \cdot 2)}}{x^{(-1 \cdot -3)} y^{(-2 \cdot -3)}} = \dfrac{x^8 y^{-6}}{x^3 y^6} = \dfrac{x^8 x^{-3}}{y^6 y^6} = \dfrac{x^{(8-3)}}{y^{(6+6)}} = \dfrac{x^5}{y^{12}}$

12. a) $\sqrt[3]{32x^4} = \sqrt[3]{2^5 x^3 x} = \sqrt[3]{2^3 2^2 x^3 x} = \sqrt[3]{2^3 x^3 4x} = 2x(\sqrt[3]{4x})$

b) $\sqrt{9x^2 y^7} = \sqrt{9x^2 y^6 y} = \sqrt{(3)^2 x^2 (y^3)^2 y} = 3|x|(y^3)\sqrt{y}$

c) $\sqrt[3]{x^3 y^8 z^4} = \sqrt[3]{x^3 y^6 y^2 z^3 z} = \sqrt[3]{x^3 (y^2)^3 y^2 z^3 z} = xy^2 z(\sqrt[3]{y^2 z})$

13. a) $3.47^2 = 12.0409$ $\qquad\qquad$ b) $1.46^3 - 3.12^4 = -91.6464$

14. a) $21.67^{4/5} = 11.714$ $\qquad\qquad$ b) $1.6783^{4/3} = 1.994$

    c) $.0007813^{-3/4} = 213.987$

15. a) $\sqrt[3]{12.78^2} = 5.4662$ b) $\sqrt[5]{18.96^4} = 10.5262$

    c) $\sqrt[3]{-2.87^5} = -5.7961$

16. a) $473,000,000 = 4.73 \times 10^8$ b) $0.0000125 = 1.25 \times 10^{-5}$

17. a) $3.475 \times 10^4 = 34750$ b) $9.473 \times 10^{-3} = 0.009473$

18. a) $(5.671 \times 10^3) + (4738.4 \times 10^{-4}) = 5671 + 0.47384 = 5671.47384$
$$= 5.67147384 \times 10^3$$

    b) $\dfrac{(3.15 \times 10^2) + (4.122 \times 10^3)}{(0.03 \times 10^{-2})} = \dfrac{315 + 4122}{(3 \times 10^{-4})} = \dfrac{(315 + 4122) \times 10^4}{3}$

    $= \dfrac{4437 \times 10^4}{3} = 1479 \times 10^4 = 1.479 \times 10^7$

19. (a) $\dfrac{23.567 \times 10^6}{2.468 \times 436.768} = 21862.9285$

    (b) $\dfrac{4987.3678 \times 10^{-2}}{0.023 \times 3.4527 \times 2^{-5}} = 20097.1602$

20. Number protons/neutrons $= \dfrac{160 \times 2.2 \times 10^3}{1.673 \times 10^{-24}} = 2.104004782 \times 10^{29}$

21. Amount $= 5000(1+.08)^{10} = \$10794.62$

22. (a) Annually: Amount $= 8700(1+.055)^7 = \$12,655.71$

    (b) Quarterly: Amount $= 8700(1+ \dfrac{.055}{4})^{(7 \times 4)} = \$12,752.15$

    (c) Monthly: Amount $= 8700(1+ \dfrac{.055}{12})^{(7 \times 12)} = \$12774.40$

23. Volume $= \dfrac{4}{3}pr^3 = \dfrac{4}{3}p(6.27)^3 = 1032.5028$ cubic inches, to 4 decimal places.

24. a) $(3x^2+4x+6) + (5x^2-7x-3) = (3x^2+5x^2) + (4x-7x) + (6-3)$
    $= 8x^2 - 3x + 3$

    b) $(2x + 3)(4x - 5) = 8x^2 + 12x - 10x - 15 = 8x^2 + 2x - 15$

    c) $(x^2 - 2x + 3)(2x^2 + 3x - 1)$
    $= x^2(2x^2 + 3x - 1) - 2x(2x^2 + 3x - 1) + 3(2x^2 + 3x - 1)$
    $= 2x^4 + 3x^3 - x^2 - 4x^3 - 6x^2 + 2x + 6x^2 + 9x - 3$

22

$= 2x^4 - x^3 - x^2 + 11x - 3$

25. a) $(2x - 5)^2 = (2x)^2 + 2(2x)(-5) + (-5)^2$
$= 4x^2 - 20x + 25$

b) $(5x^2 + y)^3 = (5x^2)^3 + 3(5x^2)^2y + 3(5x^2)y^2 + y^3$
$= 125x^6 + 75x^4y + 15x^2y^2 + y^3$

26. a) $4x^6 - 2x^3 + 12x^2 = 2x^2(2x^4 - x + 6)$

b) $3x^2(2x+1)^3 - 12x^3(2x+1)^2 = 3x^2(2x+1)^2((2x+1) - 4x)$
$= 3x^2(2x+1)^2(1 - 2x)$

c) $8x^3y^2z^9 - 4x^2yz^3 = 4x^2yz^3(2xyz^6 - 1)$

27. a) $x^2 + 6x + 5 = (x + 5)(x + 1)$
b) $x^2 - 4x + 3 = (x - 3)(x - 1)$
c) $2x^2 - x - 6 = (2x + 3)(x - 2)$
d) $15x^2 + 2x - 8 = (5x + 4)(3x - 2)$

28. a) $2x^2 + 5xy - 12y^2 = (2x - 3y)(x + 4y)$
b) $15x^2 + 4xy - 4y^2 = (5x - 2y)(3x + 2y)$
c) $4x^2 - 4xy + y^2 = (2x - y)(2x - y) = (2x - y)^2$

29. a) $2x^2 + 6xy - 4x - 12y = 2x(x + 3y) - 4(x + 3y)$
$= (x + 3y)(2x - 4) = 2(x + 3y)(x - 2)$

b) $6x^2 + 2x + 3xy + y = 2x(3x +1) + y(3x + 1)$
$= (3x + 1)(2x + y)$

c) $21x^2 + 4y - 7xy - 12x = 21x^2 - 7xy + 4y - 12x$
$= 7x(3x-y) + 4(y-3x) = 7x(3x-y) - 4(3x-y) = (3x-y)(7x-4)$

30. a) $x^2 - 4y^2 = x^2 - (2y)^2 = (x+2y)(x-2y)$

b) $9x^2 - 16y^2 = (3x)^2 - (4y)^2 = (3x+4y)(3x-4y)$

c) $4(x+2y)^2 - 9(3x-y)^2 = (2(x+2y))^2 - (3(3x-y))^2$
$= [2(x+2y) + 3(3x-y)][2(x+2y) - 3(3x-y)]$
$= (11x+y)(-7x+7y) = 7(11x+y)(y-x)$

31. a) $x^2 + 2xy + y^2 = (x+y)^2$ b) $9x^2 - 24xy + 16y^2 = (3x-4y)^2$
c) $9 - 12(x + 2y) + 4(x + 2y)^2 = (3 - 2(x+2y))^2$

32. a) $27x^3 + 8y^3 = (3x)^3 + (2y)^3$
$= (3x+2y)((3x)^2 - (3x)(2y) + (2y)^2)$
$= (3x+2y)(9x^2 -6xy + 4y^2)$

b) $64x^3 - y^3 = (4x)^3 - y^3 = (4x-y)((4x)^2 + (4x)(y) + y^2)$

23

$$= (4x-y)(16x^2 + 4xy + y^2)$$

c) $8x^3 - 125(2x-1)^3 = (2x)^3 - (5(2x-1))^3$
$$= (2x - 5(2x-1))((2x)^2 + (2x)(5(2x-1)) + (5(2x-1))^2)$$
$$= (-8x+5)(4x^2 + 10x(2x-1)) + 25(2x-1)^2)$$
$$= (5-8x)(4x^2 + 20x^2 - 10x + 100x^2 - 100x + 25)$$
$$= (5-8x)(124x^2 - 110x + 25)$$

33. a) $\dfrac{x^2 - x - 6}{x^2 + x - 2} = \dfrac{(x-3)(x+2)}{(x+2)(x-1)} = \dfrac{x-3}{x-1}$

b) $\dfrac{x^2 - 5x + 4}{x^2 -2x -8} = \dfrac{(x-4)(x-1)}{(x-4)(x+2)} = \dfrac{x-1}{x+2}$

c) $\dfrac{2x^2 + 5xy - 3y^2}{4x^2 + 17xy + 15y^2} = \dfrac{(2x-y)(x+3y)}{(4x+5y)(x+3y)} = \dfrac{2x-y}{4x+5y}$

34. a) $\dfrac{4(x-2)^5}{3(x+5y)^4} \cdot \dfrac{9(x+5y)^3}{2(x-2)^2} = \dfrac{2(x-2)^3 \cdot 3}{(x+5y)} = \dfrac{6(x-2)^3}{(x+5y)}$

b) $\dfrac{x^2-2x-3}{x^2+6x+5} \div \dfrac{2x^2-3x-9}{2x^2+13x+15} = \dfrac{x^2-2x-3}{x^2+6x+5} \cdot \dfrac{2x^2+13x+15}{2x^2-3x-9}$

$$= \dfrac{(x-3)(x+1)}{(x+5)(x+1)} \cdot \dfrac{(2x+3)(x+5)}{(2x+3)(x-3)} = 1$$

35. a) $\dfrac{4}{x+3} + \dfrac{7}{2x-1} = \dfrac{4(2x-1) + 7(x+3)}{(x+3)(2x-1)} = \dfrac{15x+17}{(x+3)(2x-1)}$

b) $\dfrac{x+2}{x+1} - \dfrac{x-3}{x+5} = \dfrac{(x+2)(x+5) - (x-3)(x+1)}{(x+1)(x+5)}$

$$= \dfrac{(x^2+7x+10) - (x^2-2x-3)}{(x+1)(x+5)} = \dfrac{9x+13}{(x+1)(x+5)}$$

c) $\dfrac{2x}{(3x - 1)(x + 2)} + \dfrac{5x}{(x^2 + 3x + 2)}$

$$= \dfrac{2x}{(3x-1)(x+2)} + \dfrac{5x}{(x+2)(x+1)} = \dfrac{2x(x+1) + 5x(3x-1)}{(3x-1)(x+2)(x+1)}$$

$$= \dfrac{17x^2 - 3x}{(3x-1)(x+2)(x+1)} = \dfrac{x(17x-3)}{(3x-1)(x+2)(x+1)}$$

36. a) $x=15$, $3x+2 = 3(15)+2 = 32 \neq 17$. Not a solution.
 b) $x=2$, $x^2 + 3x -10 = (2)^2 + 3(2) - 10 = 4+6-10 = 0$. Solution.

c) x=3, Left side = $\frac{5x-1}{2(x+4)}$ = $\frac{5(3)-1}{2(3+4)}$ = $\frac{14}{14}$ = 1.

Right side = $\frac{2x+5}{3x+2}$ = $\frac{2(3)+5}{3(3)+2}$ = $\frac{11}{11}$ = 1 = Left side. Solution.

37. a) 2x - 1 = 7
    2x - 1 + 1 = 7 + 1
    2x = 8, x = 4.

    b) 5x + 2 = 12
    5x + 2 - 2 = 12 - 2
    5x = 10, x = 2.

    c) 2x + 4 = x - 3
    2x - x = -3 - 4
    x = -7.

    d) 9x - 3 = 3x + 15
    9x - 3x = 15 + 3
    6x = 18, x = 3.

38. a) 4x + 3 = 0. Linear equation in standard form.
    b) $3x^2$ - 4x + 1. Non-linear because of the x^2 term.
    c) 5(4x - 2) = 7x + 3. Simplify and rewrite with all the terms
       on the left. 20x - 10 = 7x + 3, 13x - 13 = 0.
       Linear equation with standard form 13x - 13 = 0.

39. a) (x - 1)(x + 4) = x^2 + x + 6. x^2 + 3x - 4 = x^2 + x + 6
       2x = 10, x = 5.

    b) $\frac{3x}{2}$ - $\frac{1}{2}$ = $\frac{x}{4}$ - 4, $\frac{3x}{2}$ - $\frac{x}{4}$ = -4 + $\frac{1}{2}$, $\frac{6x}{4}$ - $\frac{x}{4}$ = -$\frac{8}{2}$ + $\frac{1}{2}$

    $\frac{5x}{4}$ = -$\frac{7}{2}$, x = $(-\frac{7}{2})(\frac{4}{5})$, x= -$\frac{14}{5}$

40. V = LWH. When V=50, L=5, W=2, then 50=(5)(2)(H), H=5.

41. A = P+Prt, A-P = Prt, r = $\frac{A-P}{Pt}$. When A=8000, P=5000, and t=10,

    r = $\frac{8000-5000}{(5000)(10)}$ = $\frac{3000}{50000}$ = .06. The interest rate would be 6%.

42. v = u+at, at = v-u, t = $\frac{v-u}{a}$.

    When a=8, u=0, and v=90, t = $\frac{90-0}{8}$ = 11.25 seconds.

43. Student has 88 average on four tests test. Let x be the
    necessary grade on the fifth test to bring average up to 90.
    Average of scores on the five tests = $\frac{4(88)+x}{5}$ = $\frac{352+x}{5}$.

    Thus $\frac{352+x}{5}$ =90, 352+x =5(90), 352+x =450, x=98. Needs a 90.

44. (a) Let P be pressure at depth d. Then P=kd. When d=4, P=200.
        Thus 200=4k, k=50. Therefore P=50d.
    (b) When P=300, d = P/50 = 300/60 = 6. Pressure is 300 pounds
        at a depth of 6 feet.

25

45. $d = kt^2$. When t=4, d=16k. When t=6, d=36k.
    36k/16k = 9/4 = 2.25. Ball rolls 2.25 times as far.

46. f(x)=3x+4. (a) f(2)=3(2)+4=10. (b) f(5)=3(5)+4=19.
    (c) f(-3)=3(-3)+4=-5.

47. $f(x)=2x^2+4x-3$. (a) $f(2)=2(2)^2+4(2)-3=13$.
    (b) $f(0)=2(0)^2+4(0)-3=-3$. (c) $f(-3)=2(-3)^2+4(-3)-3=3$.
    (d) $f(3u)=2(3u)^2+4(3u)-3=18u^2+12u-3$.
    (e) $f(2a-1) = 2(2a-1)^2+4(2a-1)-3 = 8a^2-8a+2+8a-4-3 = 8a^2-5$.

48. (a) f(x)=4x+1.Domain:Set real numbers,$(-\infty,\infty)$. Range:$(-\infty,\infty)$.
    (b) $f(x)=x^2-6$.Domain: $(-\infty,\infty)$. Range: $x^2\geq0,x^2-6\geq-6$.Set $[-6,\infty)$.
    (c) $f(x)=1-2x^2$. Domain: $(-\infty,\infty)$.
        Range: $x^2\geq0$, $-x^2\leq0$, $-2x^2\leq0$, $1-2x^2\leq1$. Set $(-\infty,1]$.
    (d) $f(x)=\sqrt{4-x}+3$. Domain: $4-x\geq0$, $4\geq x$. Set $(-\infty,4]$.
        Range: $\sqrt{4-x}\geq0$, $\sqrt{4-x}+3\geq3$. Set $[3,\infty)$.

49. $R(x)=-2x^2+650x$. (a) $R(10)=-2(10)^2+650(10)=\$6300$.
    $R(15)=-2(15)^2+650(15)=\$9300$. $R(20)=-2(20)^2+650(20)=\$12200$.
    (b) $R(0)=-2(0)^2+650(0)=0$. Expect since sales on 0 items is \$0.

50. $s(t)=-16t^2+vt$. (a) When v=440, $s(t)=-16t^2+400t$.
    $s(2)=-16(2)^2+400(2)=736$ft. $s(4)=-16(4)^2+400(4)=1344$ft.
    $s(6)=-16(6)^2+400(6)=1824$ft.(b)Table below gives s=0 when t=25.
    Object returns to earth after 25 seconds.

time (secs)	0	5	10	15	20	25
altitude(ft)	0	1600	2400	2400	1600	0

(c)Above table indicates that the object reaches its highest
point somewhere between 10 and 15 second. Note that the
symmetry of the numbers makes us suspect that it is half way
between 10 and 15 at 12.5 seconds. We investigate heights at
intervals of .1 seconds starting from 12.2 below. The results
lead us to strongly suspect that the object reaches its
highest point after 12.5 seconds, supporting the conclusion
that it takes the same time to come down as to go up. (Note
the symmetry of the numbers again.)

time (secs)	12.2	12.3	12.4	12.5	12.6	12.7
altitude(ft)	2498.6	2499.4	2499.8	2500	2499.8	2499.4

We try to fine tune the results even more on our calculator,
investigating heights at intervals of .01 second about 12.5,
but find that more accuracy using a table is not possible (see
below). Such tables of values are extremely useful in
understanding concepts and have much, but limited
computational use. We need further techniques for finding such
"highest points". This motivates us to develop further
material in the course.

X	Y1
0	0
5	1600
10	2400
15	2400
20	1600
25	0
30	-2400

X	Y1
12.2	2498.6
12.3	2499.4
12.4	2499.8
12.5	2500
12.6	2499.8
12.7	2499.4
12.8	2498.6

X	Y1
12.45	2500
12.46	2500
12.47	2500
12.48	2500
12.49	2500
12.5	2500
12.51	2500

Heights every 5 sec   Heights ever .1 sec   Heights every .01 sec
Table rounds altitudes half way between 10 and 15 sec at 12.5 sec.

51. $n(t) = 181t^2 - 131t + 73$. t=0 corresponds to 1988.

(a)	1998	1999	2000	2001	2002	2003	2004	2005
t	10	11	12	13	14	15	16	17
N(t)	16863	20533	24565	28959	33715	38833	44313	50155

(b) Viruses are likely to be a permanent part of the computer scene. MS-DOS will probably not be around in the year 2005 so that these actual numbers are not likely to be realized for MS-DOS. However the author believes that the figures will reflect in a way the increase in viruses.

X	Y1
10	16863
11	20533
12	24565
13	28959
14	33715
15	38833
16	44313

X	Y1
11	20533
12	24565
13	28959
14	33715
15	38833
16	44313
17	50155

Review Chapter Test

1. (a) natural numbers, $\{3,15\}$
   (b) integers but not natural numbers, $\{-1,0\}$.

2. (a) $(\frac{1}{4} \cdot 6) \cdot 3 = \frac{1}{4} \cdot (6 \cdot 3)$, associative property of multiplication.
   (b) $x(y + z)z = xy + xz$, distributive property.

3. $(\sqrt{9.7} + 0.26^{-1}) \div 1.73 = 4.0235$, to four decimal places.

4. $5 \boxplus 16 \boxdiv 2 \boxed{^} 3 \boxtimes 4 = 5+(\frac{16}{2^3})x4 = 5+(2x4)=5+8=13$.

5. (a) $(\sqrt[3]{\frac{27}{8}})^2 = (\frac{3}{2})^2 = \frac{9}{4} = 2.25$.

   (b) $(-27)^{5/3} = ((-27)^{1/3})^5 = (-3)^5 = -243$

6. $\frac{(x^{-1}y^2)^3}{(x^{-3}y^4)^{-2}} = \frac{x^{(-1\cdot3)}y^{(2\cdot3)}}{x^{(-3\cdot-2)}y^{(4\cdot-2)}} = \frac{x^{-3}y^6}{x^6y^{-8}} = \frac{y^6y^8}{x^3x^6} = \frac{y^{(6+8)}}{x^{(3+6)}} = \frac{y^{14}}{x^9}$

7. $\sqrt{48x^5y^2} = \sqrt{(16)(3)x^4xy^2} = \sqrt{(4)^2(3)(x^2)^2xy^2} = 4x^2|y|\sqrt{3x}$

27

8. $\sqrt[7]{41.82^4} = 8.4432$

9. $(3x - 1)(4x + 2) = 12x^2 - 4x + 6x - 2 = 12x^2 + 2x - 2$

10. (a) $6x^2 + 7x - 3 = (3x - 1)(2x + 3)$
    (b) $8x^2 + 10xy - 3y^2 = (4x - y)(2x + 3y)$
    (c) $4x^2 - 81y^2 = (2x)^2 - (9y)^2 = (2x+9y)(2x-9y)$

11. $\dfrac{3}{x-1} + \dfrac{5}{2x+1} = \dfrac{3(2x+1) + 5(x-1)}{(x-1)(2x+1)} = \dfrac{11x-2}{(x-1)(2x+1)}$

12. $3x + 1 = 5x + 3$, $3x - 5x = 3 - 1$, $-2x = 2$, $2x = -2$, $x = -1$.

13. $f(x)=3x^2+5x-1$. (a) $f(2)=3(2)^2+5(2)-1=21$.
    (b) $f(0)=3(0)^2+5(0)-1=-1$. (c) $f(-1)=3(-1)^2+5(-1)-1=-3$.
    (d) $f(2a-3) = 3(2a-3)^2+5(2a-3)-1 = 12a^2-36a+27+10a-15-1$
       $= 12a^2-26a+11$.

14. (a) $f(x)=-x^2+3$. Domain:$(-\infty,\infty)$. Range:$x^2\geq0$,$-x^2\leq0$,$-x^2+3\leq3$. $(-\infty,3]$.

    (b) $f(x)=\sqrt{x+2}-1$. Domain: $x+2\geq0$, $x\geq-2$. $[-2,\infty)$.
       Range: $\sqrt{x+2}\geq0$, $\sqrt{x+2}-1\geq-1$. $[-1,\infty)$.

15. Amount $= 10000(1+ \dfrac{.045}{12})^{(5\times12)} = \$12,517.96$

16. Student has GPA of 2.95 over 105 hours. Let x be necessary GPA
    over the last 15 hours to graduate with 3.0. Thus
    $\dfrac{105(2.95)+15x}{120} = 3.0$, $105(2.95) + 15x = 120(3.0)$,
    $309.75+15x=360$, $15x=50.25$, $x=3.35$.

17. $W=k/d^2$. When $d=3956$, $W=185$. Thus $185=k/3956^2$. $k=185(3956^2)$.
    When $d=3961$, $W=185(3956^2)/(3961^2)=184.53$ pounds (to 2 decimal
    places).

18. $s(t)=-16t^2+vt$. (a) When $v=368$, $s(t)=-16t^2+368t$.
    $s(1)=-16(1)^2+368(1)=352$ft. $s(2)=-16(2)^2+368(2)=672$ft.
    $s(3)=-16(3)^2+375(3)=960$ft. (b) Table gives $s=0$ when $t=23$.
    Object returns to earth after 23 seconds.

Chapter 1

Section 1.1

1. A(2,-1),B(4,3),
   C(6,-2),D(-5,0)

   A,quadrant IV. B,quadrant I.
   C,quadrant IV. D, x-axis.

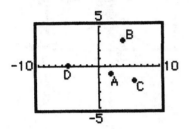

3. A(3,3), B(-3,-3),
   C(-5,-5), D(-5,5)

   A,quadrant I. B,quadrant III.
   C,quadrant III. D,quadrant II.

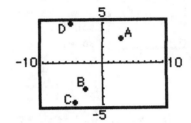

5. $y=3x-5$. (a) Point (2,1). $x=2$, $y=3(2)-5=6-5 = 1$. Yes.
   (b) Point (-3,-4). $x=-3$, $y=3(-3)-5=-9-5 =-14 \neq -4$. No.
   (c) Point (0,-5). $x=0$, $y=3(0)-5 = -5$. Yes.

7. $y=x^3$. (a) Point (-1,1). $x=-1$, $y=(-1)^3 = -1 \neq 1$. No.
   (b) Point (3,27). $x=3$, $y=(3)^3 = 27$. Yes.
   (c) Point (1.5,3.375). $x=1.5$, $y=(1.5)^3 = 3.375$. Yes.

9. Xmin=-10, Xmax=10, Xscl=2, Ymin=-4, Ymax=20, Yscl=4.

11. Xmin=-5, Xmax=10, Xscl=1, Ymin=-20, Ymax=10, Yscl=5.

13. Xmin=-12, Xmax=24, Xscl=4, Ymin=-300, Ymax=500, Yscl=100.

15. Xmin=-10, Xmax=10, Xscl=2, Ymin=-12, Ymax=10, Yscl=2.

17. Xmin=-6, Xmax=4, Xscl=1, Ymin=-6, Ymax=4, Yscl=1.

19. Xmin=-8, Xmax=4, Xscl=2, Ymin=-10, Ymax=10, Yscl=2.

21. $y=x-2$. When $x=0$,$y=-2$. y-int = -2. When $y=0$,$x=2$. x-int = 2.
    Xmin=-4, Xmax=4, Xscl=1, Ymin=-4, Ymax=4, Yscl=1.

23. $y=2x-6$. When $x=0$,$y=-6$. y-int = -6. When $y=0$,$x=3$. x-int = 3.
    Xmin=-4, Xmax=6, Xscl=1, Ymin=-10, Ymax=8, Yscl=2.

25. $y=x^2-4$.When $x=0$,$y=-4$.y-int=-4. When $y=0$,$x=-2$ or 2.x-int=-2,2.
    Xmin=-4, Xmax=4, Xscl=1, Ymin=-6, Ymax=4, Yscl=1.

27. $y=2x-16$.
When $x=0$, $y=-16$. y-int $=-16$.
When $y=0$, $x=8$. x-int $= 8$.

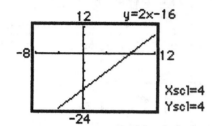

29. $y=5x-15$.
When $x=0$, $y=-15$. y-int $=-15$.
When $y=0$, $x=3$. x-int $= 3$.

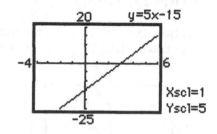

31. $y=-3x-27$.
When $x=0$, $y=-27$. y-int $=-27$.
When $y=0$, $x=-9$. x-int $= -9$.

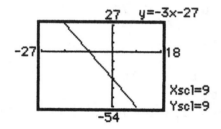

33.

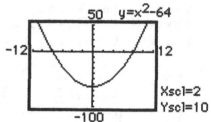

35.

37.

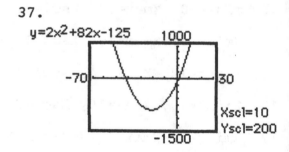

39.

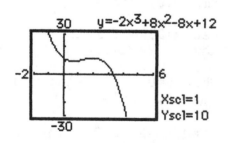

<u>Section 1.2</u>
Group Discussion: Consider line through points (x_1,y_1), (x_2,y_2).

For Highway Dept, $m=\dfrac{y_2-y_1}{\sqrt{(x_2-x_1)^2+(y_2-y_1)^2}}$ — more complicated to

work with than the math def. Let (x,y) be an arbitrary point on the line and m be the Highway Dept slope. Let us find the point/slope equation of the line. Get $m = \dfrac{y-y_1}{\sqrt{(x-x_1)^2+(y-y_1)^2}}$.

Square & cross multiply. $m^2((x-x_1)^2+(y-y_1)^2)=(y-y_1)^2$.

$(y-y_1)^2(1-m^2)=m^2((x-x_1)^2$. $y-y_1 = \dfrac{m(x-x_1)}{\sqrt{(1-m^2)}}$.

The math point/slope form, $y-y_1=m(x-x_1)$ is easier to work with. The Highway Dept def of slope is easier for the road user to appreciate - (dist road rises)/(dist motorist travels).

1. Run=2,rise=4. Slope=4/2=2.

3. Run=-5,rise=5. Slope=5/-5=-1.

5. Vertical line. Slope undef.

7. (1,1), (2,3). $m = \dfrac{y_2 - y_1}{x_2 - x_1} = \dfrac{3 - 1}{2 - 1} = 2.$

9. (1,-1), (1,-3). $x_1=x_2$. Line is vertical. Slope is undefined.

11. (4,0), (2,6). $m = \dfrac{y_2 - y_1}{x_2 - x_1} = \dfrac{6 - 0}{2 - 4} = \dfrac{6}{-2} = -3.$

13. (6,8), m=2. y-8=2(x-6), y=2x-4.

15. (0,4), m=4. y-4=4(x-0), y=4x+4.

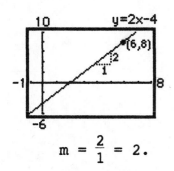

$m = \dfrac{2}{1} = 2.$

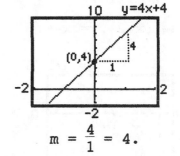

$m = \dfrac{4}{1} = 4.$

17. (0,0), m=2. y-0=2(x-0), y=2x.

19. m=2, b=5. y=2x+5.

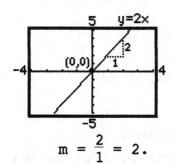

$m = \dfrac{2}{1} = 2.$

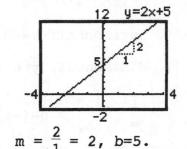

$m = \dfrac{2}{1} = 2,\ b=5.$

21. m=-1, b=-6. y=-x-6.

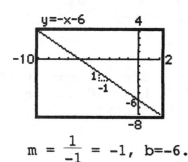

$$m = \frac{1}{-1} = -1, \; b=-6.$$

23. $m=\frac{3}{2}$, b=5. $y = \frac{3}{2}x + 5.$

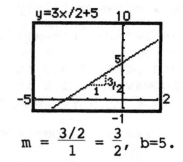

$$m = \frac{3/2}{1} = \frac{3}{2}, \; b=5.$$

25. (1,2),(2,3). $m = \frac{3-2}{2-1} = 1.$
$y-2=1(x-1), \; y=x+1.$

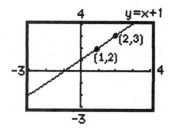

27. (2,4),(4,-2). $m= \frac{-2-4}{4-2} = -3.$
$y-4=-3(x-2), \; y=-3x+10.$

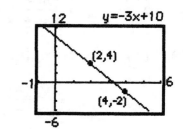

29. (2,-3), (4,0). $m= \frac{0+3}{4-2} = \frac{3}{2}.$
$y+3=\frac{3}{2}(x-2), y=\frac{3}{2}x-6.$

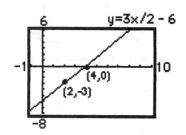

31. (1,4). Horizontal, y=4. Vertical, x=1.

33. (6,-3). Horizontal, y=-3. Vertical, x=6.

35. (-3,0). Horizontal, y=0 (x-axis). Vertical, x=-3.

37. Pts(0,0),(1,1) and (1,-1),(2,0).
$m_1 = \frac{1-0}{1-0} = 1.$ $m_2 = \frac{0-(-1)}{2-1} = 1.$ Lines parallel.
$L_1:y-0 = 1(x-0), \; y=x.$ $L_2:y+1 = 1(x-1), \; y=x-2.$ (Fig below)

39. Points $(1,-1)$, $(4,8)$ and $(6,3)$, $(9,2)$.

$m_1 = \dfrac{8-(-1)}{4-1} = \dfrac{9}{3} = 3$. $m_2 = \dfrac{2-3}{9-6} = -\dfrac{1}{3}$. Lines perpendicular.

$L_1: y+1 = 3(x-1)$, $y=3x-4$. $L_2: y-3 = -\dfrac{1}{3}(x-6)$, $y=-\dfrac{1}{3}x+5$.

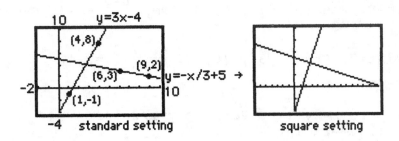

standard setting   square setting

41. Pts $(1,1),(3,3)$ and $(1,0),(2,-1)$.

$m_1 = \dfrac{3-1}{3-1} = 1$. $m_2 = \dfrac{-1-0}{2-1} = -1$. Lines perpendicular.

$L_1: y-1 = 1(x-1)$, $y=x$. $L_2: y-0 = -1(x-1)$, $y=-x+1$.

43. L:$y=2x-4$. P$(6,2)$. Parallel. $m=2$, $y-2=2(x-6)$. $y=2x-10$.

45. L:$2x-4y+8=0$. $y=\dfrac{1}{2}x+2$. P$(6,3)$. Parallel, $m=\dfrac{1}{2}$. $y-3=\dfrac{1}{2}(x-6)$. $y=\dfrac{1}{2}x$.

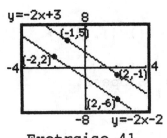

Exetrcise 41

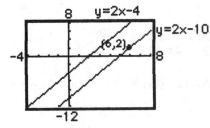

Exercise 43

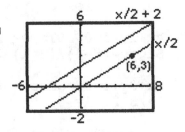

Exercise 45

47. $y=2x+8$. $m=2$, y-int=8, x-int=-4.

49. $4x-2y=-12$. $y=2x+6$, $m=2$, y-int=6, x-int=-3.

51. $3x+5y+9=0$. $y = -\dfrac{3}{5}x - \dfrac{9}{5}$. $m=-\dfrac{3}{5}$, y-int=$-\dfrac{9}{5}$, x-int=-3.

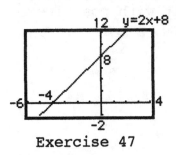

Exercise 47

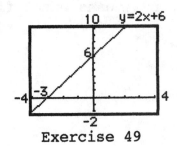

Exercise 49

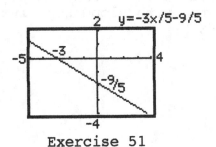

Exercise 51

53. x-int=-5, y-int=10. Pts (-5,0), (0,10). m= $\frac{10-0}{0+5}$ = 2. y=2x+10.

55. No x-int, y-int=15. Horizontal line, y=15.

57. x-int=-2, y-int=-8. Pts (-2,0), (0,-8). m= $\frac{-8-0}{0+2}$ = -4. y=-4x-8.

59. Point (1,3), y-int=7. Pts (1,3), (0,7). m= $\frac{7-3}{0-1}$ = -4.
    y-7=-4(x-0). y=-4x+7.

61. Point (5,3). Perp to x-2y+1=0. Perp to y=x/2-1/2.
    m=-2. y-3=-2(x-5), y=-2x+13.

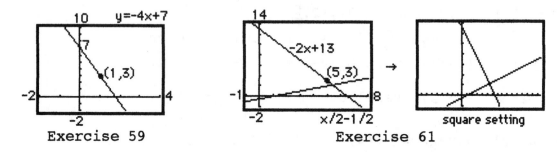

Exercise 59   Exercise 61

63. (-1,4) lies on y = mx+5. 4 = m(-1)+5, m=1.

65. Ax+2y-1=0, Slope=-A/2. Parallel to line thro (1,4),(3,8).
    Slope=(8-4)/(3-1)=2. -A/2=2. A=-4.

67. Ax+3y+4=0. Slope=-A/3. Parallel to 2x-y+1=0. Slope=2.
    -A/3=2. A=-6.

69. x+2y-3=0. Slope=-1/2. Slope of line perp = 2.
    Line of slope 2, thro (-1,6) is y-6=2(x+1), y=2x+8. x-int=-4.

Note for exercises 71 - 83 on Least Squares Lines Some calculators
will include a value for the so called coefficient of correlation
r in the window that gives the line. This coefficient lies in the
interval -1 to +1. The closer r is to -1 or +1 the better the fit.
The value r=-1 means exact fit (the line passes through all the
points) with slope negative, r=1 means exact fit with positive
slope.

71. Estimate might be y=4x+3. LSq Line y=3.9x+3.When x=5, y=22.5.

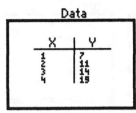
Data

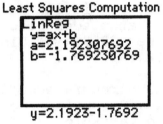
Least Squares Computation

y=3.9x+3

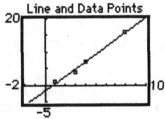
Line and Data Points

73. Estimate might be y=2x-2. LSq Line is y=2.1923x-1.7692.
    When x=7, y=13.5769.

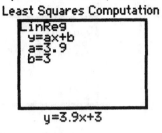
Data

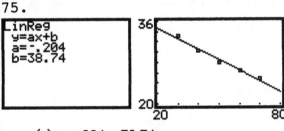
Least Squares Computation

y=2.1923-1.7692

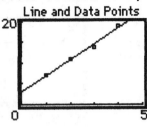
Line and Data Points

75.

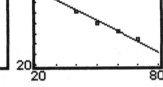

(a) y=-.204x+38.74
(b) f(55)=27.52, 27.52mpg at 55 mph

77.

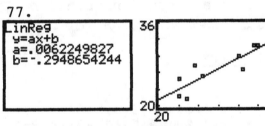

(a) y=.0062x-.2949 (to 4 dec places)
(b) f(500)=2.8, predicted GPA is 2.8.
(c) f(644)=3.7, probably had 644.

79.

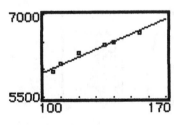

(a) y=13.7001x+4581.4966
    (to 4 dec places)
(b) f(165)=6842.0131.
    Predicted costs $6,840.

81.

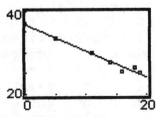

(a) y=-.6383x+36.9111 (to 4 dec places)
(b) In 2003, x=26. f(26)=18.40% smokers.

83.

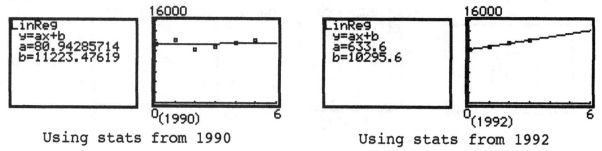

Using stats from 1990   Using stats from 1992

(a) Using stats from 1990, predicted tuitions are 2000:$12,033
2005:12,438   2010:$12,842. (b) Prediction for 2000 is below that
of 1995. Unsatisfactory. Big change took place in 1992. Let x=0
in '92. Tuition predictions then 2000(x=8):$15,364,
2005(x=13):$18,532, 2010(x=18):$21,700.

Project: When diameter is zero, circumference is zero. Thus line
    should go through the origin. Circumference intercept should be
    zero. Let C be circumference variable and d be diameter
    variable. Equation should be C=πd. Thus the slope should give
    us an approximation for π.

Section 1.3

1. f(x)=3.72x-4.3. (a) f(-3.6)= 3.72(-3.6)-4.3= -17.692.
    (b) f(4.5)=3.72(4.5)-4.3= 12.44.
    (c) f(7.89)=3.72(7.89)-4.3=25.0508.
    See also below for some calculators. (section R.7)

3. f(x)=3.5x+9.73. We see from table below that f(31)=118.23 and
    f(32)=121.73. x=32 is the first integer value of x when f
    becomes larger than 120.

X	Y₁
29	111.23
30	114.73
31	118.23
32	121.73
33	125.23
34	128.73
35	132.23

5. a) Change of 1 in x causes a constant change of -4 in f. f is
      linear. Rate of change=-4, m=-4. f(0)=3, y-int=3. f(x)=-4x+3.
      b) f(-2)-f(-3)=34. f(-1)-f(-2)=16. Change of 1 in x does not
      give constant change in f. f is not linear.
      c) Change of .5 in x causes a constant change of 1.75 in f. f
      is linear. Change of 1 in x would cause change of 3.5 in f.
      Rate of change=3.5, m=3.5. f(0)=1. y-int=1. f(x)=3.5x+1.
      d) Change of .5 in x causes a constant change of -1.15 in f. f
      is linear. Change of 1 in x would cause change of -2.3 in f.
      Rate of change=-2.3,m=-2.3. f(0)=1.7.y-int=1.7.f(x)=-2.3x+1.7.

7. (a) Initial cost of clubs = $300. Upkeep cost = $6 per month.
   Total cost function, C(t) = 6t+300, where t is time in months.

(b) Slope is monthly upkeep and y-intercept is initial cost of the clubs.

9. (a) Initial cost of dog = $470. Cost of feeding = $20 per week. Total cost function, C(t) = 20t+470, where t is time in weeks. (b) Slope is monthly feeding cost, y-intercept is initial cost of the Labrador.

11. (a) Initial cost of TV = $250. Cost of cable = $27 per month. Total cost function, C(t)=27t + 250, where t is time in months. (b) Slope is monthly cost of cable and y-intercept is initial cost of the TV.

13. (a) Amount of spent fuel in 1997= 34,000 tons. Refuse collecting at rate of 2,000 tons per year. Accumulation function A(t)=2000t+34000, where t is time in years measured from 1997. (b) In 2033, t=36. A(36)=106,000 tons.

15. (a)Markers every mile from Tampa. Starts at distance 30 miles. Then travels 65 miles per hour. In time t hrs she will be distance s(t)=65t+30 miles from Tampa. (b) Slope is speed and s-intercept is starting distance from Tampa.

17. (a) 1 meter = 1.0936 yds. f(m)=1.0936m converts m meters to yards. (b) f(100)=109.36 yds. 100 meter race is 109.36 yds. f(1500)=1640.4 yds. 1500 meter race is 1640.4 yds. (1 mile is 1,760 yds). Both races are shorter in meters.

19. (a) Cost of 50 units = $190.50. Cost of 65 units = $226.80. Let cost-volume equation be y=mx+b. Pts (50,190.5),(65,226.8). m = (226.8-190.5)/(65-50)=2.42.
    Point-slope gives y-190.5=2.42(x-50), y=2.42x+69.5.
    Fixed cost=$69.50.
    (b)

#items	70	71	72	73	74	75	76
cost	238.90	241.32	243.74	246.16	248.58	251.00	253.42

(c) y(177)=497.84, y(178)=500.26. Cost reaches $500 with the 178th item.

21. y-int=46. Graph increases from 46 to 58 in 20 years. Rise=12,run=20. Slope=12/20=3/5. W(t)=3t/5+46, where W is % women who work at time t years measured from 1975. t=25 in 2000. W(25)=61%. t=35 in 2010. W(35)=67%

23. y-int=42. Graph decreases from 42 to 23 in 30 years. Rise=-19,run=30. Slope=-19/30. S(t)=-19t/30+42, where S is % those who smoke at time t years measured from 1965. t=35 in 2000. S(35)=20%. t=45 in 2010. S(45)=14%.

25. Femle: y-int=74. Graph increases from 74 to 79 in 25 years. Rise=5,run=25. Slope=5/25=1/5. E(t)=t/5+74, where E female life expectancy at time t years measured from 1970. t=30 in 2000. E(30)=80. t=40 in 2010. E(40)=82.

Male: y-int=68. Graph increases from 68 to 73 in 25 years.
Rise=5,run=25. Slope=5/25=1/5. E(t)=t/5+68, where E
male life expectancy at time t years measured from 1970.
t=30 in 2000. E(30)=74. t=40 in 2010. E(40)=76.

Section 1.4

1. $y = x^2+6$, [-3,3].

x	-3	-2	-1	0	1	2	3
y	15	10	7	0	7	10	15

Graph below.

3. $y = x^2+2x-3$, [-4,2].

x	-4	-3	-2	-1	0	1	2
y	5	0	-3	-4	-3	0	5

5. $y = -3x^2+12x+7$, [-2,6].

x	-2	-1	0	1	2	3	4	5	6
y	-29	-8	7	16	19	16	7	-8	-29

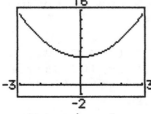
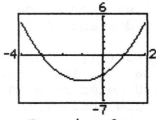
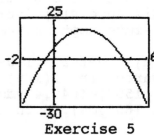

Exercise 1   Exercise 3   Exercise 5

7. $y = x^2+8x+15$. a=1,b=8,c=15. $x= \dfrac{-b}{2a} = \dfrac{-8}{2(1)} = -4$.

$y = (-4)^2 + 8(-4) + 15 = -1$. Vertex (-4,-1). Axis x=-4.
y-intercept = 15. a>0, opens up.

9. $y = 6x^2+36x-5$. a=6,b=36,c=-5. $x= \dfrac{-b}{2a} = \dfrac{-36}{2(6)} = -3$.

$y = 6(-3)^2 + 36(-3) - 5 = -59$. Vertex (-3,-59). Axis x=-3.
y-intercept = -5. a>0, opens up.

11. $y = 2x^2+5$. a=2,b=0,c=5. $x= \dfrac{-b}{2a} = \dfrac{0}{2(2)} = 0$.

$y = 2(0)^2 + 5 = 5$. Vertex (0,5). Axis x=0, y-axis.
y-intercept = 5. a>0, opens up.

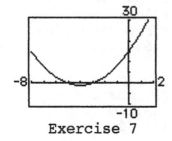
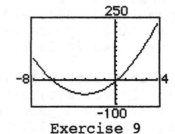
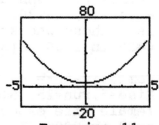

Exercise 7   Exercise 9   Exercise 11

13. $y = -2x^2+10x-1$. $a=-2,b=10,c=-1$. $x=\dfrac{-b}{2a}=\dfrac{-10}{2(-2)}=2.5$.

$y = -2(2.5)^2+10(2.5)-1 = 11.5$. Vertex $(2.5,11.5)$.
Axis $x=2.5$. y-intercept $= -1$. $a<0$, opens down.

15. 15. $y = 4x^2 - 14x - 3$. $a=4,b=-14,c=-3$. $x=\dfrac{-b}{2a}=\dfrac{14}{2(4)}=1.75$.

$y = 4(1.75)^2-14(1.75)-3 = -15.25$. Vertex $(1.75, -15.25)$.
Axis $x=1.75$. y-intercept $= -3$. $a>4$, opens up.

17. $y = 3x^2 - 7$. $a=3,b=0,c=-7$. $x=\dfrac{-b}{2a}=\dfrac{0}{2(3)}=0$.

$y = 3(0)^2-7 = -7$. Vertex $(0,-7)$.
Axis $x=0$, y-axis. y-intercept $= -7$. $a>0$, opens up.

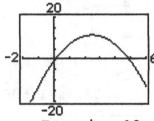

Exercise 13

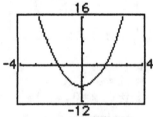

Exercise 15

Exercise 17

19. $y = 2.1x^2-6.4x-2.3$.

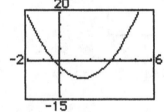

Vertex $(1.5238, -7.1762)$

21. $y = 6.2x^2+36.4x-5.3$.

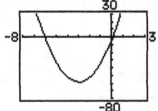

Vertex $(-2.9355, -58.7258)$

23. $y = -.1x^2+5.34x-51$.

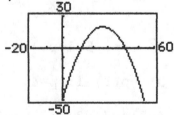

Vertex $(26.7000, 20.2890)$

25. $y = x^2-27.3x+108$.

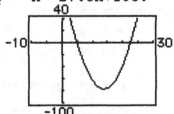

Vertex $(13.6500, -78.3225)$

27. $y=x^2+2x-15$. $y=(x+5)(x-3)$.
   x-intercepts $x=-5$, $x=3$.

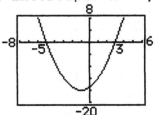

29. $y=-x^2+3x+4$. $y=(x+1)(-x+4)$.
   x-intercepts $x=-1$, $x=4$.

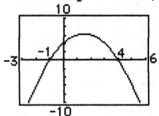

31. $y=2x^2-9x+10$. $y=(2x-5)(x-2)$.
   x-intercepts $x=5/2$, $x=2$.

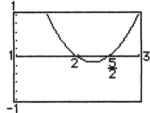

33. $y=x^2-5x+4$. $y=(x-1)(x-4)$.
   x-intercepts $x=1$, $x=4$.

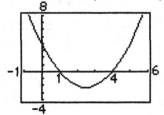

35. $y = 2x^2-13x+5$.
   x-int: $x=0.4105$, $x=6.0895$.

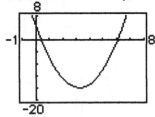

37. $y = 6x^2-7x-3$.
   x-int: $x=-.3333$, $x=1.5000$.

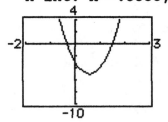

39. $y = -x^2+6.3x+5.4$.   x-int: $x=-.7644$, $x=7.0644$.

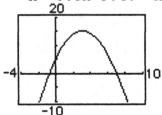

41. Point $(1,5)$ lies on $y=2x^2-x+p$. Thus $5=2(1)^2-1+p$. $5=1+p$. $p=4$.

43. Point $(3,2)$ lies on $y=-x^2+px-1$. Thus $2=-(3)^2+p(3)-1$. $p=4$.

45. Point $(-2,p)$ lies on $y=2x^2+7x-1$. Thus $p=2(-2)^2+7(-2)-1$. $p=-7$.

47. Point $(-1,-6)$ lies on $y=2x^2+px-1$. $-6=2(-1)^2+p(-1)-1$. $p=7$.

49. (a) x-intercepts, 2 and 5. Let $y=k(x-2)(x-5)$.
   y-int$=10$. Thus $10=k(-2)(-5)$. $k=1$. $y=(x-2)(x-5)=x^2-7x+10$.
   (b) $a=1,b=-7,c=10$. Vertex at $x=-b/2a=-(-7)/2=3.5$.
   Corresponding $y=(3.5)^2-7(3.5)+10=-2.25$. Vertex $(3.5,-2.25)$.

40

51. (a) x-intercepts, -5 and 2. Let y=k(x+5)(x-2).
    y-int=-30. Thus -30=k(5)(-2). k=3. y=3(x+5)(x-2)=3x^2+9x-30.
    (b) a=3,b=9,c=-30. Vertex at x=-b/2a=-(9)/6=-1.5.
    Corresp y=3(-1.5)^2+9(-1.5)-30=-36.75. Vertex (-1.5,-36.75).

53. (a) x-intercepts, -2 and 3. Let y=k(x+2)(x-3).
    y-int=12. Thus 12=k(2)(-3). k=-2. y=-2(x+2)(x-3)=-2x^2+2x+12.
    (b) a=-2,b=2,c=12. Vertex at x=-b/2a=-(-2)/(-4)=0.5.
    Corresponding y=-2(0.5)^2+2(0.5)+12=10.5. Vertex (0.5,12.5).

Section 1.5

1. s(t)=-16t^2+384t.With calculator, A(12, 2304).See below.
   Max height is 2304 ft. Reached after 12 seconds.

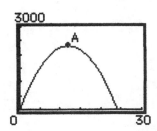

3. s(t)=-16t^2+vt. Max height after 19.5 secs. Thus vertex of the
   parabola is at x=19.5. a=-16, b=v.
   Vertex at $t=\dfrac{-b}{2a}=\dfrac{-v}{2(-16)}$ = 19.5. Thus v=32(19.5)=624.

5. R(p)=553p-p^2. Algebra: Function is a parabola opening down.
   Largest value of R will be at the vertex. a=1, b=553, c=0.
   Vertex at $p=\dfrac{-553}{2(-1)}$ =276.5.
   Table: Largest Y value of 76452 when x=276 and x=277.
   Graph: Maximum value of R at point (276.50003,76452.25).
   Charge $276.50 for the television for largest revenue.

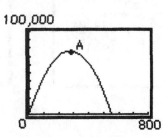

X	Y1
100	45300
150	60450
200	70600
250	75750
300	75900
350	71050
400	61200

Step size 50. Largest
Y between 250 & 300.

X	Y1
274	76446
275	76450
276	76452
277	76452
278	76450
279	76446
280	76440

Step size 1. Largest Y
between 276 & 277.

A(276.50003,76452.25)

7. (a) R(x)=(-$\dfrac{1}{38.32}$)x^2 + 90.45x. Algebra: Parabola opening down.
   Largest value of R will be at the vertex. a=-$\dfrac{1}{38.32}$, b=90.45,

c=0. Vertex x = $\dfrac{-90.45}{2(-1/38.32)}$ = (90.45)(38.32)/2 = 1733.022.
Maximum revenue when x=1,733 units.

(b) Graph below.   (c)

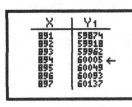

100,000

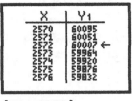

A(1733.0216,78375.92)

R≥6000 for x in [894,2572]

9. f(x)=.8x^2+x-3, g(x)=1.3x^2-3x-8. A(-1.10,-3.13), B(9.10,72.33).

11. S(p)=1.4p^2-10p+150, D(p)=0.9p^2-60p+1500. A(22.11,613.35).
Equilibrium price = \$22.11. Figure below.

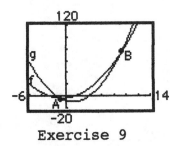

Exercise 9

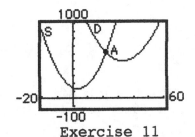

Exercise 11

13. (a) f(x)=122.1298701x^2-171.6056277x+3109.145455. Figure below.
Year 20002 corresponds to x=12. f(12)=18636.57922.
Predicted high for the year 20002 is approximately 18637.
(The Dow is more closely described by a quadratic function
than an exponential function.) (b) Plot f, y=15000 and use the
intersect finder. Intersection is (10.594771,15000). Year 2000.

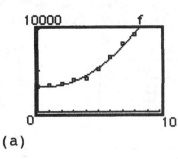

(a)

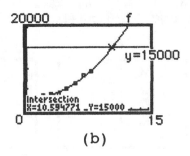

(b)

15. (a)$R(x)=-.02x^2+145x$. (b)$R(2400)=\$232800$. (c)$R(3500)=\$262500$.

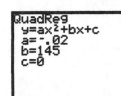

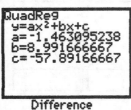

17. (a) $N(t)=6t^2+72t+9000$. (b) $t=14$ in year 2000. $N(2000)=11184$.
Number cars is 11184 hundreds, i.e. 1118400.

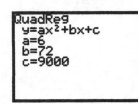

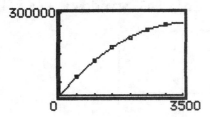

19.(a) $E(x)=-.6125x^2+7.1149x+30.3542$, to 4 dec places.
$I(x)=.8548x^2-1.8976x+88.2583$. $D(x)=-1.4631x^2+8.9917x-57.8917$.
(b) $x=13$ in year 2000. $E(13)=19.3$, $I(13)=208.1$, $D(13)=-188.3$.

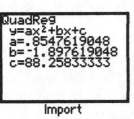

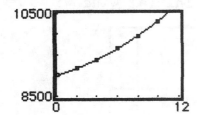

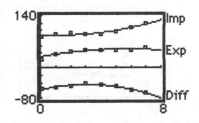

Export   Import   Difference

Section 1.6

1. $(1,3),(2,7)$. $d=\sqrt{(2-1)^2+(7-3)^2}=\sqrt{1^2+4^2}=\sqrt{17}=4.1231$

3. $(1,1),(3,-6)$. $d=\sqrt{(3-1)^2+(-6-1)^2}=\sqrt{2^2+(-7)^2}=\sqrt{53}=7.2801$

5. $(-3,0),(-2,1)$. $d=\sqrt{(-2+3)^2+(1-0)^2}=\sqrt{1^2+1^2}=\sqrt{2}=1.4142$

7. $(1,2),(3,6)$. Midpoint, $(\frac{1+3}{2},\frac{2+6}{2})=(\frac{4}{2},\frac{8}{2})=(2,4)$

9. $(3,2),(6,8)$. Midpoint, $(\frac{3+6}{2},\frac{2+8}{2})=(\frac{9}{2},\frac{10}{2})=(4.5,5)$

11. $(0,0),(5,-2)$. Midpoint, $(\frac{0+5}{2},\frac{0-2}{2})=(2.5,-1)$

13. $(x+1)^2 + (y-2)^2 = 4$. Center=$(-1,2)$, radius = $\sqrt{4}$ = 2.
$y = \pm\sqrt{4-(x+1)^2} + 2$. Initial window: $-4\le x\le4, -2\le y\le5$.

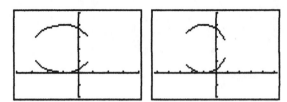

15. $(x+5)^2 + (y-2)^2 = 17$. Center=$(-5,2)$, radius=$\sqrt{17}$.
$y = \pm\sqrt{17-(x+5)^2} + 2$. Initial window: $-10\le x\le1, -3\le y\le7$.

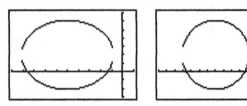

17. $(x-1)^2 + (y-3)^2 = 8$. Center=$(1,3)$, radius = $\sqrt{8}$.
$y = \pm\sqrt{8-(x-1)^2} + 3$. Window: $-3\le x\le5$, $-2\le y\le7$.

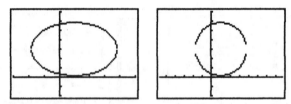

19. Center $(3,1)$, radius 4. $(x-3)^2+(y-1)^2 = 4^2$, $(x-3)^2+(y-1)^2 = 16$

21. Center $(0,4)$, radius 5. $(x-0)^2+(y-4)^2 = 5^2$, $x^2+(y-4)^2 = 25$

23. Center $(1.5,6)$, radius 3.2. $(x-1.5)^2+(y-6)^2 = 3.2^2$,
$(x-1.5)^2+(y-6)^2 = 10.24$

25. $x^2+4x+y^2-6y = 23$. $(x^2+4x+2^2) + (y^2-6y+(-3)^2) = 23+2^2+(-3)^2$.
$(x+2)^2 + (y-3)^2 = 36$. Center $(-2,3)$, radius = $\sqrt{36}$ = 6.

27. $x^2+y^2-6x-2y= -9$. $(x^2-6x+(-3)^2)+(y^2-2y+(-1)^2) = -9+(-3)^2+(-1)^2$.
$(x-3)^2 + (y-1)^2 = 1$. Center $(3,1)$, radius = $\sqrt{1}$ = 1.

29. $4x^2-8x+4y^2-24y+39 = 0$. $4(x^2-2x) + 4(y^2-6y) = -39$.
$4(x^2-2x+(-1)^2) + 4(y^2-6y+(-3)^2) = -39 + 4(-1)^2 + 4(-3)^2$.
$4(x-1)^2 + 4(y-3)^2 = 1$. $(x-1)^2 + (y-3)^2 = 1/4$.
Center $(1,3)$, radius = $\sqrt{1/4}$ = 1/2.

44

31. $9x^2+9y^2-72y+108 = 0$. $9x^2 + 9(y^2-8y) = -108$.
    $9x^2 + 9(y^2-8y+(-4)^2) = -108 + 9(-4)^2$.
    $9x^2 + 9(y-4)^2 = 36$. $x^2 + (y-4)^2 = 4$.
    Center $(0,4)$, radius $= \sqrt{4} = 2$.

33. $x^2-2x+y^2-10y-10 = 0$.
    $(x^2-2x+(-1)^2)+(y^2-10y+(-5)^2) = 10+(-1)^2+(-5)^2$.
    $(x-1)^2 + (y-5)^2 = 36$. Circle, center $(1,5)$, radius $= \sqrt{36} = 6$.

35. $x^2-2x+y^2+10y+33 = 0$. $(x^2-2x+(-1)^2)+(y^2+10y+5^2) = -33+(-1)^2+5^2$.
    $(x-1)^2 + (y+5)^2 = -7$. $r^2 = -7 < 0$, not a circle.

37. $x^2-6x+y^2+7 = 0$. $(x^2-6x+(-3)^2)+y^2 = -7+(-3)^2$.
    $(x-3)^2 + y^2 = 2$. Circle, center $(3,0)$, radius $= \sqrt{2} = 1.4142$.

39. $x^2-6x-4y^2-8y+9 = 0$. Not a circle; coefficient of x^2 not same
    as coefficient of y^2, thus cannot be written in the form
    $x^2 + y^2 + ax + by + c = 0$.

41. Both circles have center $(20,20)$. Radius of larger circle is
    10, and smaller circle is 5. Equations of circles are
    $(x-20)^2+(y-20)^2 = 100$, and $(x-20)^2+(y-20)^2 = 25$.

43. Center $(10,10)$, radius 10. Circle $(x-10)^2+(y-10)^2 = 100$.

45. Top halves of circles with centers $(10,10),(30,10),(50,10)$,
    $(70,10)$. Radius of each circle is 10. Equations of full
    circles are $(x-a)^2+(y-10)^2 = 10$, for a $= 10,30,50,70$.
    Equations top halves, $y = \sqrt{100-(x-a)^2}+10$, for a $= 10,30,50,70$.
47. Bottom halves of circles $(x-a)^2+(y-20)^2 = 10$, for a $= 10,70$.
    Equations, $y = -\sqrt{100-(x-a)^2} + 20$, for a $= 10,70$.
    Top halves of circles $(x-a)^2+(y-20)^2 = 10$, for a $= 30,50$.
    Equations, $y = \sqrt{100-(x-a)^2} + 20$, for a $= 30,50$.

49. $A(1,1),B(2,4),C(3,1)$.D is mid-pt of AB. $D(\frac{1+2}{2},\frac{1+4}{2})$, $(1.5,2.5)$.
    Let F be mid-pt of BC. $F(\frac{2+3}{2},\frac{4+1}{2})$, $(2.5,2.5)$.
    Dist(DF)=dist(FE)=1. Thus x-coord of E is 3.5.
    y-coord of E = y-coord of D = 2.5. Thus $E(3.5,2.5)$.

51. $A(1,1)$. AB=BC=3.5. $B(4.5,1)$. $C(4.5,4.5)$.
    D is mid-pt of AC. $D(\frac{1+4.5}{2},\frac{1+4.5}{2})$, $(2.75,2.75)$.

53. $A(1.8,2.8)$, $B(5.6,2.8)$. Center C is midpt AB.
    $C(\frac{1.8+5.6}{2},\frac{2.8+2.8}{2})$, $(3.7,2.8)$. Radius=AB/2=(5.6-1.8)/2=1.9.

Equation of circle is $(x-3.7)^2+(y-2.8)^2=(1.9)^2$.

55. Face: Circle center$(4,4)$, radius 2. $(x-4)^2+(y-4)^2=4$.
    Mouth: Circle center $(4,3)$, radius 0.5. $(x-4)^2+(y-3)^2=0.25$.
    Eyes: Circles centers $(3,5)$, $(5,5)$, radius 0.2.
    $(x-3)^2+(y-5)^2=0.04$ and $(x-5)^2+(y-5)^2=0.04$.
    Nose: Line $x=4$ from $(4,4)$ to $(4,5)$.

57. Points A$(1,5)$, B$(1,9)$, C$(-2,-5)$.
    $AB^2 = (1-1)^2 + (9-5)^2 = 0^2+(4)^2 = 16$.
    $AC^2 = (-2-1)^2 + (-5-5)^2 = (-3)^2+(-10)^2 = 109$.
    $BC^2 = (-2-1)^2 + (-5-9)^2 = (-3)^2+(-14)^2 = 205$.
    ABC not a right triangle.

59. Points A$(3,-1)$, B$(5,4)$, C$(-2,1)$, D$(0,6)$.
    $AB^2 = (5-3)^2 + (4+1)^2 = 2^2+5^2 = 29$.
    $AC^2 = (-2-3)^2 + (1+1)^2 = (-5)^2+2^2 = 29$.
    $AD^2 = (0-3)^2 + (6+1)^2 = (-3)^2+7^2 = 58$.
    $BC^2 = (-2-5)^2 + (1-4)^2 = (-7)^2+(-3)^2 = 58$.
    $BD^2 = (0-5)^2 + (6-4)^2 = (-5)^2+2^2 = 29$.
    $CD^2 = (0+2)^2 + (6-1)^2 = 2^2+5^2 = 29$.
    Have AB=AC=BD=CD. Square ABDC with diagonals AD=BC.

61. All points on x-axis distance 13 from point $(-4,12)$.
    Let such point be $(x,0)$. Then
    $(x+4)^2+(0-12)^2 = 13^2$. $x^2+8x+16+144 = 169$.
    $x^2+8x-9 = 0$. $(x-1)(x+9) = 0$. $x = 1$ or -9.
    Points $(1,0)$, $(-9,0)$.

63. Distance between points $(-3,4)$ and $(9,y)$ is 13.
    $(9+3)^2+(y-4)^2 = 13^2$. $144+y^2-8y+16 = 169$. $y^2-8y-9 = 0$.
    $(y-9)(y+1) = 0$. $y = 9$ or -1.

65. P$(1,2)$, Q$(5,6)$. R is three-fourths of the way from P to Q.
    Let R be (x,y). Then $(x-1)=3(5-x)$, $(y-2)=3(6-y)$.
    $x-1 = 15-3x$, $y-2 = 18-3y$. $4x=16$, $4y=20$. $x=4$, $y=5$.
    Point is $(4,5)$.

67. Circle $(x-3)^2+(y-1)^2 = 25$. At x-intercept, $y=0$. Thus
    $(x-3)^2+(0-1)^2 = 25$. $x^2-6x+9+1=25$, $x^2-6x-15=0$. $a=1,b=-6,c=-15$.
    $$x = \frac{6\pm\sqrt{(-6)^2 - 4(1)(-15)}}{2(1)} = \frac{6\pm\sqrt{96}}{2} = 7.8990, -1.8990.$$
    At y-intercept, $x=0$. Thus
    $(0-3)^2+(y-1)^2 =25$. $9+y^2-2y+1-25=0$, $y^2-2y-15=0$. $a=1,b=-2,c=-68$.
    $$y = \frac{2\pm\sqrt{(-2)^2 - 4(1)(-15)}}{2(1)} = \frac{2\pm\sqrt{64}}{2} = 5, -3.$$

69. Circle with points A(3,1) and B(3,7) as end points of a diameter. Center C is mid-point of AB. Center $(3,\frac{1+7}{2})$ = (3,4). Radius = 3. Equ, $(x-3)^2+(y-4)^2 = 9$. See figure below.

71. Circle with center (5,2), touching x-axis at point (5,0). Radius = 2. Equation, $(x-5)^2+(y-2)^2 = 4$. See figure below.

73. Circle touching x-axis at x=2 and y-axis at y=2. Center =(2,2), radius=2. Equation, $(x-2)^2+(y-2)^2=4$.

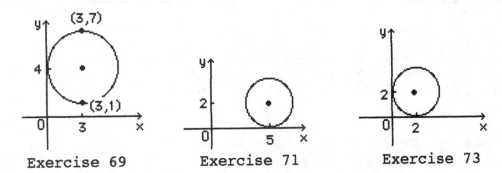

Exercise 69   Exercise 71   Exercise 73

75. Circumference and area of circle $x^2+4x+y^2-14y-11 = 0$.
$(x+2)^2+(y-7)^2 = 11+4+49 = 64$. r=8.
Circumference = $2\pi r$ = 50.2655. Area = πr^2 = 201.0619.

77. A circle that touches the lines x=5, y=4 and y=-2. Let center have y-coordinate 1, half way between y=4 and y=-2. Let radius be 3; it will touch y=4 and y=-2. Let center have x-coordinate 2; it will then touch x=5.
Equation, $(x-2)^2+(y-1)^2= 9$.See figure below.
Could also have $(x-8)^2+(y-1)^2= 9$.

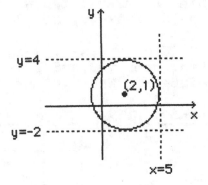

Chapter 1 Review Exercises

1. (a) A(3,5), B(-2,4),
      C(-6,-4), D(8,-1)

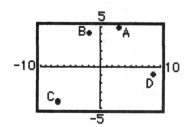

(b) P(-5,-4), Q(7,-3),
      R(-8,2), S(4,3)

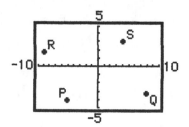

2. (a) y=2x-1.
   Point (-1,-3). x=-1, y=2(-1)-1 = -2-1 = -3. Yes.
   Point (0,-2). x=0, y=2(0)-1 = -1 ≠ -2. No.
   Point (3,5). x=3, y=2(3)-1 = 6-1 = 5. Yes.

   (b) y=3x^2-2x+1.
   Point (-2,17). x=-2, y=3(-2)^2-2(-2)+1 = 17. Yes,
   Point (1,2). x=1, y=3(1)^2-2(1)+1 = 2. Yes.
   Point (6,98). x=6, y=3(6)^2-2(6)+1 = 97 ≠ 98. No.

   (c) $y = \dfrac{1}{x-6} + 3\sqrt{x-2}$.

   Point(6,0). x=6. $y = \dfrac{1}{6-6} + 3\sqrt{6-2}$, not defined. No.

   Point $(3,\dfrac{11}{3})$. x=3. $y = \dfrac{1}{3-6} + 3\sqrt{3-2} = \dfrac{1}{3-6} + 3 = 2\dfrac{2}{3} = \dfrac{8}{3} \neq \dfrac{11}{3}$. No.

   Point $(11,\dfrac{46}{5})$. x=11. $y = \dfrac{1}{11-6} + 3\sqrt{11-2} = \dfrac{1}{5} + 9 = \dfrac{46}{5}$. Yes.

3. (a) Xmin=-5, Xmax=5,Xscl=1,Ymin=5,Ymax=25,Yscl=5.
   (b) Xmin=-10, Xmax=40,Xscl=10,Ymin=-8,Ymax=12,Yscl=2.
   (c) Xmin=-25, Xmax=50,Xscl=5,Ymin=-80,Ymax=40,Yscl=20.

4. (a) y=x-7. When x=0,y=-7.y-int=-7. When y=0,x=7. x-int = 7.
   (b) y=-2x+5. When x=0,y=5. y-int=5. When y=0,x=5/2. x-int=2.5.
   (c) y=12x-36. When x=0, y=-36.y-int=-36. When y=0,x=3. x-int=3.

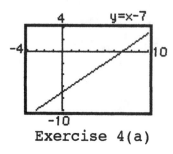

Exercise 4(a)

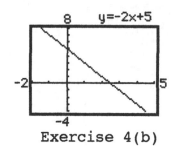

Exercise 4(b)

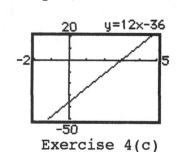

Exercise 4(c)

Chapter 1 Review Exercises

5. Slopes through points:

   (a) $(2,6),(7,1)$. $m = \dfrac{y_2 - y_1}{x_2 - x_1} = \dfrac{1 - 6}{7 - 2} = -1$.

   (b) $(-3,6),(7,-3)$. $m = \dfrac{y_2 - y_1}{x_2 - x_1} = \dfrac{-3 - 6}{7 + 3} = -\dfrac{9}{10}$.

   (c) $(-3,-8),(-1,-2)$. $m = \dfrac{y_2 - y_1}{x_2 - x_1} = \dfrac{-2 + 8}{-1 + 3} = 3$.

   (d) $(2,8),(2,-9)$. $x_1=x_2$. Line is vertical. Slope is undefined.

6. (a) $(3,5)$, $m=4$. $(y-5)=4(x-3)$, $y=4x-7$. Figure below.

   (b) $(-2,7)$, $m=-2$. $(y-7)=-2(x+2)$, $y=-2x+3$.

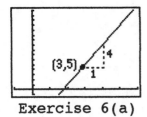

Exercise 6(a)

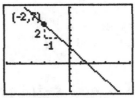

Exercise 6(b)

   (c) $(-6,-2)$, $m=0$. $(y+2)=0(x+6)$, $y=-2$. Horiz line. Fig below.

   (d) $(5,-3)$, slope not exist. Vertical line through $x=5$. $x=5$.

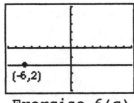

Exercise 6(c)

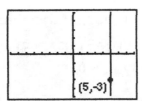

Exercise 6(d)

7. Slope of lines parallel to lines through:

   (a) $(-1,5),(7,17)$. $m = \dfrac{y_2 - y_1}{x_2 - x_1} = \dfrac{17 - 5}{7 + 1} = \dfrac{12}{8} = \dfrac{3}{2}$.

   (b) $(2,9),(-3,1)$. $m = \dfrac{y_2 - y_1}{x_2 - x_1} = \dfrac{1 - 9}{-3 - 2} = \dfrac{-8}{-5} = \dfrac{8}{5}$.

   (c) $(-5,1),(-3,-6)$. $m = \dfrac{y_2 - y_1}{x_2 - x_1} = \dfrac{-6 - 1}{-3 + 5} = \dfrac{-7}{2} = -\dfrac{7}{2}$.

8. Slope of lines perpendicular to lines through:

   (a) $(3,-6),(-1,12)$. $m_1 = \dfrac{y_2-y_1}{x_2-x_1} = \dfrac{12+6}{-1-3} = \dfrac{18}{-4} = -\dfrac{9}{2}$. $m_2 = \dfrac{2}{9}$.

   (b) $(1,1),(7,7)$. $m_1 = \dfrac{y_2-y_1}{x_2-x_1} = \dfrac{7-1}{7-1} = 1$. $m_2 = -1$.

   (c) $(-3,8),(-2,-4)$. $m_1 = \dfrac{y_2-y_1}{x_2-x_1} = \dfrac{-4-8}{-2+3} = -12$. $m_2 = \dfrac{1}{12}$.

Chapter 1 Review Exercises

9. Points $(1,-3),(3,1)$. $m = \dfrac{y_2 - y_1}{x_2 - x_1} = \dfrac{1 + 3}{3 - 1} = 2$.

   Points $(1,-3),(5,5)$. $m = \dfrac{y_2 - y_1}{x_2 - x_1} = \dfrac{5 + 3}{5 - 1} = 2$.

   The three points lie on a line.

10. (a) $(1,4),(7,-8)$. $m = \dfrac{-8-4}{7-1} = -2$. $y-4 = -2(x-1)$, $y = -2x+6$.

    (b) $(-2,-3),(2,5)$. $m = \dfrac{5+3}{2+2} = 2$. $y+3 = 2(x+2)$, $y = 2x+1$.

    (c) $(5,5),(-3,9)$. $m = \dfrac{9-5}{-3-5} = -\dfrac{1}{2}$. $y-5 = -\dfrac{1}{2}(x-5)$, $y = -\dfrac{1}{2}x + \dfrac{15}{2}$.

11. (a) $m=3,b=4$. $y = 3x + 4$.
    (b) $m=-1,b=7$. $y = -x + 7$.
    (c) $m=6,b=-2$. $y = 6x - 2$.

12. Line with slope 5 and x-intercept -2. Point $(-2,0)$.
    Point/slope form, $y-0 = 5(x+2)$. $y=5x+10$.

13. Line throught point $(2,-5)$ perpendicular to $y=-2x+4$.
    $m_1=-2$. $m_2=\dfrac{1}{2}$. Point/slope form, $y+5 = \dfrac{1}{2}(x-2)$. $y = \dfrac{1}{2}x-6$.

14. Line through $(2,7)$ perpendicular to $-6x+2y-9 = 0$.
    $2y=6x+9$. $y=3x+\dfrac{9}{2}$. $m_1=3$. $m_2=-\dfrac{1}{3}$. $y-7 = -\dfrac{1}{3}(x-2)$.
    $-3y+21 = x-2$. $x+3y-23=0$.

15. Line with x-intercept -2, ie pt$(-2,0)$, parallel to $-3x+y-7 = 0$.
    $y=3x+7$. $m_1=3$. $m_2=3$. $y-0 = 3(x+2)$. $y=3x+6$.

16. A so that $Ax+3y-7 = 0$ is parallel to $6x-2y = 5$.
    $6x-2y = 5$, $2y=6x-5$. $m=3$.
    $Ax+3y-7 = 0$, $y = (-3/A)x +7/A$. $m=-3/A$.
    Thus $-3/A = 3$. $A = -9$.

17. x-intercept of the line through $(1,5)$ perp to $4x-2y-6 =0$.
    $4x-2y-6=0$, $y=2x-3$. $m_1=2$. $m_2=-\dfrac{1}{2}$. $y-5 = -\dfrac{1}{2}(x-1)$.
    When $y=0$, $10=x-1$, $x=11$.

18. $f(x)=2.47x-3.251$. $f(-3.45)=-11.7725$, $f(2.79)=3.6403$, $f(23.72)=55.3374$.

19. (a) Change of 1 in x causes a constant change of 4 in f. f is
    linear. Rate of change=4, $m=4$. $f(0)=-2$, y-int$=-2$. $y=4x-2$.

    (b) e.g. $f(4)-f(3)=7$. $f(3)-f(2)=5$. Change of 1 in x does not
    give constant change in f. f is not linear.

(c) Change of .5 in x causes a constant change of -1.4 in f.
f is linear. Rate of change$=-2.8$, $m=-2.8$. $f(0)=4.1$, y-int$=4.1$.
$y=-2.8x+4.1$.

20. (a) Cleats cost \$105. Studs cost \$15. $f(x)=15x + 105$.
 (b) Slope is cost of a stud. y-int is initial cost of the cleats.

21. (a) Cost of 20 is \$525.50 and cost of 30 is \$560.80.
 Let $(x_1,y_1)=(20,525.50)$ and $(x_2,y_2)=(30,560.80)$.
 Then $m = \dfrac{y_2-y_1}{x_2-x_1} = \dfrac{560.8-525.5}{30-20} = \dfrac{35.3}{10} = 3.53$.
 Point/slope form gives $y-525.5 = 3.53(x-20)$.
 Cost-volume formula is $y = 3.53x+454.9$.
 (b) Cost of 50 items is $3.53(50)+454.9 = \$631.40$.
 (c) Estimates of costs of 70...76 items:

X	70	71	72	73	74	75	76
Y	702	705.53	709.06	712.59	716.12	719.65	723.18

 (d) $y(97)=797.31, y(98)=800.84$. Cost reaches \$800 with 98th item.

22. Construct table with years t measured from the base year 1988.

Years from '88	0	1	2	3	4	5
Canada	9745	9985	10255	10199	10530	10731
Japan	21280	22667	23659	24613	25539	26386
USA	116575	116573	118459	123461	123698	126728

 $t=17$ in year 2005.
 Canada: $C(t)=185.9714286t+9775.904762$.
 Japan: $J(t)=1002.857143t+21516.85714$.
 USA: $USA(t)=2204.057143t+115405.5238$.
 $C(17)=12937$ thousand, $J(17)=38565$ thousand, $U(t)=152874$ thousand.

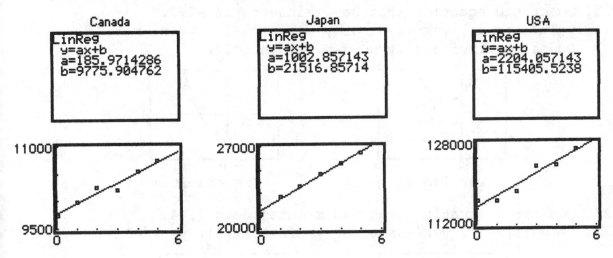

23. (a) $y=x^2-4x+9$, $a=1, b=-4, c=9$. $x= \dfrac{-b}{2a} = \dfrac{4}{2(1)} = 2$.

 $y = 2^2-4(2)+9 = 5$. Vertex $(2,5)$. Axis $x=2$.
 y-int$=9$. $a>0$, opens up. Figure below.
 Window: $Xmin=-2, Xmax=6, Xscl=1, Ymin=-2, Ymax=15, Yscl=2$.

(b) $y=x^2+6x+1$, $a=1,b=6,c=1$. $x=\dfrac{-b}{2a}=\dfrac{-6}{2(1)}=-3$.

$y=(-3)^2+6(-3)+1=-8$. Vertex $(-3,-8)$. Axis $x=-3$.
y-intercept = 1. a>0, opens up. Figure below.
Window: Xmin=-10,Xmax=2,Xscl=1,Ymin=-10,Ymax=15,Yscl=2.

(c) $y=2x^2-8x-3$, $a=2,b=-8,c=-3$. $x=\dfrac{-b}{2a}=\dfrac{8}{2(2)}=2$.

$y=2(2)^2-8(2)-3=-11$. Vertex $(2,-11)$. Axis $x=2$.
y-intercept = -3. a>0, opens up. Figure below.
Window: Xmin=-2,Xmax=8,Xscl=1,Ymin=-15,Ymax=15,Yscl=2.

(d) $y=-3x^2+18x-7$, $a=-3,b=18,c=-7$. $x=\dfrac{-b}{2a}=\dfrac{-18}{2(-3)}=3$.

$y=-3(3)^2+18(3)-7=20$. Vertex $(3,20)$. Axis $x=3$.
y-intercept = -7. a<0, opens down. Figure below.
Window: Xmin=-2,Xmax=8,Xscl=1,Ymin=-15,Ymax=25,Yscl=2.

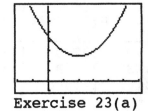 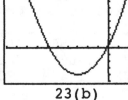 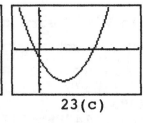 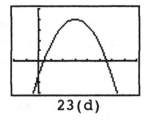

| Exercise 23(a) | 23(b) | 23(c) | 23(d) |

24. Point (1,6) lies on $y=2x^2-3x+a$. Thus $6=2(1)^2-3(1)+a$. $6=-1+a$. $a=7$. $y=2x^2-3x+7$. Graph below.

25. Quadratic equation that has x-intercepts -2,3.
$y=(x+2)(x-3)$, $y=x^2-x-6$. Or $y=-(x+2)(x-3)$, $y=-x^2+x+6$.
Opens up. Thus a>0. $y=x^2-x-6$. Graph below.

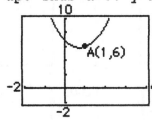 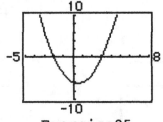

Exercise 24   Exercise25

26. Quadratic equation that has x-intercepts 1,-4.
$y=(x-1)(x+4)$. $y=x^2+3x-4$. or $y=-(x-1)(x+4)$. $y=-x^2-3x+4$.

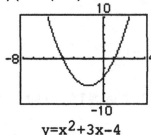 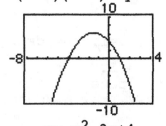

$y=x^2+3x-4$ $y=-x^2-3x+4$

27. $s=-16t^2+287t$. A(8.97,1287.02) to 2 dec places. Figure below. Maximum height is 1287.02 feet at t = 8.97 seconds.

Algebra: a=-16,b=287,c=0. Vertex at $t = \dfrac{-b}{2a} = \dfrac{-287}{2(-16)} = 8.96875$ = 8.97 to 2 decimal places. $s(8.97) = -16(8.97)^2+287(8.97) = 1287.0156$. Vertex(8.97,1287.02), confirming graph result.

28. $S(p)=0.64p^2-7p+124$, $D(p)=0.72p^2-48p+970$. A(21.54,270.15). Equilibrium price = \$21.54. Figure below.

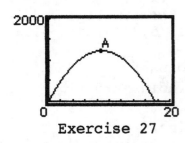

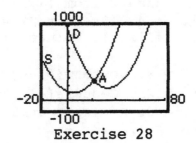

Exercise 27   Exercise 28

29. (a) $v(x)=0.001x^2+0.52x+316$. (b) v(250)=508.5feet/year.
(c) 5280 ft per ml. Length of glacier = 5280x7=36960ft.

Surface: time $= \dfrac{distance}{velocity} = \dfrac{36960}{316} = 116.9620253 = 117$ years.

Middle: Depth 125ft. v(125)=396.625ft/yr. time$=\dfrac{36960}{396.625}$ =93yrs.

Base: Av depth=250ft. time$=\dfrac{36960}{508.5}$ =73yrs.

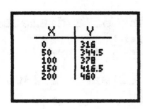

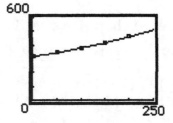

30. (a) (1,5),(3,9). d= $\sqrt{(3-1)^2+(9-5)^2} = \sqrt{2^2+4^2} = \sqrt{20} = 4.721$

(b) (-2,6),(2,5). d= $\sqrt{(2+2)^2+(5-6)^2} = \sqrt{4^2+(-1)^2} = \sqrt{17} = 4.1231$

(c) (3,-4),(-7,-1). d$=\sqrt{(-7-3)^2+(-1+4)^2}=\sqrt{(-10)^2+3^2} =\sqrt{109}=10.4403$

31. (a) (1,3),(7,9). Midpoint, $(\dfrac{1+7}{2}, \dfrac{3+9}{2}) = (\dfrac{8}{2}, \dfrac{12}{2}) = (4,6)$.

(b) (-4,-3),(-8,7). Midpoint, $(\dfrac{-4-8}{2}, \dfrac{-3+7}{2})= (\dfrac{-12}{2}, \dfrac{4}{2})= (-6,2)$.

(c) (8,0),(-9,3). Midpoint, $(\dfrac{8-9}{2}, \dfrac{0+3}{2}) = (\dfrac{-1}{2}, \dfrac{3}{2}) = (-\dfrac{1}{2}, \dfrac{3}{2})$.

Chapter 1 Review Exercises

32. Midpoint of a line segment is (2,5). One end is (-1,2). Let other end point be (x,y). Midpoint is $(\frac{x-1}{2}, \frac{y+2}{2})$.

Thus $\frac{x-1}{2}$ =2, $\frac{y+2}{2}$ =5. x-1=4,y+2=10. x=5,y=8.Other end pt (5,8).

33. Points on y-axis distance 5 from (3,-1). Let such point be (0,y). Then $(0-3)^2+(y+1)^2$ = 25. $y^2+2y-15$ = 0. (y-3)(y+5)=0. y=3 or -5. Points (0,3) and (0,-5).

34. Points (x,3x), distance 2 from (1,5).Then $(1-x)^2+(5-3x)^2 = 2^2$. $1-2x+x^2+25-30x+9x^2$ = 4. $10x^2-32x+22$ = 0. $5x^2-16x+11$ = 0. (5x-11)(x-1) = 0. x=11/5,1. Points are (11/5,33/5) and (1,3).

35. (a) $(x-3)^2+(y+4)^2$ = 9. Center=(3,-4), radius=$\sqrt{9}$ = 3.
    (b) $(x+1)^2+(y+7)^2$ = 25. Center=(-1,-7), radius=$\sqrt{25}$ = 5.
    (c) $x^2+y^2-4x-6y+9$ = 0. x^2-4x+y^2-6y+9 = 0.
    $(x^2-4x+(-2)^2)+(y^2-6y+(-3)^2)$ = $-9+(-2)^2+(-3)^2$.
    $(x-2)^2 + (y-3)^2$ = 4. Circle, center (2,3), radius = $\sqrt{4}$ = 2.
    (d) $x^2+y^2+6x+2y$ = -4. x^2+6x+y^2+2y = -4.
    $(x^2+6x+3^2)+(y^2+2y+1^2)$ = $-4+3^2+1^2$.
    $(x+3)^2 + (y+1)^2$ = 6. Circle, center(-3,-1), radius =$\sqrt{6}$=2.4495.

36. (a) Center (3,6),radius 3.$(x-3)^2+(y-6)^2 = 3^2$,
    $(x-3)^2+(y-6)^2$ = 9.
    (b) Center (-1,4), radius 6.$(x+1)^2+(y-4)^2 = 6^2$,
    $(x+1)^2+(y-4)^2$ = 36.

37. Circle with points A(1,5) and B(5,7) end points of a diameter. Center C is mid-point of AB. Center C is $(\frac{1+5}{2},\frac{5+7}{2})$= (3,6).

d(A,C)=$\sqrt{(3-1)^2+(6-5)^2}$=$\sqrt{2^2+1^2}$ = $\sqrt{5}$ = 2.2360. Radius = 2.2360. Equation, $(x-3)^2+(y-6)^2$ = 5.

38. Equation of the circle having center (-3, 1) that touches the x-axis at the point (-3, 0). Radius = 1. Equation, $(x+3)^2+(y-1)^2$ = 1.

39. Circumference and area of circle $x^2+y^2-2x-8y+13$ = 0. $(x-1)^2+(y-4)^2$ = -13+1+16 = 4. r=2. Circumference = $2\pi r$ = 4π = 12.5664. Area = πr^2 = 4π = 12.5664.

40. (a) m=-2,b=-4. y=-2x-4.
    (b) Circle, center (-8,6), radius 4. $(x+8)^2+(y-6)^2$=16. Graph as y=$\sqrt{16-(x+8)^2}$ + 6 and y=-$\sqrt{16-(x+8)^2}$ + 6.

Chapter 1 Test

1. $y=2x^2-4x+3$.
   Point $(-2,19)$. $x=-2$, $y=2(-2)^2-4(-2)+3 = 19$. Yes.
   Point $(1,1)$. $x=1$, $y=2(1)^2-4(1)+3 = 1$. Yes.
   Point $(5,31)$. $x=5$, $y=2(5)^2-4(5)+3 = 33 \neq 31$. No.

2. Xmin=-4, Xmax=16,Xscl=2,Ymin=-16,Ymax=12,Yscl=4.

3. $(-2,3)$, $m=-1$. $(y-3)=-1(x+2)$, $y=-x+1$. Figure below.

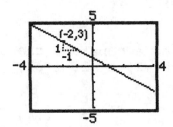

4. Slope of lines parallel to line through $(2,-3),(-1,6)$.
   $$m = \frac{y_2 - y_1}{x_2 - x_1} = \frac{6 + 3}{-1 - 2} = \frac{9}{-3} = -3.$$

5. Slope of lines perpendicular to line through $(1,1),(3,8)$.
   $$m_1 = \frac{y_2-y_1}{x_2-x_1} = \frac{8 - 1}{3 - 1} = \frac{7}{2}. \quad m_2 = -\frac{2}{7}.$$

6. $(-2,3),(-5,9)$. $m = \frac{9-3}{-5+2} = -2$. $y-3 = -2(x+2)$, $y = -2x-1$.
   See figure below.

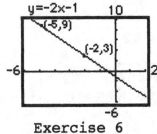

Exercise 6

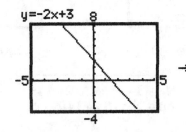 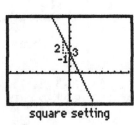

Exercise 7

7. $m=-2,b=3$. $y = -2x + 3$. See figure above.

8. Line through $(-3,4)$ perpendicular to $-2x+y-6 = 0$.
   $y=2x-6$. $m_1=2$. $m_2=-\frac{1}{2}$. $y-4 = -\frac{1}{2}(x+3)$. $y=-\frac{1}{2}x+\frac{5}{2}$.
   $2y = -x + 5$. $x+2y-5=0$. see figure below.

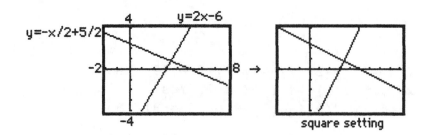

square setting

9. x-intercept of the line through (2,-8) parallel to 12x-3y-3 =0.
   12x-3y-3=0, y=4x-1. m_1=4. m_2=4. y+8 = 4(x-2).
   When y=0, 8=4(x-2), 2=x-2. x=4.

10. Change of 1 in x causes a constant change of -2 in f. f is
    linear. Rate of change=-2, m=-2. f(0)=3, y-int=3. y=-2x+3.

11. $y=x^2-6x-7$, a=1,b=-6,c=-7. $x=\dfrac{-b}{2a}=\dfrac{6}{2(1)}=3$.

    $y = 3^2-6(3)-7 = -16$. Vertex (3,-16). Axis x=3.
    y-int=-7. a>0, opens up. Figure below.

12. Quadratic equation that has x-intercepts -3,5. Consider
    y=(x+3)(x-5), $y=x^2-2x-15$; or y=-(x+3)(x-5), $y=-x^2+2x+15$.
    y-intercept is 15. Thus $y=-x^2+2x+15$. Figure below.

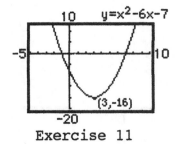

Exercise 11

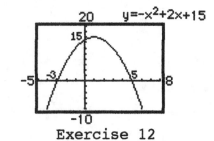

Exercise 12

13. (2,4),(-3,12). $d= \sqrt{(-3-2)^2+(12-4)^2} = \sqrt{(-5)^2+8^2} = \sqrt{89} =$
    9.433981132. Distance is 9.4340 to 4 decimal places.

14. Points of the form (x,x) distance 9 from (6,3).
    Have $(x-6)^2+(x-3)^2=(9)^2$. $x^2-12x+36+x^2-6x+9 = 81$.
    $2x^2-18x-36 = 0$. $x^2-9x-18 = 0$. x=-1.68, 10.68 to 2 dec places.
    Points are (-1.68, -1.68) and (10.68, 10.68).

15. a) $(x+5)^2+(y-1)^2 = 36$. Center=(-5,1), radius=$\sqrt{36}$ = 6.
    b) $x^2+y^2-6x-2y+6 = 0$. $x^2-6x+y^2-2y+6 = 0$.
       $(x^2-6x+(-3)^2)+(y^2-2y+(-1)^2) = -6+(-3)^2+(-1)^2$.
       $(x-3)^2 + (y-1)^2 = 4$. Circle, center (3,1), radius = $\sqrt{4}$ = 2.

16. Center (2,-7),radius 9.$(x-2)^2+(y+7)^2 = 9^2$,
    $(x-2)^2+(y+7)^2 = 81$.

17. Equation of the circle having center (2,5) that touches the y-axis at (0,5). Radius = 2. Equation, $(x-2)^2+(y-5)^2 = 4$.

18. (a) Cost of 15 is $631.50 and cost of 30 is $669.
    Let $(x_1,y_1)=(15,631.50)$ and $(x_2,y_2)=(30,669)$.
    Then $m = \dfrac{y_2-y_1}{x_2-x_1} = \dfrac{669-631.50}{30-15} = \dfrac{37.5}{15} = 2.5$.
    Point/slope form gives $y-631.5 = 2.5(x-15)$.
    Cost-volume formula is $y = 2.5x+594$.
    (b) Cost of 22 items is $2.5(22)+594 = \$649$.

19. Use table with years t measured from the base year 1900.
    $P=0.5432121212t+18.98545455$. Year 2010 corresponds to t=110.
    $P(110)=78.73878788$. Predicted pop 78.7 per square mile.

20. $S(p)=0.58p^2-6p+121$, $D(p)=0.83p^2-52p+890$. A(18.60,210.01).
    Equilibrium price = $18.60. Figure below.

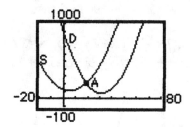

Chapter 2

Section 2.1

1. $x^2 - 3x + 2 = 0$, $(x-1)(x-2)= 0$, $x = 1,2$. Graph $y=x^2-3x+2$ below.
3. $x^2 - 4x = 0$, $x(x-4) = 0$, $x = 0, 4$. Graph of $y = x^2 - 4x$ below.
5. $-x^2 - 2x + 8=0$, $(-x+2)(x+4)=0$, $x=2,-4$. Graph $y=-x^2-2x+8$ below.

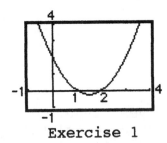
Exercise 1

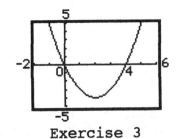
Exercise 3

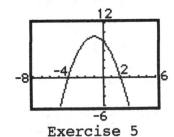
Exercise 5

7. $2x^2+5x-3=0$, $(2x-1)(x+3)=0$, $x =1/2,-3$. Graph $y=2x^2+5x-3$ below.

9. $4x^2 - 12x + 9 = 0$, $(2x-3)(2x-3)=0$, $x=3/2$. Repeated root.
   See graph of $y = 4x^2 - 12x + 9$ below.

11. $7x^2 + 13x - 2 = 0$, $(7x-1)(x+2)=0$, $x=1/7, -2$.
    See graph of $y = 7x^2 + 13x - 2$ below.

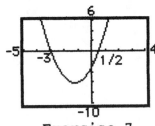

Exercise 7

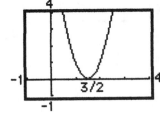
Exercise 9

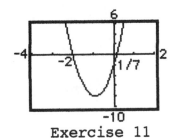
Exercise 11

13. $x^2 - 2x = 2x$. Write in standard form. $x^2 - 4x = 0$.
    $x^2 - 4x = x(x-4)=0$, $x=0,4$. See graph of $y = x^2-4x$ below.

15. $16z^2 - 8z + 1 =0$, $(4z-1)((4z-1)=0$, $z=1/4$. Repeated root.
    See graph of $y = 16z^2 - 8z + 1$ below.

17. $x^2 = 9$, $x = \pm3$. Graph of $y = x^2 - 9$ below.

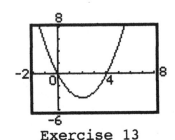
Exercise 13

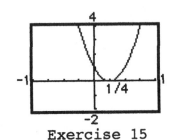
Exercise 15

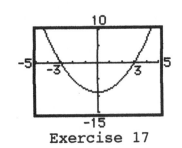
Exercise 17

19. $3x^2 - 27 = 0$, $x^2 = 9$, $x = \pm 3$. Graph of $y = 3x^2 - 27$ below.

21. $x^2 - 1/25 = 0$, $x^2 = 1/25$, $x = \pm 1/5$. Graph of $y = x^2 - 1/25$ below.

23. $(5x+2)^2 = 4$, $(5x+2) = \pm 2$, $5x = \pm 2 - 2$, $5x = 0$ or $5x = -4$, $x = 0, -4/5$.
    See graph of $y = (5x+2)^2 - 4$ below.

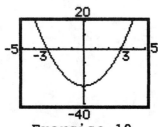
Exercise 19

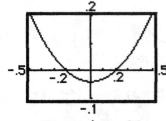
Exercise 21

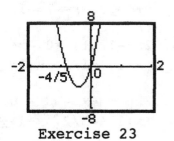
Exercise 23

25. $(3a+4)^2 = 25$, $3a+4 = \pm 5$, $3a = \pm 5 - 4$, $3a = 1$ or $3a = -9$, $a = 1/3, -3$.
    See graph of $y = (3a+4)^2 - 25$ below.

27. $x^2 = 17$, $x = \pm\sqrt{17} = \pm 4.1231$, to 4 decimal places.
    See graph of $y = x^2 - 17$ below.

29. $(x-2)^2 = 8.6$, $(x-2) = \pm\sqrt{8.6}$, $x = 2 \pm \sqrt{8.6}$,
    $x = 4.9326, -0.9326$, to 4 decimal places.
    See graph of $y = (x-2)^2 - 8.6$ below.

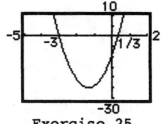
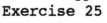
Exercise 25

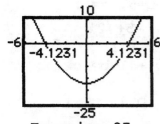
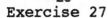
Exercise 27

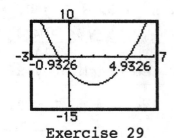
Exercise 29

31. $(3x+4)^2 = 5.6$, $3x+4 = \pm\sqrt{5.6}$, $x = (-4 \pm \sqrt{5.6})/3$,
    $x = -0.5445, -2.1221$, to 4 decimal places.
    See graph of $y = (3x+4)^2 - 5.6$ below.

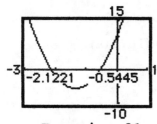
Exercise 31

33. $x^2+2x-24=0$. $x^2+2x=24$. $x^2+2x+(1)^2=24+(1)^2$. $(x+1)^2=25$. $x+1=\pm5$. $x=-6,4$.

35. $x^2-4x-12=0$. $x^2-4x=12$. $x^2-4x+(-2)^2=12+(-2)^2$. $(x-2)^2=16$. $x-2=\pm4$. $x=-2,6$.

37. $x^2+8x+7=0$. $x^2+8x=-7$. $x^2+8x+(4)^2=-7+(4)^2$. $(x+4)^2=9$. $x+4=\pm3$. $x=-7,-1$.

39. $2x^2-12x+17=0$. $x^2-6x=-17/2$. $x^2-6x+(-3)^2=-17/2+(-3)^2$. $(x-3)^2=1/2$. $x-3=\pm\sqrt{1/2}$. $x=3\pm\sqrt{1/2}$.

41. $2x^2 + 4x - 3 = 0$. $x^2+2x=3/2$. $x^2+2x+(1)^2=3/2+(1)^2$. $(x+1)^2=5/2$. $x+1=\pm\sqrt{5/2}$. $x=-1\pm\sqrt{5/2}$.

43. $3x^2+6x+2=0$. $x^2+2x=-2/3$. $x^2+2x+(1)^2=-2/3+(1)^2$. $(x+1)^2=1/3$. $x+1=\pm\sqrt{1/3}$. $x=-1\pm\sqrt{1/3}$.

45. $x^2+6x-7=0$. $(x-1)(x+7)= 0$, $x = 1,-7$.

47. $x^2+5x-3=0$. $x^2+5x=3$. $x^2+5x+(5/2)^2=3+(5/2)^2$. $(x+5/2)^2=3+(25/4)$. $x+5/2=\pm\sqrt{37/4}$. $x=-5/2\pm\sqrt{37/2}$.

49. $(2x-4)^2-11=0$. $(2x-4)^2=11$. $2x-4=\pm\sqrt{11}$. $x=2\pm\sqrt{11}/2$.

51. $x^2 + 5x + 1 = 0$. $a=1$, $b=5$, $c=1$.
$$x = \frac{-5\pm\sqrt{5^2 - 4(1)(1)}}{2(1)} = \frac{-5\pm\sqrt{21}}{2} = -0.2087, -4.7913.$$
See graph of $y=x^2 + 5x + 1$ below.

53. $2x^2 + 4x - 1 = 0$. $a=2$, $b=4$, $c=-1$.
$$x = \frac{-4\pm\sqrt{4^2 - 4(2)(-1)}}{2(2)} = \frac{-4\pm\sqrt{24}}{4} = 0.2247, -2.2247.$$
See graph of $y = 2x^2 + 4x - 1$ below.

55. $x^2 + 4x + 4 = 0$. $a=1$, $b=4$, $c=4$.
$$x = \frac{-4\pm\sqrt{4^2 - 4(1)(4)}}{2(1)} = \frac{-4\pm\sqrt{0}}{2} = -2.$$ Single solution.
See graph of $y = x^2 + 4x + 4$ below.

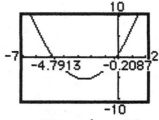
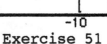

Exercise 51

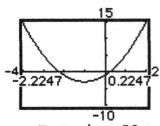

Exercise 53

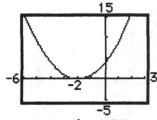

Exercise 55

57. $2s^2 - 3s + 4 = 0$. a=2, b=-3, c=4.

$$s = \frac{3 \pm \sqrt{(-3)^2 - 4(2)(4)}}{2(2)} = \frac{3 \pm \sqrt{-23}}{4}$$. No real solutions.

See graph of $y = 2x^2 - 3x + 4$ below.

59. $4x^2 - 2x - 1.1 = 0$. a=4, b=-2, c=-1.1.

$$x = \frac{2 \pm \sqrt{(-2)^2 - 4(4)(-1.1)}}{2(4)} = 0.8309, -0.3309$$

See graph of $y = 4x^2 - 2x - 1.1$ below.

61. $-2t^2 + 4t - 7 = 0$. a=-2, b=4, c=-7.

$$x = \frac{-4 \pm \sqrt{4^2 - 4(-2)(-7)}}{2(-2)} = \frac{-4 \pm \sqrt{-40}}{-4}$$. No real solutions.

See graph of $y = -2t^2 + 4t - 7$ below.

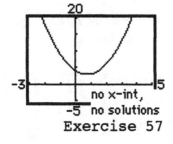

Exercise 57

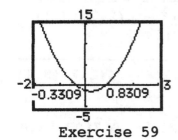

Exercise 59

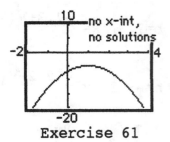

Exercise 61

63. $3x^2 - 4.8x + 1.92 = 0$. a=3, b=-4.8, c = 1.92.

$$x = \frac{4.8 \pm \sqrt{(-4.8)^2 - 4(3)(1.92)}}{2(3)} = 0.8000.$$ Single solution.

See graph of $y = 3x^2 - 4.8x + 1.92$ below.

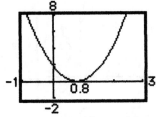

Exercise 63

65. $ax^2 - x - 2.5 = 0$. $b=-1$, $c=-2.5$.

$a=1$, $x = \dfrac{1\pm\sqrt{(-1)^2 - 4(1)(-2.5)}}{2(1)} = 2.1583, -1.1583$

$a=2.1$, (edit the above form) $x = \dfrac{1\pm\sqrt{(-1)^2 - 4(2.1)(-2.5)}}{2(2.1)} = $
$1.3549, -0.8787$

$a=4.3$, $x = \dfrac{1\pm\sqrt{(-1)^2 - 4(4.3)(-2.5)}}{2(4.3)} = 0.8876, -0.6550$

67. $2.7x^2 + bx + 3.1 = 0$. $a=2.7$, $c=3.1$.

$b=2.3$, $x = \dfrac{-2.3\pm\sqrt{(2.3)^2 - 4(2.7)(3.1)}}{2(2.7)}$

$= \dfrac{-2.3\pm\sqrt{-28.19}}{2(2.7)}$. No real solutions.

$b=7.3$, $x = \dfrac{-7.3\pm\sqrt{(7.3)^2 - 4(2.7)(3.1)}}{2(2.7)} = -2.1761, -0.5276$.

$b=-9.2$, $x = \dfrac{9.2\pm\sqrt{(-9.2)^2 - 4(2.7)(3.1)}}{2(2.7)} = 0.3791, 3.0282$.

69. $1.7x^2 - 3.8x - 4 = 0$. $x = -0.7803, 3.0156$.
71. $1.7x^2 - 4.3x - 5.1 = 0$. $x = -0.8799, 3.4093$.
73. $-5.3x^2 + 7.2x + 2.3 = 0$. $x = -0.2670, 1.6255$.

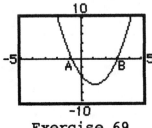

Exercise 69

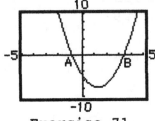

Exercise 71

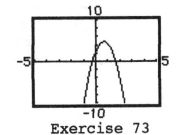

Exercise 73

75. $3.6x^2 - 42.8x + 12.7 = 0$. $x = 0.3045, 11.5844$.

77. $x^2 + x + 3 = 0$. a=1, b=1, c=3. $b^2-4ac = -11 < 0$.
    No real solutions. See graph of $y = x^2 + x + 3$ below.

79. $4x^2 + x + 1 = 0$. a=4, b=1, c=1. $b^2-4ac = -15 < 0$.
    No real solutions. See graph of $y = 4x^2 + x + 1$ below.

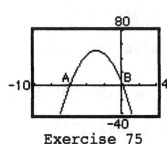

Exercise 75

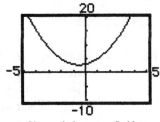

No x-int, no solution
Exercise 77

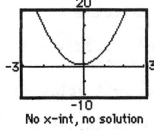

No x-int, no solution
Exercise 79

81. $2x^2 + 4x + 2 = 0$. a=2, b=4, c=2. $b^2-4ac = 0$.
    Single real solution. Graph of $y = 2x^2 + 4x + 2$ below.

83. $3x^2 - 4.8x + 1.92 = 0$. a=3, b=-4.8, c=1.92. $b^2-4ac = 0$.
    Single real solution. Graph of $y = 3x^2 - 4.8x + 1.92$ below.

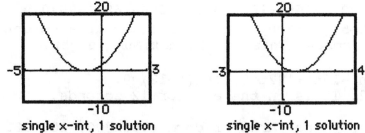

single x-int, 1 solution
Exercise 81

single x-int, 1 solution
Exercise 83

In exercises 85-89 observe that each graph has a single x-intercept. Thus the equation has a single real solution.

85. $2x^2 + kx + 8 = 0$. a=2, b=k, c=8. $b^2-4ac = k^2-4(2)(8)$
    $= k^2-64 = 0$. $k^2=64$. $k=\pm 8$. See graphs of $y=2x^2+8x+8$ (k=8)
    and $y=2x^2-8x+8$ (k=-8) below.

87. $2x^2 + 2x + k = 0$. a=2, b=2, c=k. $b^2-4ac = 2^2-4(2)(k)$
    $= 4-8k = 0$. $8k=4$. $k=1/2$.
    See graph of $y = 2x^2 + 2x + .5$ below.

89. $2x^2 + kx + k = 0$. a=2, b=k, c=k. $b^2-4ac = k^2-4(2)(k)$
    $= k^2-8k = 0$. $k(k-8)=0$. $k=0,8$.
    See graphs of $y=2x^2$ and $y=2x^2+8x+8$ below.

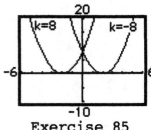
Exercise 85

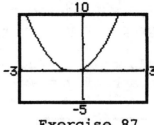
Exercise 87

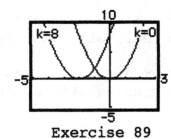
Exercise 89

91. $ax^2 - 5.7x - 6.3 = 0$. Find values of a that give two real
    solutions, a single solution, and no solutions.

    Two real solutions, $b^2 > 4ac$. $a < \dfrac{b^2}{4c}$. $a < \dfrac{(-5.7)^2}{4(-6.3)}$. $a < 1.289285714$.

    Single solution, $b^2 = 4ac$. $a = \dfrac{b^2}{4c}$. $a = \dfrac{(-5.7)^2}{4(-6.3)}$. $a = 1.289285714$.

    No solutions, $b^2 < 4ac$. $a > \dfrac{b^2}{4c}$. $a > \dfrac{(-5.7)^2}{4(-6.3)}$. $a > 1.289285714$.

Section 2.2

1. $s = -16t^2 + 320t$. When $s = 1200$, $-16t^2 + 320t = 1200$, $16t^2 - 320t + 1200 = 0$,
   $16(t^2 - 20t + 75) = 0$, $16(t-5)(t-15) = 0$, $t = 5$, 15 seconds.

3. $s = -16t^2 + 160t$. When $s = 400$, $-16t^2 + 160t = 400$, $16t^2 - 160t + 400 = 0$,
   $16(t^2 - 10t + 25) = 0$, $16(t-5)(t-5) = 0$, $t = 5$. It will be at a height
   of 400 feet after 5 seconds. This is the maximum height.

5. $s = -16t^2 + 192t$. $s = 0$ when $-16t^2 + 192t = 0$, $16t(t-12) = 0$.
   Thus $s = 0$ when $t = 0$, 12. It is in the air for 12 seconds.

7. Let x be width of new parking. Area of dorm=(500x200)=100000.
   Area of dorm+present parking=(500+50+50)(200+50)=150000.
   Thus area present parking=50,000. Area new parking=25000.
   Total new&present parking+dorm=(500+50+50+2x)(200+50+x)=
   $(600+2x)(250+x) = 2x^2 + 1100x + 150000$. New parking=$2x^2 + 1100x$.
   Thus $2x^2 + 1100x = 25000$. $2x^2 + 1100x - 25000 = 0$. $x = 21.86, -571.86$.
   Width must be positive. Width of new parking area=21.86 feet.

9. Let x be width of the grass fringe. Total area=(20+2x)(12+2x)
   Area swimming pool=(20x12)=240. Area grass=(20+2x)(12+2x)-240.
   Thus $(20+2x)(12+2x)-240=228$, $4x^2 + 64x - 228 = 0$. $4(x^2 + 16x - 57) = 0$.
   $4(x-3)(x+19) = 0$. $x = 3, -19$. Width must be positive.
   Width=3feet.

11. Let width of driveway be x. Length of driveway = x+10.
    Area of driveway=$x(x+10) = x^2 + 10x$. Concrete available =264.
    Thus $x^2 + 10x = 264$. $x^2 + 10x - 264 = 0$. $(x-12)(x+22) = 0$. $x = 12, -22$.
    Cannot have negative width. $x = 12$.
    Width of driveway=12feet, length=22feet.

64

13. Let numbers be x and 20-x. Thus x(20-x)=91, x^2-20x+91=0, (x-7)(x-13)=0, x=7,13.

15. Let numbers be x and x+2, x odd. x(x+2)=143, x^2+2x-143=0, (x-11)(x+13)=0, x=11,-13. Numbers are 11, 13 and -13, -11.

17. Let integer be n. $n-\frac{1}{n}=\frac{35}{6}$. $6n^2$-6=35n, $6n^2$-35n-6=0, (6n+1)(n-6)=0. Since n must be an integer, n=6.

19. Let x be the number. $x+5=\frac{14}{x}$. x^2+5x-14=0. (x+7)(x-2)=0, x = -7, 2. Two numbers.

21. Let planes be 200 miles apart after t hours. Planes will have flown 500t and 425t miles. $(500t)^2+(425t)^2=200^2$. $(500^2+425^2)t^2=200^2$. $t^2=200^2/(500^2+425^2)$. t=0.3047757271 hours. t=18.29 minutes approx.

23. We give answer using algebra and graph.
    Let speed of plane in still air be x mph.
    Speed with wind=x+40. Speed into wind=x-40. Dist=1170 ml.
    Time with wind=1170/(x+40). Time into wind=1170/(x-40).
    Difference in times is 1.75 hr. Thus $\frac{1170}{x-40} - \frac{1170}{x+40} = 1.75$.
    Algebra: 1170(x+40) - 1170(x-40) = 1.75(x-40)(x+40).
    1170x+46800-1170x+46800 = $1.75x^2$-2800.
    $1.75x^2$=96400. x^2=55085.71429. x=±234.7034603.
    Cannot have negative speed in still air. Thus x=234.7.
    Speed of plane in still air = 234.7 mph. Graph:
    Graph $y=\frac{1170}{x-40} - \frac{1170}{x+40}$ and y=1.75. $\frac{1170}{x-5} - \frac{1170}{x+5} = 1.75$ at A and
    B. A=(-234.7035,1.75), B(234.70346,1.75).
    The speed of the plane is 234.7 mph.

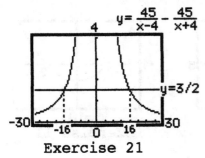

Exercise 21

Section 2.3

1. $x^4 + 2x^2 - 3 = 0$. Let $u=x^2$. u^2+2u-3=0. (u-1)(u+3)=0.
   u=1 or u=-3. x^2=1 or x^2=-3. x^2=1 gives x = ±1. x^2=-3 has no
   real roots. Two solutions, x = ±1. Graph of $y=x^4+2x^2-3$ below.

3. $x^4 - 5x^2 + 4 = 0$. Let $u=x^2$. $u^2-5u+4=0$. $(u-1)(u-4)=0$.
   u=1 or u=4. $x^2=1$ or $x^2=4$. $x^2=1$ gives $x = \pm1$.
   $x^2=4$ gives $x = \pm2$. Four solutions, $x=\pm1,\pm2$.
   See graph of $y = x^4 - 5x^2 + 4$ below.

5. $x^4 - 6x^2 + 8 = 0$. Let $u=x^2$. $u^2-6u+8=0$. $(u-2)(u-4)=0$.
   u=2 or u=4. $x^2=2$ or $x^2=4$. $x^2=2$ gives $x = \pm\sqrt{2}$.
   $x^2=4$ gives $x = \pm2$. Four solutions, $x=\pm\sqrt{2},\pm2$.
   See graph of $y = x^4 - 6x^2 + 8$ below.

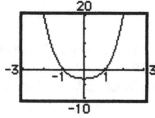
Exercise 1

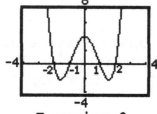
Exercise 3

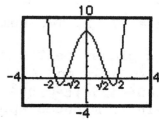
Exercise 5

7. $6x^4 + x^2 - 2 = 0$. Let $u=x^2$. $6u^2+u-2=0$. $(3u+2)(2u-1)=0$.
   u=-2/3 or u=1/2. $x^2=-2/3$ or $x^2=1/2$. $x^2=-2/3$ has no real roots.
   $x^2=1/2$ gives $x = \pm1/\sqrt{2}$. Two solutions, $x = \pm1/\sqrt{2}$.
   $x=\pm0.7071$ (to 4 dec places) Graph of $y = 6x^4+x^2-2$ below.

9. $x^4 - 2x^2 - 15 = 0$. Let $u=x^2$. $u^2-2u-15=0$. $(u-5)(u+3)=0$.
   u=5 or u=-3. $x^2=5$ or $x^2=-3$. $x^2=5$ gives $x = \pm\sqrt{5} = \pm2.2361$.
   $x^2=-3$ has no real roots. Two solutions, $x = \pm\sqrt{5}$.
   See graph of $y = x^4 - 2x^2 - 15$ below.

11. $(x-1)^2+2(x-1)-3 =0$. Let $u=x-1$. $u^2+2u-3=0$. $(u-1)(u+3)=0$.
    u=1 or u=-3. $x-1=1$ or $x-1=-3$. x=2 or x=-2.
    See graph of $y = (x-1)^2+2(x-1)-3$ below.

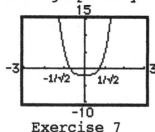
Exercise 7

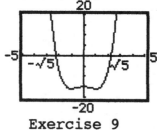
Exercise 9

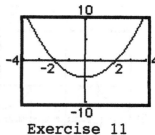
Exercise 11

13. $(p-3)^2-5(p-3)+4=0$. Let $u=p-3$. $u^2-5u+4=0$. $(u-1)(u-4)=0$.
    u=1 or u=4. $p-3=1$ or $p-3=4$. p=4 or p=7.
    See graph of $y = (p-3)^2-5(p-3)+4$ below.

15. $x^6-9x^3+8=0$. Let $u=x^3$. $u^2-9u+8=0$. $(u-1)(u-8)=0$.
    u=1 or u=8. $x^3=1$ or $x^3=8$. x=1 or x=2.
    See graph of $y = x^6-9x^3+8$ below.

17. $x^{-2}-4x^{-1}+3=0$. Let $u=x^{-1}$. $u^2-4u+3=0$. $(u-1)(u-3)=0$.
    $u=1$ or $u=3$. $x^{-1}=1$ or $x^{-1}=3$. $x=1$ or $x=1/3$.
    See graph of $y = x^{-2}-4x^{-1}+3$ below.

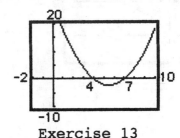
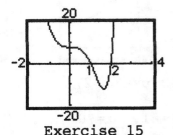
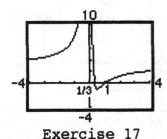

Exercise 13   Exercise 15   Exercise 17

19. $x^{-2}-3x^{-1}-10=0$. Let $u=x^{-1}$. $u^2-3u-10=0$. $(u-5)(u+2)=0$.
    $u=5$ or $u=-2$. $x^{-1}=5$ or $x^{-1}=-2$. $x=1/5$ or $x=-1/2$.
    See graph of $y = x^{-2}-3x^{-1}-10$ below.

21. $8(x-1)^2-6(x-1)+1=0$. Let $u=x-1$. $8u^2-6u+1=0$. $(2u-1)(4u-1)=0$
    $2u=1$ or $4u=1$. $u=1/2$ or $u=1/4$. $x-1=1/2$ or $x-1=1/4$.
    $x=3/2$ or $x=5/4$. See graph of $y = 8(x-1)^2-6(x-1)+1$ below.

23. $x-\sqrt{x}-6=0$. Let $u=\sqrt{x}$. $u^2-u-6=0$. $(u-3)(u+2)=0$.
    $u=3$ or $u=-2$. $\sqrt{x}=3$ or $\sqrt{x}=-2$. $\sqrt{x}=3$ gives $x=9$.
    $\sqrt{x}=-2$ has no solutions. Single solution, $x=9$.
    See graph of $y = x-\sqrt{x}-6$ below.

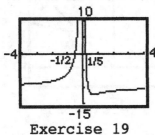
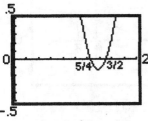
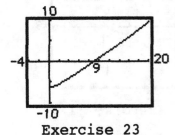

Exercise 19   Exercise 21   Exercise 23

25. $x^{1/2}+3x^{1/4}-4=0$. Let $u=x^{1/4}$. $u^2+3u-4=0$. $(u-1)(u+4)=0$.
    $u=1$ or $u=-4$. $x^{1/4}=1$ or $x^{1/4}=-4$. $x^{1/4}=1$ gives $x=1$.
    $x^{1/4}=-4$ has no solutions. Single solution, $x=1$.
    See graph of $y = x^{1/2}+3x^{1/4}-4$ below.

27. $\dfrac{1}{(x+1)^2} + \dfrac{3}{(x+1)} + 2 = 0$. $(x+1)^{-2}+3(x+1)^{-1}+2=0$.
    Let $u=(x+1)^{-1}$. $u^2+3u+2=0$. $(u+1)(u+2)=0$. $u=-1$ or $u=-2$.
    $(x+1)^{-1}=-1$ or $(x+1)^{-1}=-2$. $x+1=-1$ or $x+1=-1/2$.
    $x=-2$ or $-3/2$. See graph of $y = \dfrac{1}{(x+1)^2} + \dfrac{3}{(x+1)} + 2$ below.

29. $\sqrt{2x-1}=3$. Square. $2x-1=9$. $2x=10$. $x=5$.

67

Check: $\sqrt{2x-1} = \sqrt{2(5)-1} = \sqrt{9} = 3$. Equ is satisfied. x=5.
See graphs of y = $\sqrt{2x-1}$ and y=3 below.

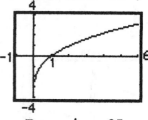

Exercise 25

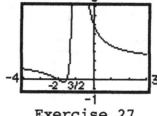

Exercise 27

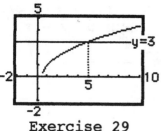

Exercise 29

31. $\sqrt{7x+11}=9$. Square. 7x+11=81. 7x=70. x=10.
    Check: $\sqrt{7x+11}= \sqrt{7(10)+11}= \sqrt{81}= 9$. Equ is satisfied. x=10.
    See graphs of y = $\sqrt{7x+11}$ and y=9 below.

33. $x=\sqrt{x+6}$. Square. $x^2=x+6$. $x^2-x-6=0$. (x-3)(x+2)=0. x=3 or x=-2.
    Check: When x=3,$\sqrt{x+6}=\sqrt{3+6}=\sqrt{9}=3=x$. Equ is satisfied.
    When x=-2,$\sqrt{x+6}=\sqrt{-2+6}=\sqrt{4}=2\neq x$. Equ is not satisfied.
    One solution, x=3. See graphs of y=x and y=$\sqrt{x+6}$ below.

35. $\sqrt{2t+8}=t$. Square. $2t+8=t^2$. $t^2-2t-8=0$. (t-4)(t+2)=0. t=4 or -2.
    Check: When t=4, $\sqrt{2t+8}=\sqrt{2(4)+8}=\sqrt{16}=4=t$. Equ is satisfied.
    When t=-2, $\sqrt{2t+8}=\sqrt{2(-2)+8}=\sqrt{4}=2\neq t$. Equ not satisfied.
    One solution, t=4. See graphs of y=$\sqrt{2t+8}$ and y=t below.

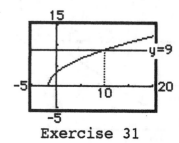

Exercise 31

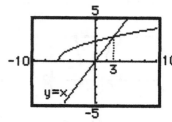

Exercise 33

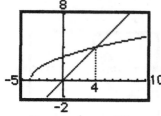

Exercise 35

37. $\sqrt[3]{x+1}=2$. Cube. x+1=8. x=7. Check: $\sqrt[3]{x+1}=\sqrt[3]{7+1}=\sqrt[3]{8}=2$.
    Equ is satisfied. One solution, x=7.
    See graphs of y=$\sqrt[3]{x+1}$ and y=2 below.

39. $\sqrt{2x-3} = \sqrt{x+3}$. Square. 2x-3=x=3. x=6.
    Check: $\sqrt{2x-3}=\sqrt{2(6)-3}=\sqrt{9}=3$. $\sqrt{x+3}=\sqrt{6+3}=\sqrt{9}=3$.
    Equ is satisfied. One solution, x=6.
    See graphs of y=$\sqrt{2x-3}$ (1) and y=$\sqrt{x+3}$ (2) below.

41. $2\sqrt{x+1} = \sqrt{2x+3}+1$. Square. $4(x+1)=(2x+3) + 2\sqrt{2x+3} + 1$.
    $4x+4-2x-3-1=2\sqrt{2x+3}$. $2x=2\sqrt{2x+3}$. Square. $4x^2=4(2x+3)$.
    $4x^2=8x+12$. $4x^2-8x-12=0$. $x^2-2x-3=0$. (x+1)(x-3)=0.
    x=-1 or x=3. Check:

When x=-1. $2\sqrt{x+1}=0$. $\sqrt{2x+3}+1=1+1=2\neq-1$. Equ not satisfied.
When x=3. $2\sqrt{x+1}=4$. $\sqrt{2x+3}+1=3+1=4$. Equ is satisfied.
One solution, x=3.
See graphs of $y=2\sqrt{x+1}$ (1) and $y=\sqrt{2x+3}+1$ (2) below.

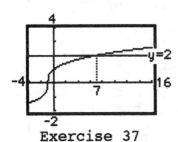
Exercise 37

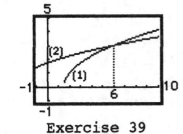
Exercise 39

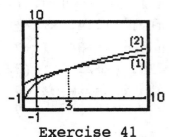
Exercise 41

43. $\sqrt{2x+1}+\sqrt{x-3}=4$. $\sqrt{2x+1}=-\sqrt{x-3}+4$. Square. $2x+1=(x-3)-8\sqrt{x-3}+16$.
$2x+1-x+3-16=-8\sqrt{x-3}$. $x-12=-8\sqrt{x-3}$. Square. $(x-12)^2=64(x-3)$.
$x^2-24x+144=64x-192$. $x^2-88x+336=0$. $(x-4)(x-84)=0$.
x=4 or x=84. Check:
When x=4, $\sqrt{2x+1}+\sqrt{x-3}=\sqrt{9}+\sqrt{1}=4$. Equ is satisfied.
When x=84, $\sqrt{2x+1}+\sqrt{x-3}=\sqrt{169}+\sqrt{81}=13+9=22\neq4$. Equ not
satisfied. One solution, x=4.
See graphs of $y=\sqrt{2x+1}+\sqrt{x-3}$ and y=4 below.

45. If the numerator of the exponent is odd then can take power
and root in one step as follows.
$x^{3/5}=27$. $(x^{3/5})^{5/3}=27^{5/3}$. $x=(27^{1/3})^5=3^5=243$.
Check: $x^{3/5}=243^{3/5}=(243^{1/5})^3=3^3=27$.
Equation is satisfied. One solution, x=243.
See graphs of $y=x^{3/5}$ and y=27 below.

47. $(x-1)^{1/3}=-2$. $((x-1)^{1/3})^3=(-2)^3$. $x-1=-8$. $x=-7$.
Check: $(x-1)^{1/3}=(-7-1)^{1/3}=(-8)^{1/3}=-2$. Solution is x=-7.
See graphs of $y=(x-1)^{1/3}$ and y=-2 below.

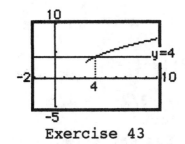
Exercise 43

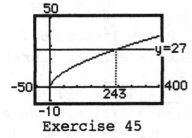
Exercise 45

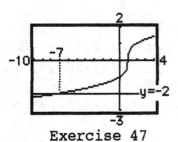
Exercise 47

49. If numerator of the exponent is even then must take two steps
otherwise we lose one solution. (do not get the ±).
$(x+2)^{2/3}=4$. $((x+2)^{2/3})^3=4^3$. $(x+2)^2=64$.
$x+2=\pm\sqrt{64}$. $x+2=\pm8$. $x=-10$ or 6.
Check: x=-10. $(x+2)^{2/3}=(-8)^{2/3}=((-8)^{1/3})^2=(-2)^2=4$.

$x=6$. $(x+2)^{2/3} = (8)^{2/3} = (8^{1/3})^2 = (2)^2 = 4$.
Solutions, $x=-10$, 6. See graphs of $y=(x+2)^{2/3}$ and $y=4$ below.
TI-82 does not display the left branch.

51. $(x+2)^{-2/5} = 1/4$. $((x+2)^{-2/5})^{-5} = (1/4)^{-5}$. $(x+2)^2=(4^{-1})^{-5}$.
    $(x+2)^2=1024$. $x+2=\pm\sqrt{1024}$. $x+2=\pm32$. $x=-34$ or 30.
    Check: $x=-34$. $(x+2)^{-2/5} =(-32)^{-2/5} =((-32)^{1/5})^{-2} =(-2)^{-2} =1/4$.
    $x=30$. $(x+2)^{-2/5}=(32)^{-2/5} =(32^{1/5})^{-2} = (2)^{-2}=1/4$.
    Solutions $x=-34,30$. See graphs of $y=(x+2)^{-2/5}$ and $y=1/4$ below.
    Note that the TI-82 does not display the left branch.

53. $(x+5)^{-5/3}=32$. $((x+5)^{-5/3})^{-3/5}=32^{-3/5}$. $x+5=(32^{1/5})^{-3}$.
     $x+5=(2)^{-3}$. $x= 1/8 - 5$. $x=-39/8$.
    Check: $(-39/8+5)^{-5/3}=(1/8)^{-5/3}=(8)^{5/3}=((8)^{1/3})^5=2^5=32$.
    Solution $x=-39/8$. See graphs of $y=(x+5)^{-5/3}$ and $y=32$ below.

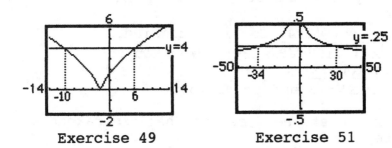

Exercise 49   Exercise 51

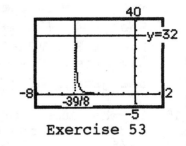

Exercise 53

55. $2x^4+3x^2-5=0$. Let $u=x^2$. $2u^2+3u-5=0$. $(2u+5)(u-1)=0$.
    $u=-5/2$ or $u=1$. $x^2=-5/2$ or $x^2=1$. $x^2=-5/2$ has no real roots.
    $x^2=1$ gives $x=\pm1$. Two solutions, $x=\pm1$.
    See graph of $y=2x^4+3x^2-5$ below.

57. $x^{-2}-8x^{-1}+12=0$. Let $u=x^{-1}$. $u^2-8u+12=0$. $(u-6)(u-2)=0$.
    $u=6$ or $u=2$. $x=1/6$ or $x=1/2$. See graph of $y=x^{-2}-8x^{-1}+12$ below.

59. $\sqrt{5x+6} - 1 = \sqrt{x+1}$. Square. $5x+6 - 2\sqrt{5x+6} + 1 = x+1$.
    $4x+6 = 2\sqrt{5x+6}$. Square. $(4x+6)^2=4(5x+6)$.
    $16x^2+48x+36=20x+24$. $16x^2+28x+12=0$. $(16x+12)(x+1)=0$.
    $x=-3/4$ or $x=-1$. Check:
    When $x=-3/4$, $\sqrt{5x+6}-1 = \sqrt{-15/4 + 6}-1 = \sqrt{9/4}-1=3/2-1 = 1/2$.
             $\sqrt{x+1} = \sqrt{1/4} = 1/2$. Equ satisfied.
    When $x=-1$, $\sqrt{5x+6}-1=\sqrt{1}-1=0$. $\sqrt{x+1} =\sqrt{-1+1}=0$. Equ satisfied.
    Two solutions, $x=-3/4$ and $x=-1$.
    See graphs of $y=\sqrt{5x+6} - 1$ (1) and $y=\sqrt{x+1}$ (2) below.

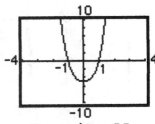

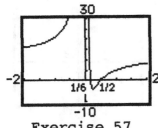

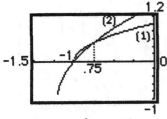

Exercise 55   Exercise 57   Exercise 59

61. $x^{2/3}-4 = 0$. $x^{2/3}=4$. $(x^{2/3})^3=4^3$. $x^2=64$. $x=\pm 8$.

See graph of $y=x^{2/3}-4$ below. Note that the TI-82 does not display the left branch.

63. $x^6-4x^3+3 = 0$. Let $u=x^3$. $u^2-4u+3=0$. $(u-1)(u-3)=0$.

   $u=1$ or $u=3$. $x^3=1$ or $x^3=3$. Two solutions, $x=1$ and
   $x=\sqrt[3]{3} = 1.4422$ (to 2 dec places). Graph $y=x^6-4x^3+3$ below.

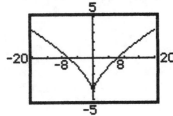

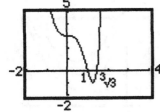

Exercise 61   Exercise 63

65. $x^2 + 8x - 2 = 0$. A: $x=-8.24$. B: $x=0.24$.
67. $-3x^2 + 7x + 15 = 0$. A: $x=-1.36$. B: $x=3.69$.
69. $5.76x^2 + 17.28x - 51.84 = 0$. A: $x=-4.85$. B: $x=1.85$.

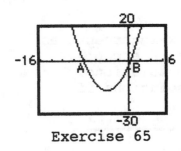

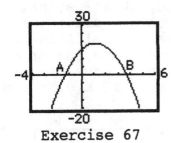

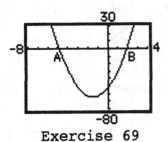

Exercise 65   Exercise 67   Exercise 69

71. $10x^3+17x^2-35x-12=0$. A:$x=-2.7977$, B:$x=-0.3056$, C:$x=1.4033$.
73. $5x^3-2x^2-17x+7=0$. A:$x=-1.8487$, B:$x=0.4124$, C:$x=1.8363$.
75. $x^4+x^3-14x^2-x+12=0$. A:$x=-4.1485$, B:$x=-0.9609$, C:$x=0.9540$.
   D: $x=3.1554$.

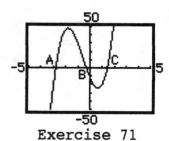
Exercise 71

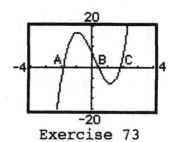
Exercise 73

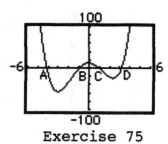
Exercise 75

77. $-x^3+2x^2+3x=5$. Write as $-x^3+2x^2+3x-5=0$.
A:$x=-1.6511$, B:$x=1.2739$, C:$x=2.3772$.

79. $y=x^2-10x+k$. $a=1,b=-10,c=k$. disc,$D=b^2-4ac=(-10)^2-4(1)(k)$.
$D=100-4k$. (a) Two real solutions, $100-4k>0$, $100>4k$, $k<25$.
(b) One real solution, $100-4k=0$. $k=25$.
(c) No real solutions, $100-4k<0$, $-4k<-100$, $k>25$.
Graphs below for $y=x^2-10x+k$ for $k=30$ (1), $k=25$ (2), $k=20$ (3).

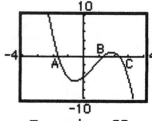
Exercise 77

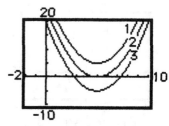
Exercise 79

81. Equation $x^4-8x^2+3x+k=0$. Consider graphs of $y=x^4-8x^2+3x+k$,
below. We see that:
$k=5$: 4 x-intercepts, thus 4 real solutions to equation.
$k=-0.25$: 3 x-intercepts, thus 3 solutions.
$k=-5$: 2 x-intercepts, thus 2 solutions.
These represent all the possibilities for x-intercepts as
k varies. Graph can never have one or no x-intercepts.
Thus the equation can never have one or no solutions.
In general: four solutions if $k>0$, three solutions if
$k=0$, and two if $k<0$.

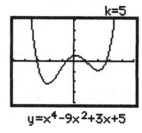
$y=x^4-9x^2+3x+5$

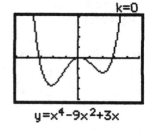
$y=x^4-9x^2+3x$

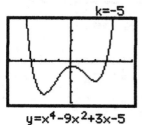
$y=x^4-9x^2+3x-5$

Windows: $-5 \le x \le 5$, $-40 \le y \le 40$.

Section 2.4

1. $x+2 \le 7$, $x \le 7-2$, $x \le 5$. Interval $(-\infty,5]$. See graph below.

. $-2x+1 < 11$, $-2x < 10$, $x > -5$. Interval $(-5,\infty)$.

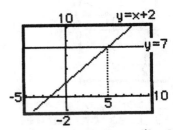

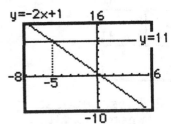

Line $y=x+2$ lies on/below
$y=7$ when $x\leq5$.
Exercise 1

Line $y=-2x+1$ lies below
$y=11$ when $x>-5$.
Exercise 3

. $5x \leq x+12$, $4x \leq 12$, $x \leq 3$. Interval $(-\infty,3]$.
. $5x-2 \leq x+2$, $4x \leq 4$, $x\leq 1$. Interval $(-\infty,1]$.

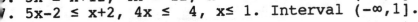

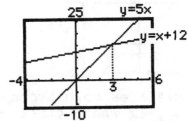

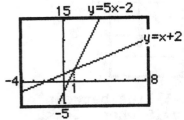

Line $y=5x$ lies on/below
$y=x+12$ when $x\leq3$.
Exercise 5

Line $y=5x-2$ lies on/below
$y=x+2$ when $x\leq1$.
Exercise 7

9. $19-3x > 5x+1$, $-8x > -18$, $x < \frac{9}{4}$. Interval $(-\infty,\frac{9}{4})$.

11. $2z+5 > -3z + 2$, $5z > -3$, $z>-\frac{3}{5}$. Interval $(-\frac{3}{5},\infty)$.

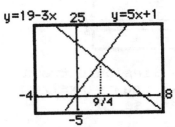

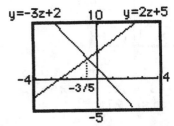

Line $y=19-3x$ lies above
$y=5x+1$ when $x<9/4$.
Exercise 9

Line $y=2z+5$ lies above
$y=-3z+2$ when $x>-3/5$.
Exercise 11

13. $1 < 3x-2 < 7$, $3 < 3x < 9$, $1 < x < 3$. Interval $(1,3)$.

15. $5 < x+3 \leq 7$, $2 < x \leq 4$. Interval $(2,4]$.

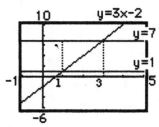

Line y=3x-2 lies between
y=1 and y=7 when 1<x<3.
Exercise 13

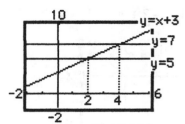

Line y=2x+1 lies on/between
Line y=x+3 lies on/between
Exercise 15

17. -9 < 4x-5 < 15, -4 < 4x < 20, -1 < x < 5. Interval (-1,5).
19. 2 < -2x + 4 ≤ 12, -2 < -2x ≤ 8, 1>x≥-4 or -4≤x<1. [-4,1).

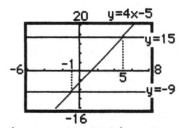

Line y=4x-5 lies between
y=-9 and y=15 when -1<x<5.
Exercise 17

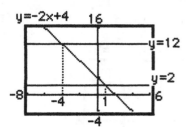

Line -2x+4 lies on/between
y=2 and y=12 when -4≤x<1
Exercise 19

21. 13 ≥ -3x + 1 ≥ 4, 12 ≥ -3x ≥ 3, -4 ≤ x ≤ -1. [-4,-1].
23. 1 ≤ 2x + 3 ≤ 5, -2 ≤ 2x ≤ 2, -1 ≤ x ≤ 1. [-1,1].

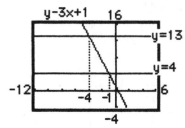

Line y=-3x+1 lies on/between
y=4 and y=13 when -4≤x≤-1.
Exercise 21

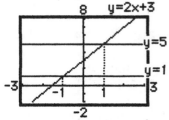

Line y=2x+3 lies on/between
y=1 and y=5 when -1≤x≤1.
Exercise 23

25. -4<x+2<8. Display gives -6<x<6.
    Algebra: -4-2<x<8-2, -6<x<6. (-6,6).

27. 2<-x+4<8. Display give -4<x<2.
    Algebra: 2-4<-x<8-4, -2<-x<4, 2>x>-4 or -4<x<2. (-4,2).

29. x^2+x-2 < 0. (x-1)(x+2) = 0. Critical values x = 1 & -2.
    (x-1): x-1<0 when x<1, x-1=0 when x=1, x-1>0 when x>1.
    (x+2): x+2<0 when x<-2, x+2=0 when x=-2, x+2>0 when x>-2.
    Get following sign diagram. Solution (-2,1).
    Graph below x-axis over (-2,1).

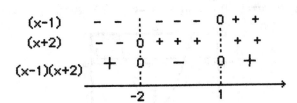
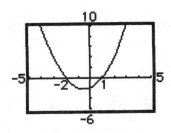

31. $-x^2-x+6 < 0$. $(-x+2)(x+3) = 0$. Critical values x = 2 & -3.
    $(-x+2)$: $-x+2<0$ when x>2, $-x+2=0$ when x=2, $-x+2>0$ when x<2.
    $(x+3)$: x+3<0 when x<-3, x+3=0 when x=-3, x+3>0 when x>-3.
    Get following sign diagram. Solution $(-\infty,-3) \cup (2,\infty)$.
    Graph below x-axis over $(-\infty,-3)$ and $(2,\infty)$.

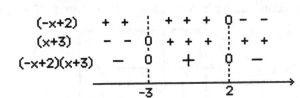
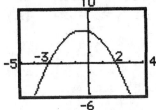

33. $x^2-1 > 0$. $(x+1)(x-1) = 0$. Critical values x = -1 & 1.
    $(x+1)$: x+1<0 when x<-1, x+1=0 when x=-1, x+1>0 when x>-1.
    $(x-1)$: x-1<0 when x<1, x-1=0 when x=1, x-1>0 when x>1.
    Get following sign diagram. Solution $(-\infty,-1) \cup (1,\infty)$.
    Graph above x-axis over $(-\infty,-1)$ and $(1,\infty)$.

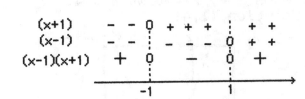
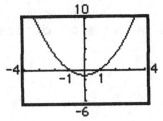

35. $x^2-9x-7<3$. $x^2-9x-10<0$. $(x+1)(x-10)=0$. Critical #s x=-1,10.
    $(x+1)$: x+1<0 when x<-1, x+1=0 when x=-1, x+1>0 when x>-1.
    $(x-10)$: x-10<0 when x<10, x-10=0 when x=10, x-10>0 when x>10.
    Get following sign diagram. Solution $(-1,10)$.
    Graph $y=x^2-9x-7$ (A) is below graph y=3 (B) over $(-1,10)$.

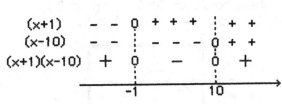
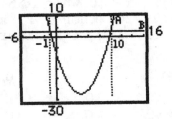

37. $x^2-8>1$. $x^2-9>0$. $(x+3)(x-3)=0$. Critical values x=-3,3.
    $(x+3)$: x+3<0 when x<-3, x+3=0 when x=-3, x+3>0 when x>-3.
    $(x-3)$: x-3<0 when x<3, x-3=0 when x=3, x-3>0 when x>3.
    Get following sign diagram. Soln $(-\infty,-3) \cup (3,\infty)$.

Graph of $y=x^2-8$(A) above graph $y=1$(B) over $(-\infty,-3)$ and $(3,\infty)$.

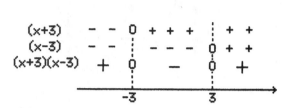

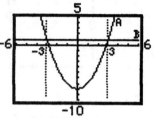

39. $x^2-5x-9\le-2x+1$. $x^2-3x-10\le0$. $(x+2)(x-5)=0$. Critical #s $x=-2,5$.
    $(x+2)$: $x+2<0$ when $x<-2$, $x+2=0$ when $x=-2$, $x+2>0$ when $x>-2$.
    $(x-5)$: $x-5<0$ when $x<5$, $x-5=0$ when $x=5$, $x-5>0$ when $x>5$.
    Get following sign diagram. Solution $[-2,5]$.
    Graph of $y=x^2-5x-9$ (A) on/below graph $y=-2x+1$ (B) over $[-2,5]$.

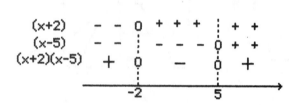

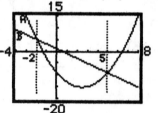

41. $x^2-5x+4<0$. Display gives graph of y below x-axis for x in
    interval $(1,4)$. Solutions $(1,4)$. Algebra:
    $(x-1)(x-4)=0$. Critical values $x=1$ and $x=4$.
    $(x-1)$: $x-1<0$ when $x<1$, $x-1=0$ when $x=1$, $x-1>0$ when $x>1$.
    $(x-4)$: $x-4<0$ when $x<4$, $x-4=0$ when $x=4$, $x-4>0$ when $x>4$.
    Get following sign diagram. Solution $(1,4)$.

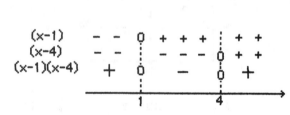

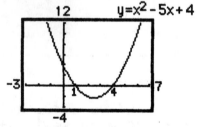

43. $-x^2+8x-12<0$. Display gives graph of y below x-axis for x on
    interval $(-\infty,2)$ and $(6,\infty)$. Solns $(-\infty,2)\cup(6,\infty)$.
    Algebra: $(-x+6)(x-2)=0$. Critical values $x=2$ and $x=6$.
    $(-x+6)$: $-x+6<0$ when $x>6$, $-x+6=0$ when $x=6$, $-x+6>0$ when $x<6$.
    $(x-2)$: $x-2<0$ when $x<2$, $x-2=0$ when $x=2$, $x-2>0$ when $x>2$.
    Get following sign diagram. Solution $(-\infty,2)\cup(6,\infty)$.

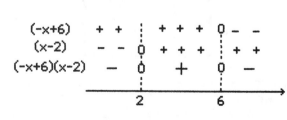

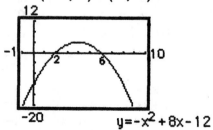

45. $\dfrac{3x+1}{x-1} < 2$, $\dfrac{3x+1}{x-1} - 2 < 0$, $\dfrac{(3x+1) - 2(x-1)}{x-1} < 0$, $\dfrac{x+3}{x-1} < 0$.
Critical values are x = -3, 1.
(x+3): x+3<0 when x<-3, x+3=0 when x=-3, x+3>0 when x>-3.
(x-1): x-1<0 when x<1, x-1=0 when x=1, x-1>0 when x>1.
Get following sign diagram. Solution (-3,1).
Graph of y=(3x+1)/(x-1) is below line y=2 over (-3,1).

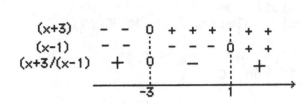 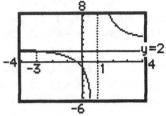

47. $\dfrac{-x+8}{x-2} < 1$, $\dfrac{-x+8}{x-2}-1<0$, $\dfrac{-x+8-(x-2)}{x-2} <0$, $\dfrac{-2x+10}{x-2} <0$, $\dfrac{2(-x+5)}{x-2} <0$,
$\dfrac{-x+5}{x-2} <0$. Critical values are x = 2, 5.
(-x+5): -x+5<0 when x>5, -x+5=0 when x=5, -x+5>0 when x<5.
(x-2): x-2<0 when x<2, x-2>0 when x>2.
Get following sign diagram. Solution (-∞,2)∪(5,∞).
Graph of y=(-x+8)/(x-2) is below line y=1 over (-∞,2) and (5,∞).

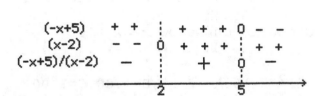 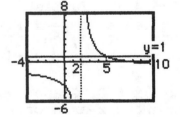

49. $\dfrac{6x}{x-4} >-6$, $\dfrac{6x}{x-4} +6 >0$, $\dfrac{6x+6(x-4)}{x-4} >0$, $\dfrac{12x-24}{x-4} >0$, $\dfrac{12(x-2)}{x-4} >0$,
$\dfrac{x-2}{x-4} >0$. Critical values are x = 2, 4.
(x-2): x-2<0 when x<2, x-2=0 when x=2, x-2>0 when x>2.
(x-4): x-4<0 when x<4, x-4>0 when x>4.
Get following sign diagram. Solution (-∞,2)∪(4,∞).
Graph of y=6x/(x-4) is above line y=-6 on (-∞,2) & (4,∞).

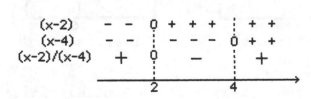 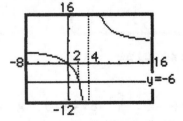

51. $x^3-2x^2-5x+6>0$. The graph is above the x-axis over the intervals (-2,1) and (3,∞). Solution (-2,1)∪(3,∞).

53. $x^4+5x^3+5x^2-5x-6\geq0$. The graph is on/above the x-axis over the intervals $(-\infty,-3]$, $[-2,-1]$ and $[1,\infty)$. Solutions $(-\infty,-3]\cup[-2,-1]\cup[1,\infty)$.

55. $x^2-4x-6<0$. Graph below. A=-1.16, B=5.16 (to 2 dec places) Graph is below x-axis on (A,B). Solution (-1.16,5.16).

57. $-x^2+6x+15>0$. Graph below. A=-1.90, B=7.90. Graph is above x-axis on (A,B). Solution (-1.90,7.90).

59. $x^3-12x+9>0$. Graph below. A=-3.79, B=0.79, C=3.00. Graph is above x-axis on (A,B) and (C,∞). Solution (-3.79,0.79)$\cup$(3,∞).

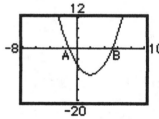

Exercise 55

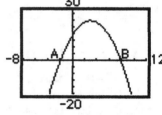

Exercise 57

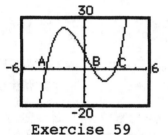

Exercise 59

61. $-x^3+3x^2+4x\leq7$. Easiest to write as $-x^3+3x^2+4x-7\leq0$ and find x-intercepts of $y=-x^3+3x^2+4x-7$. Graph below. A=-1.71, B=1.14, C=3.57. Graph is on/below x-axis on [A,B] and [C,∞). Solution [-1.71,1.14]$\cup$[3.57,∞).

63. $x^3-2x^2-5x+6>0$. Graph $y=x^3-2x^2-5x+6$. A=-2.00,B=1.00,C=3.00. Graph above x-axis on (A,B) and (C,∞). Solution (-2,1)$\cup$(3,∞).

65. $\dfrac{6x+2}{x-2}<3$. Graph $y=\dfrac{6x+2}{x-2}-3$, below, A=-2.67, B=2. Graph is below x-axis on (A,B). Solution (-2.67,2).

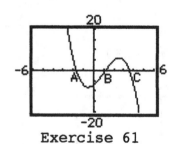

Exercise 61

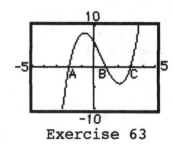

Exercise 63

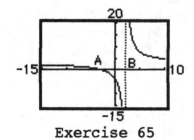

Exercise 65

67. $x^2>6x-9$. $x^2-6x-9>0$. $x^2-6x-9=0$. (x-3)(x-3)=0. Single critical value x=3.
x<3: (x-3)<0, (x-3)(x-3)>0. Solutions.
x>3: (x-3)>0, (x-3)(x-3)>0. Solutions.
x=3 does not satisfy the inequality. Not a solution.

78

Thus solution set is $(-\infty, 3) \cup (3, \infty)$. $x=3$ is the only real number that does not satisfy the inequality. See graphs below. The line touches the curve at the point $(3,9)$. We say that it is the tangent to the curve at $(3,9)$. The curve lies above the line for all x values other than 3.

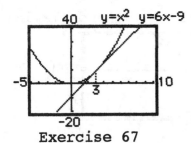

Exercise 67

69. $F = \frac{9}{5}C + 32$. $F \geq 59$, $\frac{9}{5}C + 32 \geq 59$, $\frac{9}{5}C \geq 27$, $C \geq \frac{5}{9}(27)$, $C \geq 15°$.

71. $C = \frac{5}{9}(F-32)$. $C > 20$, $\frac{5}{9}(F-32) > 20$, $F-32 > (\frac{9}{5})20$, $F > 36+32$, $F > 68°$.

73. $C = 16n + 1200$. $32000 \leq C \leq 38000$. $32000 \leq 16n+1200 \leq 38000$.
$30800 \leq 16n \leq 36800$. $1925 \leq n \leq 2300$. The number of items produced should be kept between 1,925 and 2,300.
Graph of $C = 16n+1200$ lies on/between $C=3800$ and $C=3200$ for values of n on/between 1925 and 2300. See graph below.

75. (a) $A = P + Prt$. $P = 5000$, $r = .04$. Thus $A = 5000 + (5000)(.04)t$,
$A = 5000 + 200t$. Want $A > 10000$. $5000 + 200t > 10000$. $200t > 5000$, $t > 25$.
Money should be left in the account for over 25 years.
(b) $A = P + Prt$. $r = .06$, $t = 10$. Thus $A = P + P(.06)(10)$, $A = P + .6P$,
$A = 1.6P$. For $A > 6000$, $1.6P > 6000$, $P > 6000/1.6$, $P > 3750$.
The initial investment should be over \$3,750.

77. $A = (1/2)bh$. $b = 6$. Thus $A = 3h$. Want $A > 24$. Thus $3h > 24$, $h > 8$.
The height must be greater than 8 inches.
Graph of $A = 3h$ lies above line $A = 24$ when $h > 8$.

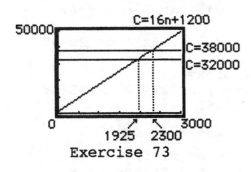

Exercise 73

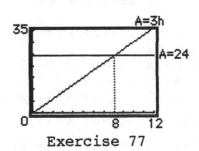

Exercise 77

79. $s = -16t^2 + vt$. $v = 240$. $s = -16t^2 + 240t$.
$s < 576$ when $-16t^2 + 240t < 576$, $-16t^2 + 240t - 576 < 0$.
$16(-t^2 + 15t - 36) < 0$, $16(-t+3)(t-12) < 0$, $(-t+3)(t-12) < 0$.
Corresponding equation is $(-t+3)(t-12) = 0$.

Critical values are t=3 and t=12.
(-t+3): -t+3<0 when t>3, -t+3=0 when t=3, -t+3>0 when t<3.
(t-12): t-12<0 when t<12, t-12=0 when t=12, t-12>0 when t>12.
Get following sign diagram. Solution $(-\infty,3)\cup(12,\infty)$.
Physically however, s=0 when $-16t^2+240t=0$, 16t(t-15) =0.
Object is at ground level at t=0,15.
Object is below 576' during time intervals [0,3) and (12,15].

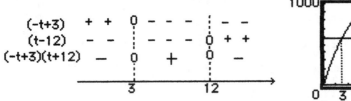
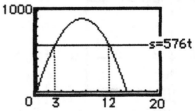

81. $D=900P-P^2$. Want D>100000. Thus $900P-P^2>100000$.
    $-P^2+900P-100000>0$. Graph of $y=-P^2+900P-100000$ is above
    P-axis from A to B. A=129.84379, B=770.15621.
    Thus price should lie between $129.84 and $770.16.

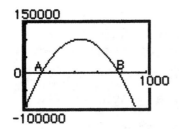

Section 2.5

1. $|x|=5$. x=5 or x=-5.
3. $|x-4|=6$. x-4=6 or x-4=-6. x=10 or -2.
5. $|3x+4|=9$. 3x+4=9 or 3x+4=-9. 3x=5 or 3x=-13. x=5/3 or
    x=-13/3.
7. $|3z-1|=4$. 3z-1=4 or 3z-1=-4. 3z=5 or 3z=-3. z=5/3 or -1.
9. $|2x-4|=|x-2|$. 2x-4=x-2 or 2x-4=-(x-2). x=2 or 3x=6. x=2.
11. $|3x+2|=|x-6|$. 3x+2=x-6 or 3x+2=-(x-6). 2x=-8 or 4x=4.
    x=-4 or x=1.
13. $|5x-2|=|2-3x|$. 5x-2=2-3x or 5x-2=-(2-3x). 8x=4 or 2x=0.
    x=1/2 or x=0.
15. $|-5x|=20$. -5x=20 or -5x=-20. x=-4 or x=4.
17. $|2x-3|=|-9|$. 2x-3=-9 or 2x-3=9. 2x=-6 or 2x=12. x=-3 or 6.
19. $|3x|=15$. 3x=15 or 3x=-15. x=5 or x=-5.
21. $\dfrac{3}{|2x-1|}=\dfrac{6}{7}$. 21=6|2x-1|. 7/2=|2x-1|. 2x-1=7/2 or
    2x-1=-7/2. 2x=9/2 or 2x=-5/2. x=9/4 or x=-5/4.
    Original equation is satisfied for both values of x.

23. $|x|<4$. -4<x<4. Set of solutions is (-4,4). Graph below.
    Graph of y=|x| is below y=4 over (-4,4).

25. |x|>5. x<-5 or x>5. (-∞,-5)U(5,∞). Graph of y=|x| lies above
    y=5 over the intervals (-∞,-5) and (5,∞).

27. |x-3|>8. x-3<-8 or x-3>8. x<-5 or x>11. (-∞,-5)U(11,∞).
    Graph of y=|x-3| lies above y=8 over (-∞,-5) and (11,∞).

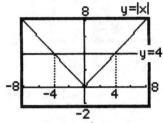
Exercise 23

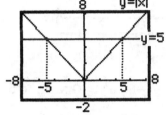
Exercise 25

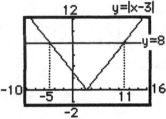
Exercise 27

29. |2x-3|≥7. 2x-3≤-7 or 2x-3≥7. 2x≤-4 or 2x≥10. x≤-2 or x≥5. Set
    of solutions is (-∞,-2]U[5,∞).
    Graph of y=2x-3 is on/above y=7 over (-∞,-2] and [5,∞).

31. |1-2x|>7. 1-2x<-7 or 1-2x>7. -2x<-8 or -2x>6.
    x>4 or x<-3. (-∞,-3)U(4,∞).
    Graph of y=1-2x is above y=7 over (-∞,-3) and (4,∞).

33. |9-3t|≥6. 9-3t≤-6 or 9-3t≥6. -3t≤-15 or -3t≥-3.

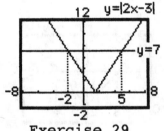
Exercise 29

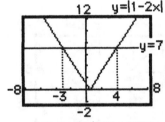
Exercise 31

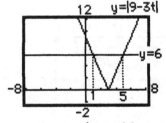
Exercise 33

35. |2x+1|<7. Display shows that the graph of y=|2x+1| lies below
    line y=7 over the interval (-4,3). Soln is (-4,3).
    Algebra: -7<2x+1<7. -8<2x<6. -4<x<3.

37. |2x-5|≤7. Display: Graph of y=|2x-5| lies on/below y=7 over
    [-1,6]. Solution is [-1,6].
    Algebra: -7≤2x-5≤7. -2≤2x≤12. -1≤x≤6.

39. |x+1|<5. Display shows that graph of y=|x+1| lies below y=5
    over (-6,4). Solution (-6,4).
    Algebra: -5<x+1<5. -6<x<4.

41. |2x-7|≤7. Display shows that graph of y=|2x-7| lies on/below
    y=7 over [0,7]. Solution [0,7].
    Algebra: -7≤2x-7≤7. 0≤2x≤14. 0≤x≤7.

43. |5x-7|≥|-12|. |5x-7|≥12. 5x-7≤-12 or 5x-7≥12.
    5x≤-5 or 5x≥19. x≤-1 or x≥19/5. (-∞,-1]U[19/5,∞).

Graph of y=|5x-7| is on/above y=|-12| over (-∞,-1] & [19/5,∞).

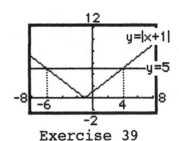
Exercise 39

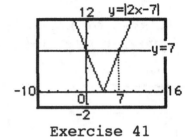
Exercise 41

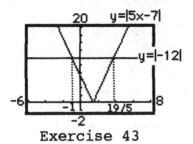
Exercise 43

45. |x|+2<8. |x|<6. -6<x<6. (-6,6). Graph of y=|x|+2 is below y=8 over (-6,6).

47. |$\frac{x-1}{3}$|≤5. -5 ≤ $\frac{x-1}{3}$ ≤ 5. -15≤x-1≤15. -14≤x≤16. [-14,16].

    y=|$\frac{x-1}{3}$| is on/below y=5 over [-14,16].

49. 2<|x|<5. Two cases:
    (i) 2<|x|. x<-2 or x>2.x must lie in (-∞,-2)∪(2,∞).
    (ii) |x|<5. -5<x<5. x must lie in (-5,5).
    Combine (i) and (ii). x must lie in (-5,-2) or (2,5).
    Solution set is (-5,-2)∪(2,5).
    Graph y=|x| lies between y=2 and y=5 on (-5,-2) & (2,5).

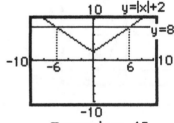
Exercise 45

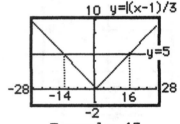
Example 47

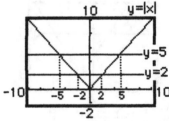
Exercise 49

51. 9>|2x-5|>0. (i) 9>|2x-5|. -9<2x-5<9. -4<2x<14. -2<x<7.
    x must lie in (-2,7).
    (ii) |2x-5|>0. 2x-5<0 or 2x-5>0. 2x<5 or 2x>5. x<5/2 or x>5/2.
    x must lie in (-∞,5/2)∪(5/2,∞). i.e. x≠5/2.
    Combine (i) and (ii). Solution set is (-2,5/2)∪(5/2,7).
    Graph y=|2x-5| between y=9 & x-axis over (-2,5/2) & (5/2,7).

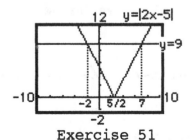
Exercise 51

82

Chapter 2 Project

1. Square root: (a) Let us use 10 = 2x5. Thus a=2 and b=5.

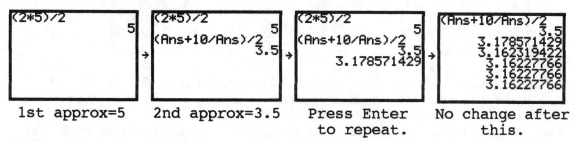

1st approx=5   2nd approx=3.5   Press Enter   No change after
                                      to repeat.   this.

   Best approximation is 3.16227766. Calculator gives same.

(b) Why the method works: (a+b)/2 is the average of a and b.
   Ans=(a+b)/2 lies between a and b. At the next step we use
   (Ans+n/Ans)/2. Note that n = Ansx(n/Ans). Thus (Ans+n/Ans)/2
   repeats taking averages, but with n now lying between Ans and
   n/Ans, two numbers that are closer together than a and b. We
   get closer and closer to Ans being equal to n/Ans. If Ans=n/Ans
   then of course n=Ans^2, and Ans = $\sqrt{n}$. Thus the sequence of
   averages gets closer and closer to $\sqrt{n}$.

2. Quadratic Equation: (a) Construct a quadratic with solutions -3
   and 1. Let (x+3)(x-1)=0. The equation is x^2+2x-3=0.
   Try x=8 as first guess, below. We get the solution x=1 closest
   to 8. Next try x=-6, a negative number to search for a second
   solution. We arrive at x=-3, below.

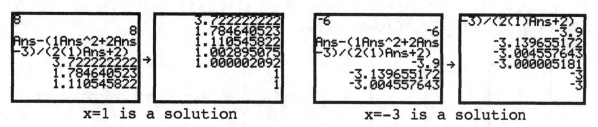

      x=1 is a solution   x=-3 is a solution

(b) Single solution: Use (x-4)(x-4)=0. x^2-8x+16=0 has the single
solution x=4. Use x=2 as the first guess. The best approximation
we can get to this single solution is is 3.999999215

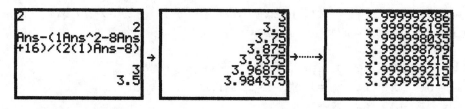

(c) No solution: Use 2x^2+3x+5=0. a=2,b=3,c=5. b^2-4ac=-31<0,
thus no solution. Let first guess be 1. It can be seen below
that the sequence of numbers does not converge to any specific
number, revealing that the equation has no real solutions.

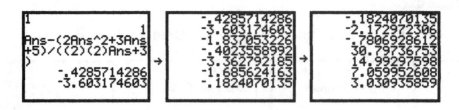

(d) Use $x^2+2x-3=0$ of (a), which has solutions -3 and 1. $a=1$, $b=2$. Try $-b/2a=-1$ as initial guess. We get an error with message that divide by zero has occured. On inspection we see that the denominator of algorithmic expression is zero when Ans$=-b/2a$.

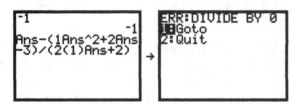

Chapter 2 Review Exercises

1. (a) $x^2+x-2 = 0$, $(x+2)(x-1) = 0$. $x = -2,1$.
   See graph of $y=x^2+x-2$ below.

   (b) $x^2+10x+21 = 0$, $(x+7)(x+3) = 0$, $x =-7,-3$.
   See graph of $y=x^2+10x+21$ below.

   (c) $2x^2+9x-5 = 0$, $(2x-1)(x+5) = 0$, $x = 1/2, -5$
   See graph of $y=2x^2+9x-5$ below.

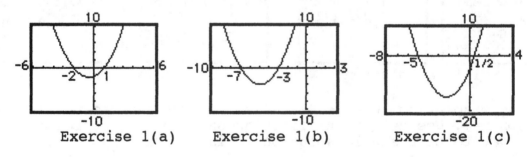

Exercise 1(a)   Exercise 1(b)   Exercise 1(c)

1. (d) $6x^2-17x-14 = 0$, $(3x+2)(2x-7) = 0$, $x = -2/3, 7/2$.
   See graph of $y=6x^2-17x-14$ below.

2. (a) $x^2=16$. $x=\pm4$. See graph of $y=x^2-16$ below.
   (b) $9x^2-1=0$. $9x^2=1$. $x^2=1/9$. $x=\pm1/3$.
   See graph of $y=9x^2-1$ below.

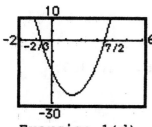

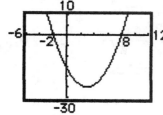

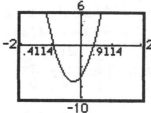

Exercise 1(d)	Exercise 2(a)	Exercise 2(b)

2. (c) $2x^2-8=0$. $2x^2=8$. $x^2=4$. $x=\pm2$. See graph of $y=2x^2-8$ below.

(d) $(x-3)^2=25$. $(x-3)=\pm5$. $x=\pm5+3$. $x=8$ or $x=-2$.
See graph of $y=(x-3)^2-25$ below.

(e) $(4x+1)^2=7$. $(4x+1)=\pm\sqrt{7}$. $4x=\pm\sqrt{7}-1$. $x=\pm\sqrt{7}/4-1/4$.
$x=0.4114$ or $x=-0.9114$ (to 4 decimal places).
See graph of $y=(4x+1)^2-7$ below.

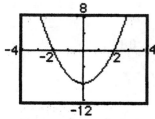

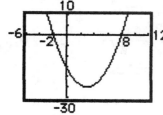

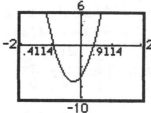

Exercise 2(c)	Exercise 2(d)	Exercise 2(e)

3. (a) $x^2-10x+21=0$. $x^2-10x=-21$. $x^2-10x+(-5)^2=-21+(-5)^2$.
$(x-5)^2=4$. $x-5=\pm2$. $x=3,7$.

(b) $x^2+14x+24=0$. $x^2+14x=-24$. $x^2+14x+(7)^2=-24+(7)^2$.
$(x+7)^2=25$. $x+7=\pm5$. $x=-12,-2$.

(c) $2x^2-16x+31=0$. $x^2-8x=-31/2$. $x^2-8x+(-4)^2=-31/2+(-4)^2$.
$(x-4)^2=1/2$. $x-4=\pm\sqrt{1/2}$. $x=4\pm\sqrt{1/2}$.

(d) $3x^2-6x-8=0$. $x^2-2x=8/3$. $x^2-2x+(-1)^2=8/3+(-1)^2$.
$(x-1)^2=1/2$. $x-1=\pm\sqrt{11/3}$. $x=1\pm\sqrt{11/3}$.

4. (a) $x^2+4x-3 = 0$, $a=1$, $b=4$, $c=-3$.
$$x = \frac{-4\pm\sqrt{4^2 - 4(1)(-3)}}{2(1)} = \frac{-4\pm\sqrt{28}}{2} = -2\pm\sqrt{7} = 0.6458, -4.6458.$$
See graph of $y=x^2+4x-3$ below.

(b) $x^2-x-9 = 0$, $a=1$, $b=-1$, $c=-9$.
$$x = \frac{1\pm\sqrt{(-1)^2 - 4(1)(-9)}}{2(1)} = \frac{1\pm\sqrt{37}}{2} = 3.5414, -2.5414.$$
See graph of $y=x^2-x-9$ below.

(c) $2x^2+8x-7 = 0$, $a=2$, $b=8$, $c=-7$.

$$x = \frac{-8\pm\sqrt{(8)^2 - 4(2)(-7)}}{2(2)} = \frac{-8\pm\sqrt{120}}{4} = 0.7386, -4.7386.$$

See graph of $y=2x^2+8x-7$ below.

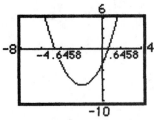
Exercise 4(a)

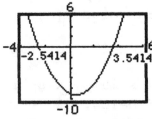
Exercise 4(b)

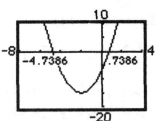
Exercise 4(c)

(d) $3x^2+9x+1 = 0$, $a=3$, $b=9$, $c=1$.

$$x = \frac{-9\pm\sqrt{9^2 - 4(3)(1)}}{2(3)} = \frac{-9\pm\sqrt{69}}{6} = -0.1156, -2.8844.$$

See graph of $y=3x^2+9x+1$ below.

(e) $4x^2+x+2 = 0$, $a=4$, $b=1$, $c=2$.

$$x = \frac{-1\pm\sqrt{1^2 - 4(4)(2)}}{2(2)} = \frac{-1\pm\sqrt{-31}}{4}. \quad \sqrt{-31} \text{ not a real number.}$$

No solutions. See graph of $y= 4x^2+x+2$ below.

(f) $x^2+6x+9 = 0$, $a=1$, $b=6$, $c=9$.

$$x = \frac{-6\pm\sqrt{6^2 - 4(1)(9)}}{2(1)} = \frac{-6\pm\sqrt{0}}{2} = -3. \text{ Single solution.}$$

See graph of $y=x^2+6x+9$ below.

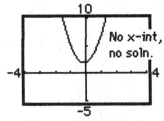
Exercise 4(d)

No x-int, no soln.

Exercise 4(e)

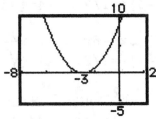
Exercise 4(f)

5. For the instructor's convenience we give the quadratic formula form of the solutions for this exercise since zero finders vary from calculator to calculator, and the graph. $2.3x^2-3.5x-6.8=0$. $a=2.3$, $b=-3.5$, $c=-6.8$.

$$x = \frac{3.5 \pm \sqrt{(-3.5)^2 - 4(2.3)(-6.8)}}{2(2.3)} = 2.6411 \text{ or } -1.1194.$$

6. (a) $x^2+3x-5 = 0$. a=1, b=3, c=-5.
   $b^2-4ac = (3)^2 - 4(1)(-5) = 9+20 = 29 > 0$.
   Two real solutions. See graph of $y=x^2+3x-5$ below.
   Two x-int, two real solutions.

   (b) $3x^2+x-2 = 0$. a=3, b=1, c=-2.
   $b^2-4ac = (1)^2 - 4(3)(-2) = 1+24 = 25 > 0$.
   Two real solutions. See graph of $y=3x^2+x-2$ below.
   Two x-int, two real solutions.

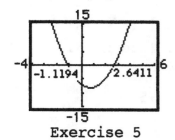
Exercise 5

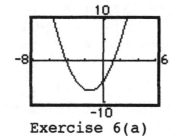
Exercise 6(a)

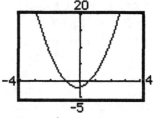
Exercise 6(b)

6. (c) $7x^2+2x+3 = 0$. a=7, b=2, c=3.
   $b^2-4ac = (2)^2 - 4(7)(3) = 4-84 = -80 < 0$.
   No real solutions. See graph of $y=7x^2+2x+3$ below.
   No x-intercepts, no real solutions.

   (d) $9x^2-6x+1 = 0$. a=9, b=-6, c=1.
   $b^2-4ac = (-6)^2 - 4(9)(1) = 36-36 = 0$.
   One real solution. See graph of $y=9x^2-6x+1$ below.
   One x-intercept, one real solution.

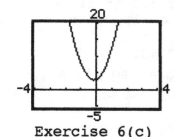
Exercise 6(c)

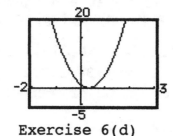
Exercise 6(d)

7. Any quadratic equation whose solutions are x=1 and x=-4 must have factors (x-1) and (x+4). Thus let the equation be (x-1)(x+4)=0. That is $x^2+3x-4=0$. See graph of $y=x^2+3x-4$ below. Any other quadratic equation of the form p(x-1)(x+4)=0 for p≠0 also has solutions x=1 and x=-4. That is $px^2+3px-4p=0$.

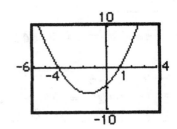

8. Let x mph be speed of boat in still water.
   Speed of boat upstream = x-5 and speed downstream = x+5.
   Distance=40 miles. Time upstream= $\frac{40}{x-5}$, &downstream= $\frac{40}{x+5}$.

   Upstream time is 2 hrs more than downstream.Thus $\frac{40}{x-5} - \frac{40}{x+5}$ =2.

   40(x+5) - 40(x-5) = 2(x-5)(x+5).
   40x+200-40x+200=2x²-50. 2x²=450. x²=225. x=±15.
   Cannot have negative speed in still water. Thus x=15.
   Speed of boat in still water is 15 mph. Graph:
   Graph y= $\frac{40}{x-5} - \frac{40}{x+5}$ and y=2 (below). $\frac{40}{x-5} - \frac{40}{x+5}$ = 2 when x=-15
   and x=15. The speed of the boat is 15 mph.

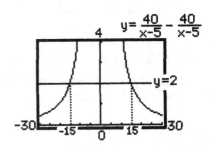

9. s=-16t²+vt. v=224. Thus s=-16t²+224t. When s=720,
   720=-16t²+224t. -16t²+224t-720=0. (-16t+144)(t-5)=0.
   t=144/16=9 or t=5. It will be at a height of 720 ft at 5
   seconds and 9 seconds; once on way up, once on way down.

10.(a) x⁴-x²-6=0. Let u=x². u²-u-6=0. (u-3)(u+2)=0.
   u=3 or u=-2.x²=3 or x²=-2. x=±√3=±1.7321 (to 4 dec places)
   x²=-2, no real solns. See graph of y=x⁴-x²-6 below.

   (b) 2x⁴+5x²-3=0. Let u=x². 2u²+5u-3=0. (2u-1)(u+3)=0.
   u=1/2 or u=-3. x²=1/2 or x²=-3. x=±1/√2=±0.7071.
   x²=-3, no solns. See graph of y=2x⁴+5x²-3 below.

   (c) 15(x+3)⁴-7(x+3)²-2=0. Let u=(x+3)². 15u²-7u-2=0.
   (5u+1)(3u-2)=0. u=-1/5 or u=2/3. (x+3)²=-1/5 or (x+3)²=2/3
   (x+3)²=-1/5 has no solution. x+3=±√2/3. x=-3±√2/3.
   x=-3.8165,-2.1835.See of graph y=15(x+3)⁴-7(x+3)²-2 below.

Chapter 2 Review Exercises

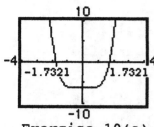

Exercise 10(a)

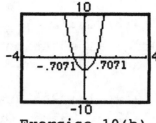

Exercise 10(b)

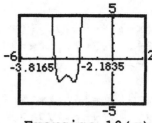

Exercise 10(c)

10. (d) $x-x^{1/2}-12=0$. Let $u=x^{1/2}$. $u^2-u-12=0$. $(u-4)(u+3)=0$.
$u=4$ or $u=-3$. $x^{1/2}=4$ or $x^{1/2}=-3$. $x^{1/2}=4$ gives $x=16$.
$x^{1/2}=-3$ has no solution since $x^{1/2}$ cannot be negative.
See graph of $y=x+x^{1/2}-12$ below.

11. (a) $\sqrt{4x-3}=1$. Square, $4x-3=1$. $4x=4$, $x=1$. Original
equation is satisfied. See graphs $y=\sqrt{4x-3}$ & $y=1$ below.

(b) $\sqrt{3x+7}=7$. Square, $3x+7=49$. $3x=42$, $x=14$. Original
equation is satisfied. See graphs $y=\sqrt{4x-3}$ & $y=7$ below.

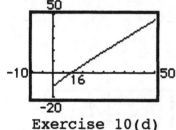

Exercise 10(d)

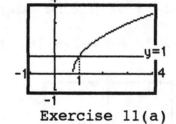

Exercise 11(a)

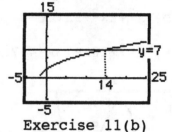

Exercise 11(b)

(c) $\sqrt{3x}=\sqrt{6x-1}$. Square, $3x=6x-1$. $3x=1$, $x=1/3$. Original
equation satisfied. Graphs $y=\sqrt{3x}$ (1), $y=\sqrt{6x-1}$ (2) below.

(d) $\sqrt{4x+3}=\sqrt{2x-1}$. Square. $4x+3=2x-1$. $2x=-4$, $x=-2$. Check
in original equ. $\sqrt{4x+3}=\sqrt{-5}$ and $\sqrt{2x-1}=\sqrt{-5}$. Not a real
number. No solutions. See graphs $y=\sqrt{4x+3}$ (1) & $y=\sqrt{2x-1}$
(2) below. No pt. of intersection, thus no solution.

(e) $\sqrt{x+2}+\sqrt{x-3}=5$. $\sqrt{x+2}=5-\sqrt{x-3}$. Square.
$x+2=25-10\sqrt{x-3}+(x-3)$. $-20=-10\sqrt{x-3}$. $2=\sqrt{x-3}$. Square.
$4=x-3$, $x=7$. Original equation is satisfied.
See graphs $y=\sqrt{x+2}+\sqrt{x-3}$ & $y=5$ below.

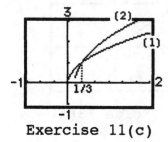

Exercise 11(c)

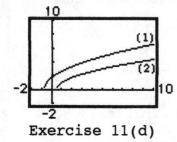

Exercise 11(d)

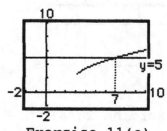

Exercise 11(e)

89

(f) $(x+3)^{2/3} = 9$. $((x+3)^{2/3})^{3/2} = 9^{3/2}$. $((x+3) = (9^{1/2})^3$. $x+3 = 3^3$. $x+3 = 27$, $x = 24$. Original equation is satisfied. See graphs of $y = (x+3)^{2/3}$ and $y = 9$ below.

(g) $x^{-4/5} = 16$. $(x^{-4/5})^{-5/4} = (16)^{-5/4}$. $x = ((16)^{1/4})^{-5}$. $x = (\pm 2)^{-5}$, $x = (\pm 1/2)^5$, $x = \pm 1/32 = \pm 0.03125$. Original equation is satisfied. See graphs of $y = x^{-4/5}$ and $y = 16$ below.

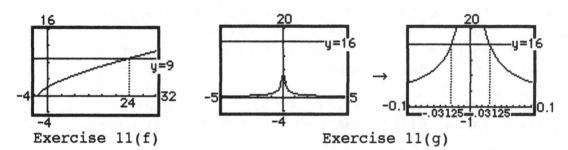

Exercise 11(f)   Exercise 11(g)

12. (a) $3x^3 - 7x^2 - 22x + 19 = 0$. A:$x = -2.2296$, B:$x = 0.7437$, C:$x = 3.8192$.

(b) $x^3 + 2x^2 + 5x + 16 = 0$. A:$x = -2.5270$. One solution.

(c) $4x^4 + 20x^3 + 3x^2 - 47x + 40 = 0$. A:$x = -3.8176$, B:$x = -2.8356$.

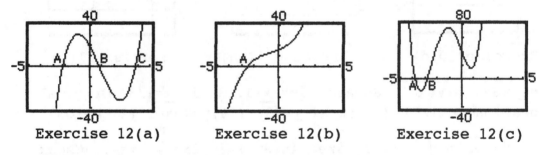

Exercise 12(a)   Exercise 12(b)   Exercise 12(c)

13. (a) $x - 3 \le 7$. $x \le 7 + 3$, $x \le 10$. $(-\infty, 10]$. Graph of $y = x - 3$ lies on/below line $y = 7$ when $x \le 10$.

(b) $2x - 7 < 15$. $2x < 22$, $x < 11$. $(-\infty, 11)$. Graph of $y = 2x - 7$ lies below line $y = 15$ when $x < 11$.

(c) $3x + 2 \ge -x$. $4x \ge -2$, $x \ge -1/2$. $[-1/2, \infty)$. Graph of $y = 3x + 2$ lies on/above $y = -x$ when $x \ge -1/2$.

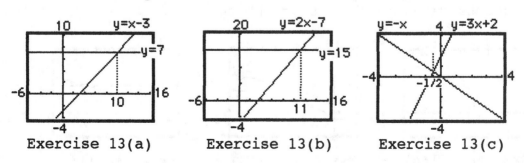

Exercise 13(a)   Exercise 13(b)   Exercise 13(c)

13. (d) 4x-5<7x+1. -3x<6, x>-2. (-2,∞). Graph of y=4x-5 lies below y=7x+1 when x>-2.

14. (a) 5≤2x+1≤7. 4≤2x≤6, 2≤x≤3. [2,3]. Graph of y=2x+1 lies on/between lines y=5 and y=7 when 2≤x≤3.

(b) -3<4x-3<9. 0<4x<12. 0<x<3. (0,3). Graph of y=4x-3 lies between lines y=-3 and y=9 when 0<x<3.

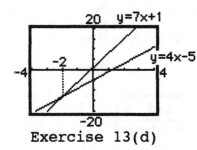
Exercise 13(d)

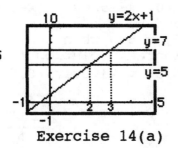
Exercise 14(a)

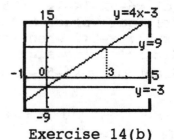
Exercise 14(b)

14. (c) 22≥3x+1≥5. 21≥3x≥4, 7≥x≥4/3, 4/3≤x≤7. [4/3,7]. Graph of y=3x+1 lies on/between y=22 and y=5 when 4/3≤x≤7.

(d) -3<4x+1≤7. -4<4x≤6, -1<x≤3/2. (-1,3/2]. Graph of y=4x+1 lies on/between y=-3 and y=7 when -1<x≤3/2.

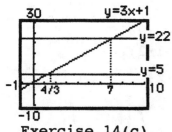
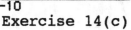
Exercise 14(c)

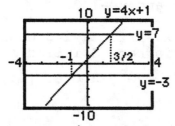
Exercise 14(d)

15. (a) x²-x-6≤0. (x+2)(x-3)=0. Critical values x=-2 and 3.
(x+2): x+2<0 when x<-2, x+2=0 when x=-2, x+2>0 when x>-2.
(x-3): x-3<0 when x<3, x-3=0 when x=3, x-3>0 when x>3.
Get following sign diagram. Solution [-2,3].
Graph on or below x-axis over [-2,3].

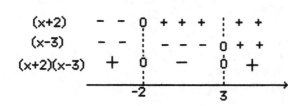
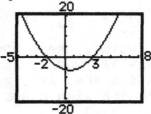

(b) x²-5x+4>0. (x-1)(x-4)=0. Critical values x=1 and 4.
(x-1): x-1<0 when x<1, x-1=0 when x=1, x-1>0 when x>1.
(x-4): x-4<0 when x<4, x-4=0 when x=4, x-4>0 when x>4.
Get following sign diagram. Solution (-∞,1)∪(4,∞).
Graph above x-axis over (-∞,1)∪(4,∞).

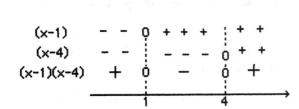
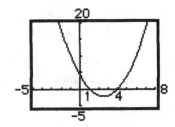

(c) $x^2-5x-1<-1$. $x^2-5x<0$. $x^2-5x=0$. $x(x-5)=0$.
Critical values $x=0$ and 5.
$(x-5)$: $x-5<0$ when $x<5$, $x-5=0$ when $x=5$, $x-5>0$ when $x>5$.
Get following sign diagram. Solution $(0,5)$.
Graph of $y=x^2-5x-1$ below $y=-1$ over $(0,5)$.

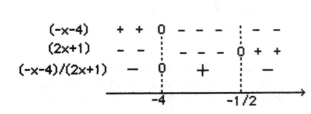
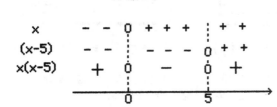

16. (a) $\dfrac{5x-1}{2x+1} \geq 3$. $\dfrac{5x-1}{2x+1} -3 \geq 0$. $\dfrac{(5x-1)-3(2x+1)}{2x+1} \geq 0$. $\dfrac{-x-4}{2x+1} \geq 0$.
Critical vales, $x=-4$ and $x=-1/2$.
$(-x-4)$: $-x-4<0$ when $x>-4$, $-x-4=0$ when $x=-4$, $-x-4>0$ when $x<-4$.
$(2x+1)$: $2x+1<0$ when $x<-1/2$, $2x+1>0$ when $x>-1/2$.
Get following sign diagram. Solution $[-4,-1/2)$.

Graph of $y= \dfrac{5x-1}{2x+1}$ on/above $y=3$ over $[-4,-1/2)$.

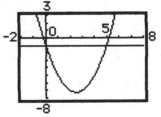
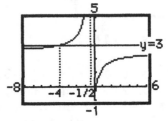

(b) $\dfrac{6x+3}{x+4} \leq 5$. $\dfrac{6x+3}{x+4} -5 \leq 0$. $\dfrac{(6x+3)-5(x+4)}{x+4} \leq 0$. $\dfrac{x-17}{x+4} \leq 0$.
Critical values, $x=17$ and $x=-4$.
$(x-17)$: $x-17<0$ when $x<17$, $x-17=0$ when $x=17$, $x-17>0$ when $x>17$.
$(x+4)$: $x-4<0$ when $x<4$, $x-4>0$ when $x>4$.
Get following sign diagram. Solution $(-4,17]$.
Graph of $y=\dfrac{6x+3}{x+4}$ on/below $y=5$ over $(-4,17]$.

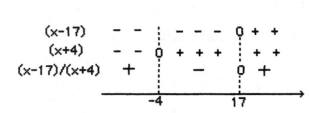
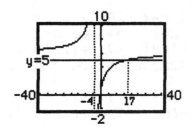

(c) $\dfrac{-x+1}{x-5}>-3$. $\dfrac{-x+1}{x-5}+3>0$. $\dfrac{(-x+1)+3(x-5)}{x-5}>0$. $\dfrac{2x-14}{x-5}>0$.

$\dfrac{2(x-7)}{x-5}>0$. $\dfrac{(x-7)}{x-5}>0$. Critical values x=7, 5.

(x-7): x-7<0 when x<7, x-7=0 when x=7, x-7>0 when x>7.
(x-5): x-5<0 when x<5, x-5>0 when x>5.
Get following sign diagram. Solution $(-\infty,5)\cup(7,\infty)$.

Graph of $y=\dfrac{-x+1}{x-5}$ above y=-3 over$(-\infty,5)$ and$(7,\infty)$.

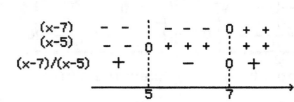
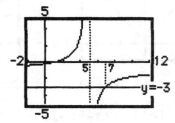

17. $s(t)=-16t^2+vt$. v=288. $s(t)=-16t^2+288t$.
    s>1232 when $-16t^2+288t>1232$. $-16t^2+288t-1232>0$.
    $16(-t^2+18t-77)>0$. $(-t+7)(t-11)>0$. Critical vals, t=7,11.
    (-t+7): -t+7<0 when t>7, -t+7=0 when t=7, -t+7>0 when t<7.
    (t-11): t-11<0 when t<11, t-11=0 when t=11, t-11>0 when t>11.
    Get following sign diagram. Solution (7,11).
    Graph of $s=-16t^2+288t$ is above s=1232 over (7,11).
    Object is higher than 1232 feet between 7 and 11 seconds.

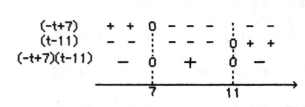
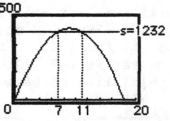

18. A=P+Prt. P=1000, r=.06. A=1000+(1000)(.06)t. A=1000+60t.
    A>1480 when 1000+60t>1480. 60t>480, t>8. It will be above
    $1,480 after more than 8 years.

19. (a) $|x+1|=7$. x+1=7 or x+1=-7. x=6 or x=-8.
    (b) $|x-4|=9$. x-4=9 or x-4=-9. x=13 or x=-5.
    (c) $|2x-3|=5$. 2x-3=5 or 2x-3=-5. 2x=8 or 2x=-2.
        x=4 or x=-1.
    (d) $|2x+1|=|x-3|$. 2x+1=x-3 or 2x+1=-(x-3). x=-4 or 3x=2.
        x=-4 or x=2/3.

93

(e) $|3x-2|=|4x+1|$. $3x-2=4x+1$ or $3x-2=-(4x+1)$. $-x=3$ or $7x=1$. $x=-3$ or $x=1/7$.

(f) $-|2x|+4=7$. $-|2x|=3$. $|2x|=-3$. No solution since $|2x|$ cannot be negative.

20. (a) $|x|\le7$. $-7\le x\le7$. Solution $[-7,7]$. Graph of $y=|x|$ lies on/below $y=7$ over $[-7,7]$.

(b) $|x+3|\le4$. $-4\le x+3\le4$, $-7\le x\le1$. Solution $[-7,1]$. Graph of $y=|x+3|$ lies on/below $y=4$ over $[-7,1]$.

(c) $|3x-5|<6$. $-6<3x-5<6$, $-1<3x<11$. Solution $(-1/3,11/3)$, $(-.3333,3.6667)$. Graph of $y=|3x-5|$ lies below $y=6$ over $(-.3333,3.6667)$.

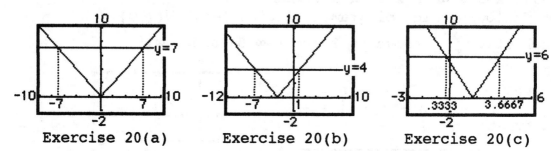

Exercise 20(a)   Exercise 20(b)   Exercise 20(c)

(d) $|2x-5|\ge13$. $2x-5\le-13$ or $2x-5\ge13$, $2x\le-8$ or $2x\ge18$, $x\le-4$ or $x\ge9$. Solution $(-\infty,-4]\cup[9,\infty)$. Graph of $|2x-5|$ lies on/above $y=13$ over $(-\infty,-4]$ and $[9,\infty)$.

(e) $|7x-3|>6$. $7x-3<-6$ or $7x-3>6$, $7x<-3$ or $7x>9$, $x<-3/7$ or $x>9/7$. Solution $(-\infty,-3/7)\cup(9/7,\infty)$, $(-\infty,-.4286)\cup(1.2857,\infty)$. Graph of $y=|7x-3|$ lies above $y=6$ over $(-\infty,-.4286)$ and $(1.2857,\infty)$

(f) $1\le|x|\le7$. Two cases:
(i) $1\le|x|$. $x\le-1$ or $x\ge1$. x must lie in $(-\infty,-1]\cup[1,\infty)$.
(ii) $|x|\le7$. $-7\le x\le7$. x must lie in $[-7,7]$.
Combine (i) and (ii). x must lie in $[-7,-1]$ or $[1,7]$.
Solution set is $[-7,-1]\cup[1,7]$.
Graph $y=|x|$ lies on/between $y=1$ & $y=7$ on $[-7,-1]$ & $[1,7]$.

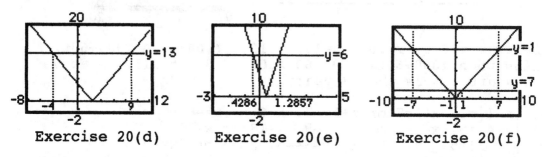

Exercise 20(d)   Exercise 20(e)   Exercise 20(f)

(g) $-3<|x+4|<5$. Two cases:
(i) $-3<|x+4|$. True for all x.

(ii) $|x+4|<5$. $-5<x+4<5$. $-9<x<1$. x must lie in $(-9,1)$.
Combine (i) and (ii). x must lie in $(-9,1)$.
Solution set is $(-9,1)$.
Graph $y=|x+4|$ lies between $y=-3$ & $y=5$ on $(-9,1)$.

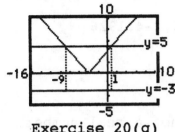

Exercise 20(g)

21.a) $5<4x+1<21$. Display gives $1<x<5$. Algebra: $4<4x<20,1<x<5$.
Solution set is $(1,5)$.

b) $x^2+2x-24\geq0$. Graph appears to be on or above x-axis on
$(-\infty,6]$ and $[4,\infty)$. Algebra: $(x+6)(x-4)\geq0$, $(x+6)(x-4)=0$.
Critical values are $x=-6$ and $x=4$.
$(x+6)$: $x+6<0$ when $x<-6$, $x+6=0$ when $x=-6$, $x+6>0$ when $x>-6$.
$(x-4)$: $x-4<0$ when $x<4$, $x-4=0$ when $x=4$, $x-4>0$ when $x>4$.
Get following sign diagram. Solution is $(-\infty,-6]\cup[4,\infty)$.

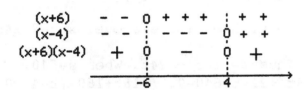

c) $|2x+3|<9$. Display suggests that graph of $y=|2x+3|$ lies
below line $y=9$ over the interval $(-6,3)$. Soln is $(-6,3)$.
Algebra: $-9<2x+3<9$. $-12<2x<6$. $-6<x<3$.

Chapter 2 Test

1. $2x^2-17x+35 = 0$, $(x-5)(2x-7) = 0$. $x = 5$, $7/2$.
See graph of $y=2x^2-17x+35$ below.

2. $6x^2-54=0$. $6x^2=54$. $x^2=9$. $x=\pm3$.See graph of $y=6x^2-54$ below.

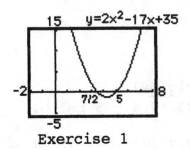

Exercise 1

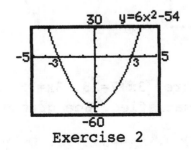

Exercise 2

3. $2x^2-8x+3=0$. $x^2-4x=-3/2$. $x^2-4x+(-2)^2=-3/2+(-2)^2$. $(x-2)^2=5/2$. $x-2=\pm\sqrt{5/2}$. $x=2\pm\sqrt{5/2}$.

4. (a) $x^2+3x-5 = 0$, $a=1$, $b=3$, $c=-5$.

$x = \dfrac{-3\pm\sqrt{3^2 - 4(1)(-5)}}{2(1)} = \dfrac{-3\pm\sqrt{29}}{2} = 1.1926, -4.1926$

See graph of $y=x^2+3x-5$ below.

(b) $x^2-2x+7 = 0$, $a=1$, $b=-2$, $c=7$.

$x = \dfrac{2\pm\sqrt{(-2)^2 - 4(1)(7)}}{2(1)} = \dfrac{2\pm\sqrt{-24}}{2}$. $\sqrt{-24}$ not a real number.

No solutions. See graph of $y=x^2-2x+7$ below.

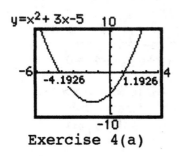

Exercise 4(a)

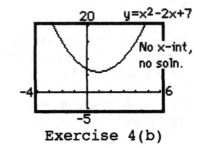

Exercise 4(b)

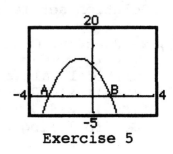

Exercise 5

5. $-2.8x^2 - 4.7x + 9.3 = 0$. See figure above. $x_A=-2.8457, x_B=1.1672$

6. $s=-16t^2+vt$. $v=224$. Thus $s=-16t^2+224t$. When $s=640$, $640=-16t^2+224t$. $-16t^2+224t-640=0$. $(-16t+160)(t-4)=0$. $t=160/16=10$ or $t=4$. It will be at a height of 640 ft at 4 seconds and 10 seconds; once on way up, once on way down.

7. $x^4-2x^2-15=0$. Let $u=x^2$. $u^2-2u-15=0$. $(u+3)(u-5)=0$. $u=-3$ or $u=5$. $x^2=-3$ or $x^2=5$. $x^2=-3$, no solns. $x^2=5$, $x=\pm\sqrt{5}$. $x_A=\sqrt{5}=2.2361$, $x_B=-\sqrt{5}=-2.2361$. See graph of $y=x^4-2x^2-15$ below.

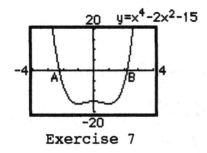

Exercise 7

8. $\sqrt{3x+7} =5$. Square, $3x+7=25$. $3x=18$, $x=6$. Original equation is satisfied. See graphs of $y=\sqrt{3x+7}$ & $y=5$ below.

9. $2x^3-5x^2-19x+62=0$. $x_A=-3.2794$. Only one solution. See graph of $y=2x^3-5x^2-19x+62$ below.

10. 3x-4<7x+8. -4x<12, x>-3. (-3,∞). Graph of y=3x-4 lies
    below y=7x+8 when x>-3. See graphs below.

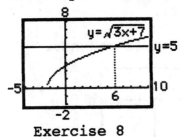
Exercise 8

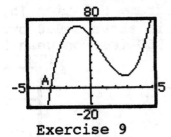
Exercise 9

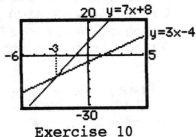
Exercise 10

11. $x^2-6x+5<0$. (x-1)(x-5)=0. Critical values x=1 and 5.
    x<1: (x-1)<0 & (x-5)<0. (x-1)(x-5)>0. Not solutions.
    (x-1): x-1<0 when x<1, x-1=0 when x=1, x-1>0 when x>1.
    (x-5): x-5<0 when x<5, x-5=0 when x=5, x-5>0 when x>5.
    Get following sign diagram. Solution (1,5).
    Graph below x-axis over (1,5).

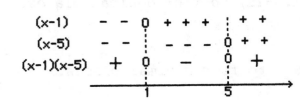

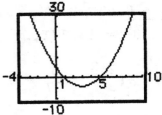

12. |2x-1|=|3x+7|. 2x-1=3x+7 or 2x-1=-(3x+7). -x=8 or
    5x=-6. x=-8 or x=-6/5.

13. -2<|x+3|<7. Two cases:
    (i) -2<|x+3|. True for all x.
    (ii) |x+3|<7. -7<x+3<7. -10<x<4. x must lie in (-10,4).
    Combine (i) and (ii). x must lie in (-10,4).
    Solution set is (-10,4).
    Graph y=|x+3| lies between y=-2 & y=7 on (-10,4). See below.

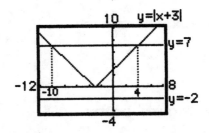

14. Let width be x. Length=x+25. Area=x(x+25)=x^2+25x. 350 sq feet
    of concrete is available. Thus $x^2+25x=350$. $x^2+25x-350=0$.
    (x-10)(x+35)=0. x=10,-35. Width cannot be negative. x=10.
    Width of region is 10 feet, length is 35 feet.

15. s(t)=$-16t^2$+vt. v=272. s(t)=$-16t^2$+272t.
    s>960 when $-16t^2$+272t>960. $-16t^2$+272t-960>0.

16$(-t^2+17t-60)$>0. $(-t+5)(t-12)$>0. Critical vals, t=5,12.
$(-t+5)$: -t+5<0 when t>5, -t+5=0 when t=5, -t+5>0 when t<5.
$(t-12)$: t-12<0 when t<12, t-12=0 when t=12, t-12>0 when t>12.
Get following sign diagram. Solution (5,12).
Graph of s=$-16t^2+272t$ is above s=960 over (5,12).
Object is higher than 960 feet between 5 and 12 seconds.

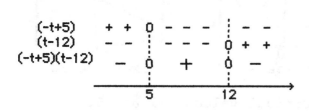
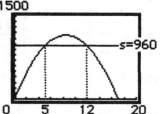

Cumulative Test   Chapters R,1,2

1. (a) $(\sqrt{8.3} + 0.57^{-1}) \div 2.91 = 1.5929$, to four decimal places.
   (b) $\sqrt[6]{27.38^5} = 15.7711$.

2. a) $(\sqrt[4]{\frac{16}{81}})^3 = (\frac{2}{3})^3 = \frac{8}{27} = 0.2963$, to four decimal places.

   b) $(-125)^{2/3} = ((-125)^{1/3})^2 = (-5)^2 = 25$

3. $\frac{(x^{-2}y^3)^4}{(x^{-2}y^5)^{-2}} = \frac{x^{(-2\cdot4)}y^{(3\cdot4)}}{x^{(-2\cdot-2)}y^{(5\cdot-2)}} = \frac{x^{-8}y^{12}}{x^4y^{-10}} = \frac{y^{12}y^{10}}{x^4x^8} = \frac{y^{(12+10)}}{x^{(4+8)}} = \frac{y^{22}}{x^{12}}$

4. $(2x - 3)(3x + 5) = 6x^2 - 9x + 10x - 15 = 6x^2 + x - 15$

5. a) $12x^2 - 5x - 2 = (3x - 2)(4x + 1)$
   b) $9x^2 - 25y^2 = (3x)^2 - (5y)^2 = (3x+5y)(3x-5y)$

6. a) $f(x)=-2x^2+4$. Domain:$(-\infty,\infty)$.Range:$x^2\geq0,-2x^2\leq0,-x^2+4\leq4$.
      $(-\infty,4]$.
   b) $f(x)=\sqrt{x-1}+3$. Domain: x-1≥0, x≥1. [1,∞).
      Range: $\sqrt{x-1}\geq0$, $\sqrt{x-1}+3\geq3$. [3,∞).

7. (3,-1), m=2. (y-(-1))=2(x-3), y=2x-7. See figure below.

8. (-1,5),(4,-5). m = $\frac{-5-5}{4+1}$ = -2. y-5 = -2(x+1), y = -2x+3.
   See figure below.

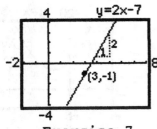
Exercise 7

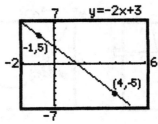
Exercise 8

9. x-intercept of the line through (-1,5) parallel to 10x-2y+4 =0.
10x-2y+4=0, y=5x+2. m_1=5. m_2=5. y-5 = 5(x+1).
When y=0, -5=5(x+1), -5=5x+5. x=-2.

10. $y=x^2-8x-4$, a=1,b=-8,c=-4. x= $\dfrac{-b}{2a}$ = $\dfrac{8}{2(1)}$ = 4.

y = $4^2-8(4)-4$ = -20. Vertex (4,-20). Axis x=4.
y-int=-4. a>0, opens up. See figure below.

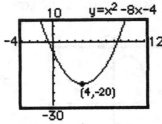
Exercise 10

11. (1,5),(-2,8). d= $\sqrt{(-2-1)^2+(8-5)^2}$ = $\sqrt{(-3)^2+3^2}$ = $\sqrt{18}$ = 4.242640687. Distance is 4.2426 to 4 decimal places.

12. a) $(x-3)^2+(y+7)^2$ = 25. Center=(3,-7), radius=$\sqrt{25}$ = 5.
b) $x^2+y^2-6x-2y+6$ = 0. x^2+2x+y^2-6y+1 = 0.
$(x^2+2x+(1)^2)+(y^2-6y+(-3)^2)$ = $-1+(1)^2+(-3)^2$.
$(x+1)^2 + (y-3)^2$ = 9. Circle, center (-1,3), radius = $\sqrt{9}$ = 3.

13. $2x^2+5x-12$ = 0, (2x-3)(x+4) = 0. x = 3/2, -4.
See graph of $y=2x^2+5x-12$ below.

14. (a) x^2+4x-7 = 0, a=1, b=4, c=-7.

x = $\dfrac{-4\pm\sqrt{4^2 - 4(1)(-7)}}{2(1)}$ = $\dfrac{-4\pm\sqrt{44}}{2}$ = $-2 \pm \dfrac{\sqrt{44}}{2}$ = 1.3166, -5.3166
See graph of $y=x^2+4x-7$ below.

(b) x^2-3x+8 = 0, a=1, b=-3, c=8.

x = $\dfrac{3\pm\sqrt{(-3)^2 - 4(1)(8)}}{2(1)}$ = $\dfrac{3\pm\sqrt{-23}}{2}$. $\sqrt{-23}$ not a real number.
No solutions. See graph of $y=x^2-3x+8$ below.

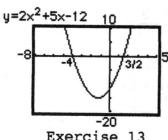

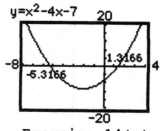

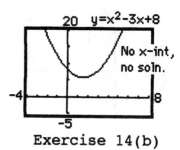

Exercise 13   Exercise 14(a)   Exercise 14(b)

15. $x^4-2x^2-8=0$. Let $u=x^2$. $u^2-2u-8=0$. $(u+2)(u-4)=0$.
    $u=-2$ or $u=4$. $x^2=-2$ or $x^2=4$. $x^2=-2$, no solns. $x^2=4$, $x=\pm2$.
    See graph of $y=x^4-2x^2-8$ below.

16. $2x^3-4x^2-18x+57=0$. $x_A=-3.3130$. Only one solution. See graph of
    $y=2x^3-4x^2-18x+57$ below.

17. $x-4<3x+2$. $-2x<6$, $x>-3$. $(-3,\infty)$. Graph of $y=x-4$ lies
    below $y=3x+2$ when $x>-3$. See graphs below.

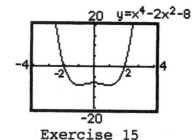

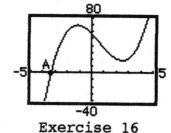

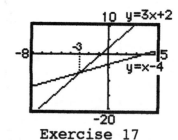

Exercise 15   Exercise 16   Exercise 17

18. Average of 78 after three tests. Let x be necessary grade
    on fourth test to bring average to 80. Thus
    $\frac{3(78)+1x}{4}$ = 80, 234 + x = 320, x = 86. She needs a grade of 86.

19. Cost of 10 is \$520 and cost of 20 is \$550.
    Let $(x_1,y_1)=(10,520)$ and $(x_2,y_2)=(20,550)$.
    Then m = $\frac{y_2-y_1}{x_2-x_1}$ = $\frac{550-520}{20-10}$ = $\frac{30}{10}$ = 3.
    Point/slope form gives y-520 = 3(x-10).
    Cost-volume formula is y = 3x+490.
    Cost of 17 items is 3(17)+490 = \$541.

20. $s=-16t^2+vt$. $v=256$. Thus $s=-16t^2+256t$. When s=880,
    $880=-16t^2+256t$. $-16t^2+256t-880=0$. $(-16t+176)(t-5)=0$.
    t=176/16=11 or t=5. It will be at a height of 6400 ft at 5
    seconds and 11 seconds; once on way up, once on way down.

100

Chapter 3

1.

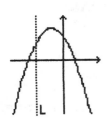

Function. Every vertical line L cuts the graph in one point.

3.

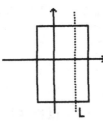

Not a function. There is a vertical line L that cuts the graph in two points.

5.

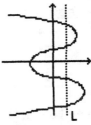

Not a function. L cuts the graph in four points. (There is another vertical line that cuts it in 2 points.)

7. $f(x)=(x-3)(x+5) = x^2+2x-15$. Polynomial of degree 2.
9. $f(x)=(x+1)(x-2)(x+5)=(x+1)(x^2+3x-10) = x^3+4x^2-7x-10$. Degree 3.
11. $f(x)=x^2-2x-3 = (x+1)(x-3)$. $x = -1,3$.
13. $f(x)=2x^2-4x-6 = (2x+2)(x-3)$. $x = -1,3$.
15. $f(x)=x^2-4x-5 = (x+1)(x-5)$. $x = -1,5$.
17. $f(x)=x^3-3x^2+2x = x(x^2-3x+2) = x(x-1)(x-2)$. $x = 0,1,2$.
19. $f(x)=x(x-2)(x+5)$. $x = -5, 0, 2$.
21. $f(x)=x^2(x+3)(3x-7)$. $x = -3, 0, 7/3$.

23. $f(x)=(x-1)(x-3)(x-6)$. x-intercepts=1,3,6. y-int=$f(0)$=-18.

Interval	Test Pt	$f(x)=(x-1)(x-3)(x-6)$	above/below x-axis
$(-\infty,1)$	x=0	f(0)=-18<0	below
(1,3)	x=2	f(2)=4>0	above
(3,6)	x=4	f(4)=-6<0	below
$(6,\infty)$	x=7	f(7)=24>0	above

25. $f(x)=(x+4)(x+1)(2-x)$. x-intercepts=-4,-1,2. y-int=8.

Interval	Test Pt	$f(x)=(x+4)(x+1)(2-x)$	above/below x-axis
$(-\infty,-4)$	x=-5	f(-5)=28>0	above
(-4,-1)	x=-2	f(-2)=-8<0	below
(-1,2)	x=0	f(0)=8>0	above
$(2,\infty)$	x=3	f(3)=-28<0	below

27. $f(x)=x^2(x-2)^2(x+3)$. x-intercepts=-3,0,2. y-int=0.

Interval	Test Pt	$f(x)=x^2(x-2)^2(x+3)$	above/below x-axis
$(-\infty,-3)$	x=-4	f(-4)=-5760<0	below
(-3,0)	x=-1	f(-1)=18>0	above
(0,2)	x=1	f(1)=4>0	above
$(2,\infty)$	x=3	f(3)=54>0	above

Exercise 23

Exercise 25

Exercise 27

29. $f(x)=x^4-4x^2=x^2(x^2-4)=x^2(x+2)(x-2)$. x-int=-2,0,2. y-int=0.

Interval	Test Pt	$f(x)=x^4-4x^2$	above/below x-axis
$(-\infty,-2)$	x=-3	f(-3)=45>0	above
$(-2,0)$	x=-1	f(-1)=-3<0	below
$(0,2)$	x=1	f(1)=-3<0	below
$(2,\infty)$	x=3	f(3)=45>0	above

31. $f(x)=x^3-7x^2+10x=x(x^3-7x+10)=x(x-5))(x-2)$. x-int=0,2,5. y-int=0.

Interval	Test Pt	$f(x)=x^3-7x^2+10x$	above/below x-axis
$(-\infty,0)$	x=-1	f(-1)=-18<0	below
$(0,2)$	x=1	f(1)=4>0	above
$(2,5)$	x=3	f(3)=-6<0	below
$(5,\infty)$	x=6	f(6)=24>0	above

33. $f(x)=x(x^2-1)(x+3)=x(x-1)(x+1)(x+3)$. x-int=-1,-3,0,1. y-int=0.

Interval	Test Pt	$f(x)=x(x^2-1)(x+3)$	above/below x-axis
$(-\infty,-3)$	x=-4	f(-4)=60>0	above
$(-3,-1)$	x=-2	f(-2)=-6<0	below
$(-1,0)$	x=-.5	f(-.5)=.9375>0	above
$(0,1)$	x=.5	f(.5)=-1.3125<0	below
$(1,\infty)$	x=2	f(1)=30>0	above

Exercise 29

Exercise 31

Exercise 33

35. x-int=-3,1,5. y-int=45. Let $f(x)=k(x-a)(x-b)(x-c)$.
    x-int are a,b,c. Thus a=-3,b=1,c=5. $f(x)=k(x+3)(x-1)(x-5)$.
    y-int=f(0)=15k. Thus 15k=45, k=3. $f(x)=3(x+3)(x-1)(x-5)$.

37. x-int=-3,2,6. y-int=-72. Let $f(x)=k(x-a)(x-b)(x-c)$.
    x-int are a,b,c. Thus a=-3,b=2,c=6. $f(x)=k(x+3)(x-2)(x-6)$.
    y-int=f(0)=36k. Thus 36k=-72, k=-2. $f(x)=-2(x+3)(x-2)(x-6)$.

39. x-int=-2,3. Try $f(x)=(x+2)(x-3)$.

Interval	Test Pt	$f(x)=(x+2)(x-3)$	above/below x-axis
$(-\infty,-2)$	x=-3	f(-3)=6>0	above
$(-2,3)$	x=0	f(0)=-6<0	below
$(3,\infty)$	x=4	f(4)=6>0	above

f(x)=(x+2)(x-3) satisfies the conditions (graph below).
f(x)=k(x+2)(x-3) where k>0 will also satisfy the conditions.

41. x-int=-2,1,4. Try f(x)=(x+2)(x-1)(x-4).

Interval	Test Pt	f(x)=(x+2)(x-1)(x-4)	above/below x-axis
(-∞,-2)	x=-3	f(-3)=-28<0	below
(-2,1)	x=0	f(0)=8>0	above
(1,4)	x=2	f(2)=-8<0	below
(4,∞)	x=5	f(5)=28>0	above

Thus f(x)=(x+2)(x-1)(x-4) satisfies the conditions (graph
below). f(x)=k(x+2)(x-1)(x-4), k>0 also satisfies conditions.

43. x-int: y=0 when x=-5,2,4. y-int: x=0 when y=40.

Interval	Test Pt	f(x) from table	above/below x-axis
(-∞,-5)	x=-6	f(-6)=-80<0	below
(-5,2)	x=0	f(0)=40>0	above
(2,4)	x=3	f(3)=-8<0	below
(4,∞)	x=5	f(5)=30>0	above

Let f(x)=k(x+5)(x-2)(x-4). f(0)=40k=40 when k=1.
Thus f(x)=(x+5)(x-2)(x-4) has correct x, y-intercepts.
Graph below. Gives correct table values.

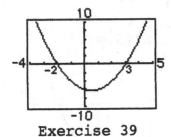

Exercise 39

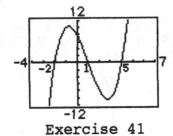

Exercise 41

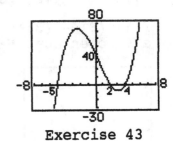

Exercise 43

45. x-int: y=0 when x=-2,1,5. y-int: x=0 when y=30.

Interval	Test Pt	f(x) from table	above/below x-axis
(-∞,-2)	x=-3	f(-3)=-96<0	below
(-2,1)	x=0	f(0)=30>0	above
(1,5)	x=2	f(2)=-36<0	below
(5,∞)	x=6	f(6)=120>0	above

Let f(x)=k(x+2)(x-1)(x-5). f(0)=10k=30 when k=3.
Thus f(x)=3(x+2)(x-1)(x-5) has correct x, y-intercepts.
Graph below. Gives correct table values.

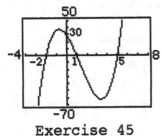

Exercise 45

Section 3.2

1. Increasing: [b,∞). Decreasing [a,b]. Constant (-∞,a].
   Local max: none. Local min: at x=b.

3. Increasing: $(-\infty,a]$. Decreasing: $[a,b]$. Constant: $[b,\infty)$.
    Local max: at x=a. Local min: none.

5. Increasing: $[a,b]$, $[c,\infty)$. Decreasing: $(-\infty,a]$, $[b,c]$.
    Local max: at x=b. Local min: at x=a, x=c.

<u>In the windows of exercises 7-11 Xscl=1, Yscl=1.</u>

7. $f(x)= x^2+8x+10$. a=1,b=8,c=10. $x=\dfrac{-b}{2a} = \dfrac{-8}{2(1)} = -4$.

    $f(-4)=(-4)^2+8(-4)+10 = -6$. Vertex $(-4,-6)$. y-intercept = 12.
    a>0, opens up. Min at $(-4,-6)$. Dec$(-\infty,-6]$, Inc$[-6,\infty)$.

9. $f(x)= -4x^2+8x+7$. a=-4,b=8,c=7. $x=\dfrac{-b}{2a} = \dfrac{-8}{2(-4)} = 1$.

    $f(2)=-4(1)^2+8(1)+7 = 11$. Vertex $(1,11)$. y-int=7.
    a<0, opens down. Max at $(1,11)$. Inc$(-\infty,1]$. Dec$[1,\infty)$.

11. $f(x)= x^2-6x+4$. a=1,b=-6,c=4. $x=\dfrac{-b}{2a} = \dfrac{6}{2(1)} = 3$.

    $f(1)=(3)^2-6(3)+4 = -5$. Vertex $(3,-5)$. y-int=4.
    a>0, opens up. Max at $(3,-5)$. Dec$(-\infty,3]$. Inc$[3,\infty)$.

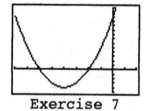
Exercise 7

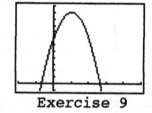
Exercise 9

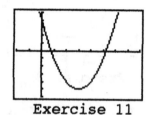
Exercise 11

13. $f(x)= 4x^3-3x^2-8x+5$. Max$(-0.6039,7.8562)$, Min$(1.1039,-2.1062)$.
    Inc$(-\infty,-0.6039]$, $[1.1039,\infty)$. Dec$[-0.6039,1.1039]$.

15. $f(x)= -5x^3+7x^2+3x+9$. Max$(1.1130,14.1166)$, Min$(-0.1797,8.7160)$.
    Dec$(-\infty,-.1797]$, $[1.1130,\infty)$. Inc$[-0.1797,1.1130]$.

17. $f(x)=-2x^3+7x^2+9x-4$. Max$(2.8581,32.2099)$, Min$(-0.5248,-6.5062)$.
    Dec$(-\infty,-0.5248]$, $[2.8581,\infty)$. Inc$[-0.5248,2.8581]$.

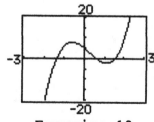
Exercise 13

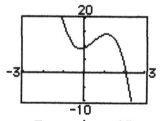
Exercise 15

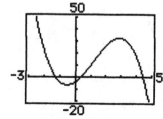
Exercise 17

19. $f(x)=x^3+34x^2+40x-3000$. Max$(-22.0623,1928.1248)$,
    Min$(-0.6043,-3011.9770)$.
    Inc$(-\infty,-22.0623]$, $[-0.6043,\infty)$. Dec$[-22.0623,-0.6043]$.

Section 3.2

21. $N(t)=2t^3-27t^2+84t+416$. Max(2,492). Min(7,367).
Population peaked with 492 ducks in 1992. Was at its lowest in 1997 with 367 ducks.

23. $P(N)=-\dfrac{N^3}{15} + \dfrac{4N^2}{5} + 4N$. Max(10,53.3333). Publisher should produce 10 thousand copies in order to maximize profit.

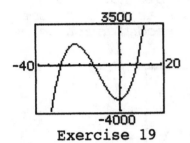
Exercise 19

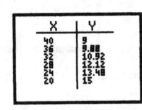
Exercise 21

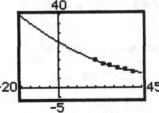
Exercise 23

25. $f(x)=-0.0004044x^4+0.01745x^3-0.1505x^2-1.4929x+82.19$.
Min(13.4803,64.1086). The interest in brand names decreased from 1978 to mid 1991 (corresp to x=13.5), but has been on the increase ever since.

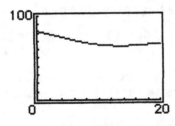

27. Data was entered, equation found using quadratic regression, below, and data and graph plotted. Graph closely fits the data.

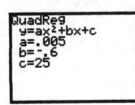

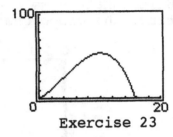

29. Path of asteroid is $y=-0.02x^2+2.6x-34.5$. (exercise 28)
Dist between Earth (0,0) & asteroid $(x,-0.02x^2+2.6x-34.5)$ is d
$= \sqrt{(x-0)^2+((-0.02x^2+2.6x-34.5)-0)^2} = \sqrt{x^2+(-0.02x^2+2.6x-34.5)^2}$.
Graph this function, below. Min A(12.192686,13.490013).
Closest asteroid gets to earth is approx.13.49 million miles.

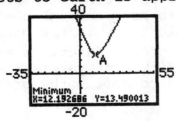

31. (a) $N(t)=-0.075t^4+3.8t^3+11.65t^2$. t is time from 1945. Max at A(39.94,69841). Number peaked when t=40, in 1985 (numbers are approximate, thus round). (b) In 1995 t=50. N(50)=35,375 nuclear warheads. (c) Observe how the graph drops rapidly. According to this function, nuclear warheads should be completely abolished by the year corresonding to B. B(53.57,0), t=54. No nuclear weapons by the year 1999! Did not happen. Unfortunately we shall have nuclear weapons around for some time, the graph changing to look like the second graph below.

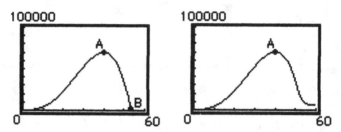

33. Let y yd under river, x yd along the bank (below). Total cost C=20x+50y. $PQ^2=PR^2+RQ^2$. $PQ=\sqrt{PR^2+RQ^2}$. PQ=y, PR=500-x, RQ=100. $y=\sqrt{(500-x)^2+100^2}$. $y=\sqrt{x^2-1000x+260000}$.
Thus $C=20x+50\sqrt{x^2-1000x+260000}$. Graph below.
Min A(456.36,14582.58) Cost is min of $14,582.58 when x=456.4.
$y=\sqrt{(456.4)^2-1000(456.4)+260000}$ = 109.1.
Thus 456yd along the bank and 109yd under the river.

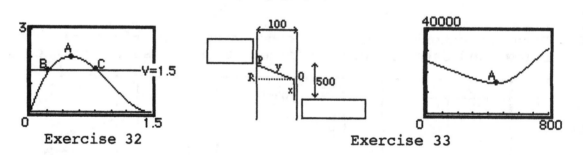

Exercise 32   Exercise 33

Section 3.3

Group Discussion It is not possible for a function to have a graph that is symmetric with respect to the x-axis. Such a graph would have a vertical line that cuts it at more than one point. See graph below. It is possible for an equation that does not represent a function to have a graph that is symmetric about the x-axis. The example below is the graph of the equation $x=y^2$. It is the graph of two functions $x=\sqrt{y}$ and $x=-\sqrt{y}$. A circle is also symmetric about y-axis, x-axis, and the origin. It needs two functions to define the graph.

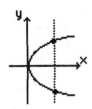

1. f(x)=2x-4.
    y=2x-4, straight line.
    y-int=-4, x-int=2.
    Domain:$(-\infty,\infty)$.Range:$(-\infty,\infty)$.

3. f(x)=x.
    y=x, straight line.
    Through origin & (2,2).
    Domain:$(-\infty,\infty)$.Range:$(-\infty,\infty)$.

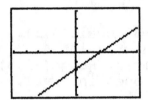

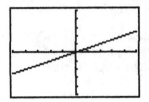

5. $f(x)=x^2+2x+3$. $y=x^2+2x+3$. a=1,b=2,c=3. $x=\dfrac{-b}{2a}=\dfrac{-2}{2(1)}=-1$.

    $y=(-1)^2+2(-1)+3=2$. Vertex=(-1,2). a>0,opens up. y-int=3.
    Domain:$(-\infty,\infty)$. Range:$[2,\infty)$. See graph below.

7. $g(x)=2x^2-8x+5$. $y=2x^2-8x+5$. a=2,b=-8,c=5. $x=\dfrac{-b}{2a}=\dfrac{8}{2(2)}=2$.

    $y=2(2)^2-8(2)+5=-3$. Vertex=(2,-3). a>0,opens up. y-int=5.
    Domain:$(-\infty,\infty)$. Range:$[-3,\infty)$. See graph below.

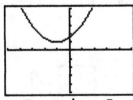

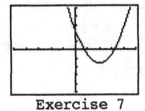

Exercise 5   Exercise 7

9. $g(x)=x^2+2$. $y=x^2+2$.
    Vertex (0,2). y-int=2.
    Or use graph of $y=x^2$ and
    slide up distance 2.
    (prepare for next section)
    Domain:$(-\infty,\infty)$.Range:$[2,\infty)$.

11. $f(x)=x^4$. $y=x^4$.
    Case $f(x)=x^n$, n pos even.
    Domain:$(-\infty,\infty)$.Range:$[0,\infty)$.

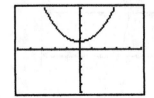

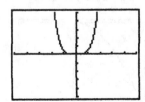

13. $g(x)=-x^2$.
    Domain:$(-\infty,\infty)$.Range:$(-\infty,0]$.

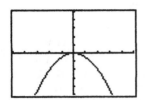

15. $f(x)=ax^2$, for a=2,3,-5.
     $y_1=2x^2$, $y_2=3x^2$, $y_3=-5x^2$.
     Domains:$(-\infty,\infty)$.
     Range(for a=2,3):$[0,\infty)$.
     Range(a=-5):$(-\infty,0]$.

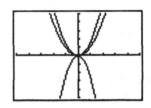

17. $f(x)=2x-1$.Domain$[-2,2]$.$y=2x-1$.
    Line, y-int=-1,x-int=1/2.
    $f(-2)=-5$, $f(2)=3$. Range$[-5,3]$.

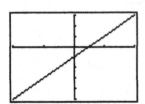

19. $g(x)=-2x+8$,domain$(0,5)$.
    Line, y-int=8,x-int=4.
    $f(0)=8$, $f(5)=-2$. Range$(-2,8)$.
    Range is an open interval.

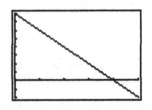

21. $f(x)=2x^2+4x+1$,domain $[-3,2]$.
    Parabola,x= $\dfrac{-b}{2a}$ = $\dfrac{-4}{2(2)}$ = -1.
    $f(-1)=-1$. Vertex(1,-1).
    $f(-3)=7$, $f(2)=17$. Range$[-1,17]$.

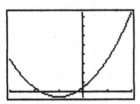

23. $f(x)=x^3$, domain $[-2,2]$.
    $f(-2)=-8$,$f(2)=8$.Range$[-8,8]$

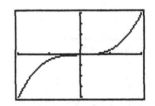

25. $h(x)=-x^2$, domain $[-4,3]$.
    Down parabola,vertex (0,0).
    $f(-4)=-16$,$f(3)=-9$.Range$[-16,0]$

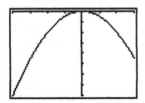

27. $f(x)=x^2-4$. $f(-x)=(-x)^2-4$.
    $f(-x)=f(x)$, $f(-x)\neq-f(x)$.
    Even function.

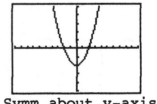

Symm about y-axis.

29. $f(x)=x+4$. $f(-x)=(-x)+4$.
$f(-x)\neq f(x)$, $f(-x)\neq -f(x)$.
Neither symmetry.

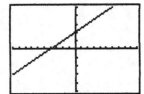

Not symm about y-axis
or origin.

31. $f(x)=4x^3+2$. $f(-x)=4(-x)^3+2$.
$f(-x)\neq f(x)$, $f(-x)\neq -f(x)$.
Neither symmetry.

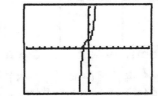

Not symm about y-axis
or origin

33. $f(x)=\dfrac{1}{x^2-4}$. $f(-x)=\dfrac{1}{(-x)^2-4}$
$f(-x)=f(x)$, $f(-x)\neq -f(x)$.
Even function.

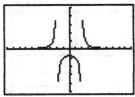

Symm about y-axis.

35. $f(x)=4$. $f(-x)=4$.

$f(-x)=f(x)$, $f(-x)\neq -f(x)$.
Even function.

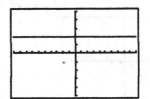

Symm about y-axis.

37.

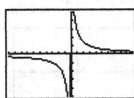

Symm about origin. Odd.
$f(x)=\dfrac{6}{x}$. $f(-x)=\dfrac{6}{(-x)}$.
$f(-x)\neq f(x)$, $f(-x)=-f(x)$. Odd.

39.

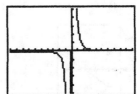

Symm about origin. Odd.
$f(x)=\dfrac{8}{x^3}$. $f(-x)=\dfrac{8}{(-x)^3}$.
$f(-x)\neq f(x)$, $f(-x)=-f(x)$. Odd.

41.

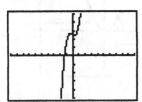

Not symm about y-axis
or origin. Neither
$f(x)=3x^3-x+5$.
$f(-x)=3(-x)^3-(-x)+5$.
$f(-x)\neq f(x)$, $f(-x)\neq -f(x)$. Neither.

43.

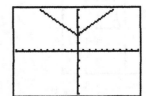

Symm about y-axis. Even.
$f(x)=|x| + 4$.
$f(-x)=|(-x)| + 4$.
$f(-x)=f(x)$, $f(-x)\neq -f(x)$. Even.

45.

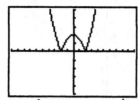

Symm about y-axis. Even.
$f(x)=|x^2-4|. f(-x)=|(-x)^2-4|.$
$f(-x)=f(x), f(-x)\neq-f(x).$ Even.

47. Even. $f(-x)=f(x)$. e.g. $f(-6)=288=f(6)$.
49. Neither. e.g. $f(-6)=-45$ but $f(6)=9$. $f(-6)\neq f(6)$ & $f(-6)\neq-f(6)$.
51. Even. $f(-x)=f(x)$. e.g. $f(-9)=-1320=f(9)$.

53. $f(x) = \begin{cases} 2x+1 & \text{if } x \leq 4 \\ 2 & \text{if } x > 4 \end{cases}$

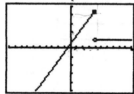

55. $f(x) = \begin{cases} -4x+5 & \text{if } x \leq 3 \\ -7 & \text{if } x > 3 \end{cases}$

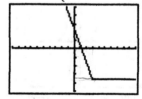

57. $f(x) = \begin{cases} 6-x & \text{if } x < 2 \\ x-6 & \text{if } x \geq 2 \end{cases}$

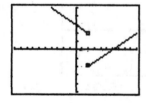

59. $f(x) = \begin{cases} x^2 & \text{if } x \leq 0 \\ \dfrac{x}{2} & \text{if } x > 0 \end{cases}$

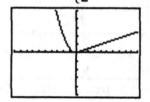

61. $f(x) = \begin{cases} -x+1 & \text{if } x \leq -2 \\ x^3 & \text{if } x > -2 \end{cases}$

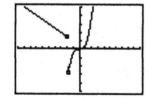

63. $f(x) = \begin{cases} -3 & \text{if } x \leq -2 \\ x^2 & \text{if } -2<x\leq 3 \\ 2 & \text{if } x > 3 \end{cases}$

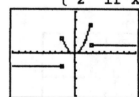

65. $f(x) = \begin{cases} x + 1 & \text{if } x \leq 3 \\ -x - k & \text{if } x > 3 \end{cases}$

Using $f(x)=x+1$ when $x\leq 3$ we get $f(3)=4$ when $x=3$. For continuity we need $-x-k=4$ when $x=3$. Thus $k=-7$.
Enter the above function for various values of k and observe the coresponding graphs. We get continuity when $k=-7$.

110

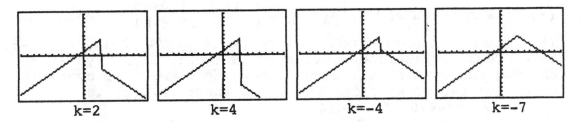

k=2   k=4   k=-4   k=-7

67. $f(x)=x^3-9x^2+15x$, with domain [0, 8]. Min is (5,-25) and $f(8)=56$. All the y values for x in [0, 8] lie in the interval [-25, 56]. See graph below. Range is [-25, 56].

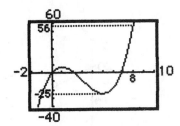

69. $N(t)=t^2-10t-20$ for $12\le t\le 24$. (a) N(12)=4, N(24)=316. A 12 month old child would know about 4 words, a 24 month old child about 316 words! (b) N(18)=124. Increase from 12/18 months=124-4=120 words. Increase from 18/24 months=316-124=192 words.
(c) Number of words per year would gradually increase and then start decreasing giving the curve below. This is a typical learning curve where one starts learning slowly, then rate increases, and then the rate decreases.

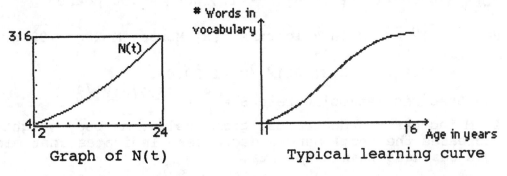

Graph of N(t)   Typical learning curve

71. (a) Print shop:5¢ 1st 50 pages,3¢ next 50,2¢ over 100 pages. C(50)=5(5)=250¢. C(100)= 1st 50 plus 2nd 50 =5(50)+3(50)=400¢. Let x pages be printed. If x≤50 total cost is 5x.
If 50<x≤100, total cost is 250 for 1st 50 pages plus 3(x-50) for the next x-50 pages.
If x>100, total cost is 400 for 1st 100 pages plus 2(x-100) for the next x-100 pages.

$$C(x) = \begin{cases} 5x & \text{if } 0 \le x \le 50 \\ 250+3(x-50) & \text{if } 50<x\le100 \\ 400+2(x-100) & \text{if } x > 100 \end{cases}$$

(b) C(75)=325=$3.25. C(120)=440=$4.40.

111

73. (a) Consumer awareness, $A(t) = \begin{cases} -t/4+20 & \text{if } 0 \le t \le 80 \\ 0 & \text{if } t > 80 \end{cases}$ .Graph below.

(b) Value of A should not fall below 14. $-t/4+20=14$, $-t/4=-6$, $t=24$. Consumer awareness drops to 14 twenty four days after showing the commercial. Should show the commercial every 24 days. Desired advertising awareness function $D(t)$ is

$$D(t) = \begin{cases} -t/4+20 & \text{if } 0 \le t < 24 \\ -t/4 + 26 & \text{if } 24 \le t < 48 \\ -t/4 + 32 & \text{if } 48 \le t < 72 \\ \quad \cdots \end{cases}$$

The function $A(t)$ is given a boost of 6 to bring it up from 14 to 20 every 24 days.

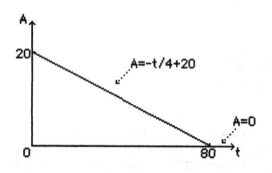

75. Humans: $M_{brain}=0.085(M_{body})^{0.66}$. (1 gram=2.2046 pounds)
Other mammals $M_{brain}=0.01(M_{body})^{0.70}$. (a) Elephant:
$M_{body}=6000$kgm. $M_{brain}=0.01(6000)^{0.70}=4.41$kg=9.72 pounds.
(b) Human: $M_{body}=140$ pounds$=140/2.2046$kg.
$M_{brain}=0.085(140/2.2046)^{0.66}=1.3160$kg $=2.90$ pounds.

77. $R=73.3(M)^{0.74}$, M in kilograms. (a) Mouse: M=20gm=.02kg.
$R=73.3(.02)^{0.74}=4.05$.
Cow: M=300Kg. $R=73.3(300)^{0.74}=4990.84$.

(b) Specific metabolic rate $S = \dfrac{R}{M} = \dfrac{73.3(M)^{0.74}}{M}$.

As M increases S decreases, graph below. As body weight increases the restlessness decreases. Indicates that mice are more restless than elephants.

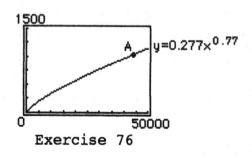

Exercise 76

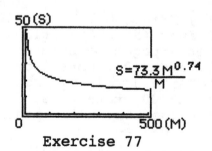

Exercise 77

Project: (a) Graphs of $f(x)=x^n$, $g(x)=x^m$, n,m even positive integers with n<m below. Graphs cross at (1,1) and (-1, 1). Symmetry about y-axis.

(b) Graphs of $f(x)=x^n$, $g(x)=x^m$, n,m odd positive integers with n<m below. Graphs cross at (1, 1) and (-1, -1). Symmetry about origin.

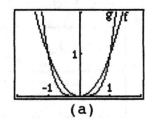

(a)

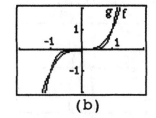

(b)

(c) & (d) $f(x)=x^{1/n}$. n even: graphs cross at (1,1). No symmetry. n odd:cross at (1,1) & (1,-1). Symmetry about origin.

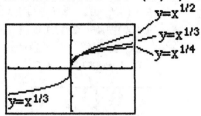

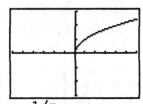

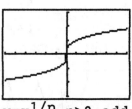

$y=x^{1/n}$,n>0, even. $y=x^{1/n}$,n>0,odd.
Larger n, 'lower' the graph in each case.

Section 3.4

1. $g(x)=x^2+4$
    Shift $f(x)=x^2$
    up 4.

3. $g(x)=(x+3)^2$
   Shift $f(x)=x^2$
   left 3.

5. $g(x)=x^3-4$
   Shift $f(x)=x^3$
   down 4.

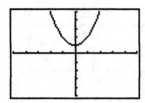

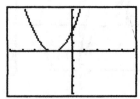

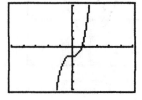

7. $g(x)=-x^3+6$
    Reflect $f(x)=x^3$
    in x-axis, up 6.

9. $g(x)=-(x-4)^2$
   Shift $f(x)=x^2$
   right 4, reflect in
   x-axis.

11. $g(x)=(x-3)^3-8$
    Shift $f(x)=x^3$
    right 3, down 8.

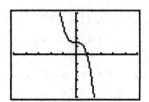

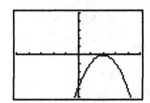

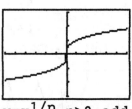

13. $g(x)=|x+3|-5$
Shift $f(x)=|x|$
left 3, down 5.

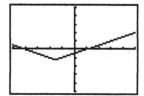

15. $g(x)=-2x^4+10$
Reflect $f(x)=2x^4$
in x-axis, up 10.

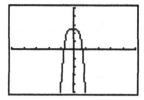

17. $g(x)=\sqrt{x+6}-8$
Shift $f(x)=\sqrt{x}$
left 6, down 8.

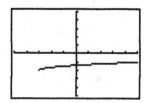

19. $g(x)=x^2-4x+2$.
$g(x)=(x^2-4x)+2$
$=(x^2-4x+2^2)+2-2^2$
$=(x-2)^2-2$.
$f(x)=x^2$.Shift f
right 2,down 2.

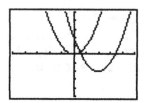

21. $g(x)=x^2+2x+4$.
$g(x)=(x^2+2x)+4$
$=(x^2+2x+1^2)+4-1^2$
$=(x+1)^2+3$
$f(x)=x^2$.Shift f
left 1,up 3.

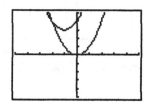

23. $g(x)=-x^2+4x-1$.
$g(x)=-(x^2-4x)-1$
$=-(x^2-4x+2^2)-1+2^2$
$f(x)=x^2$.Shift f
right 2, reflect
in x-axis, up 3.

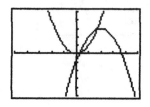

25. $g(x)=x^2+c$,
$c=-2,0,2$.

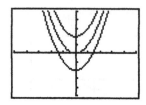

27. $g(x)=-(x-2)^3+c$,
$c=-3,0,3$.

29. $g(x)=-\sqrt{x+c}+c$,
$c=-3,0,3$.

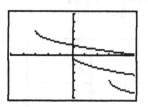

In exercises 31-41 use D for shift down, U for up, L for left, R for right, Rx-axis for reflect in x-axis.

31. Shift D5
$g(x)=x^2-5$.

33. Rx-axis,L6,U5.
$g(x)=-(x-6)^2+5$.

35. Rx-axis,Up5.
$g(x)=-x^3+5$.

37. Up3.
$g(x)=|x|+3$.

39. Rx-axis,R2,U6.
$g(x)=-|x-2|+6$.

41. Rx-axis,D1.
$g(x)=-\sqrt{x}-1$.

43. $f(x)=(x+5)^2-10$,
$f(x)=x^2$,
$f(x)=(x-5)^2-10$.

45. $f(x)=x^2$,
$f(x)=-x^2$,

47. $f(x)=|x-4|$,
$f(x)=|x|$,
$f(x)=|x+4|$.

49. (a) $f(x)=x^2$ and $g(x)=x^2+2$. Table shows that the value of g is 2 more than the value of f at each of given x values.
    (b) $f(x)=x^2$. Value of h is 3 less than the value of f at each of given x values. Let $h(x)=x^2-3$.
    (c) $f(x)=x^3$. Value of k is 5 more than the value of f at each of given x values. $k(x)=x^3+5$.

51. (i) In table (a) f values are square of x values. This suggests $f(x)=x^2$. This fits graph (f). (a)<->(f).
    (ii) In table (b) g values are the negative of the square of the x values. This suggests $g(x)=-x^2$. This fits graph (d). (b)<->(d).
    (iii) Process of elimination gives (c)<->(e). Let us see that this is so. The function in (e) looks as if it could be $h(x)=-x^2$. This fits values of h in (c).

53. (a) $25=(-5)^2, 16=(-4)^2, \ldots, 0=0^2, 1=1^2$. $f(x)$ is square of (x-2). $f(x)=(x-2)^2$.
    (b) $4=2^2, 9=3^2, \ldots, 49=7^2, 64=8^2$. $f(x)$ is the square of (x+5). $f(x)=(x+5)^2$.
    (c) $-36=-6^2, -25=-5^2, \ldots, -1=-1^2, 0=-0^2$. $f(x)$ is the negative of the square of(x-3). $f(x)=-(x-3)^2$.

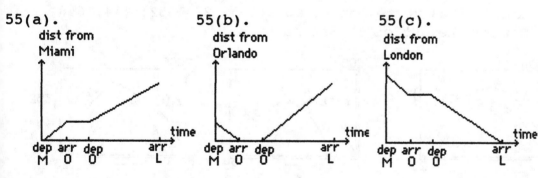

55(a). dist from Miami

55(b). dist from Orlando

55(c). dist from London

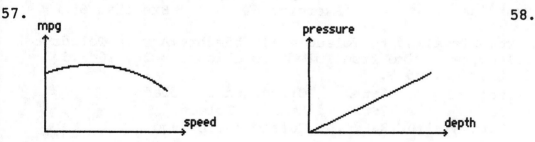

57. mpg

58.

pressure

59.

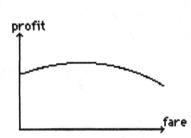

profit

fare

61.

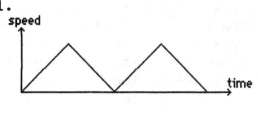

speed

time

63. (a) $s(t)=-16t^2+vt+h$. $s_1(t)=-16t^2+288t$ and $s_2(t)=-16t^2+288t+352$.
(b) Motion 1: Graph is a parabola opening down, with vertex at
$t=\frac{-b}{2a}$. a=-16, b=288. Vertex at $t=\frac{-b}{2a}=\frac{-288}{2(-16)}$ =9. $s_1(9)$=1296.
Vertex A(9,1296),(graph below) Max height 1296ft at t=9.
(c) Motion 2: Shift graph of s_1 up 352 to get graph of s_2.
Thus vertex of s_2 at B(9,1296+352)=(9,1648).
Max height of 1648ft at t=9.

65. (a) $s(t)=-16t^2+vt+h$. Let t be the time measured from instant
at which the 1st object was projected. Initial vel both
objects is 288. $s_1(t)=-16t^2+288t$. At time t 2nd object will
have been traveling for time t-5 from initial height of 352.
Thus $s_2(t)=-16(t-5)^2+288(t-5)+352$.
(b) Motion 1: Vertex A(9,1296) (see ex 65). Max ht 1296ft at
t=9. (c) Motion 2: Shift graph of s_1 right 5 up 352 to get
graph of s_2.Thus vertex of s_2 at B(9+5,1296+352)=(14,1648).
Thus maximum height of 1648ft at t=14.

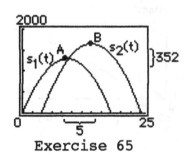

Exercise 63   Exercise 64   Exercise 65

67. Time from Cleveland to Toledo = 110/55=2hrs.Arr in Toledo,t=2.
Dep Toledo,t=3. Time from Toledo to Chicago = 245/70=3.5.
Arr in Chicago,t=6.5.

$$d(t)=\begin{cases}55t & 0\leq t\leq 2 & \text{Cleveland to Toledo} \\ 110 & 2\leq t\leq 3 & \text{Meal in Toledo} \\ 70(t-3)+110 & 3\leq t\leq 6.5 & \text{Toledo to Chicago}\end{cases}$$

69. (a) Dist=speedxtime.Dist=1 mile,time=3min 59.4sec=239.4sec.
Speed=dist/time=1/239.4 ml/sec=(60x60/239.4)mph = 15.04mph.
His speed would have been 15.04mph, to two decimal places.
(b) Since the distance of the race is 1 mile and the duration
is less than 4 minutes let us construct a function using

116

velocity in miles per minute. Speed = 15.04/60 miles per minute. Thus d(t)=15.04t/60 miles, where t is in minutes. (c) Graph below.(d) Probable race is that runners jostle for first 1/4 mile (1st lap), run at a constant pace for the middle 1/2 mile or so, and then sprint the last stretch.

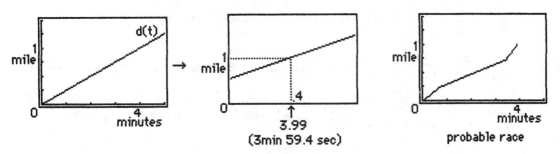

71. Field goal from opponents' 40yd line. Use m=1,v=75,(x-180).
(Shift 60yds, 180ft from left). $f(x)=-\dfrac{32}{75^2}(x-180)^2 +(x-180)$. The goal is 110yd (330ft) from left of screen. Thus choose window 0≤x≤330, 0≤y≤100. Graphs below. The ball passes over goal posts at a height of 22 feet, thus 3 points.

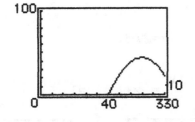
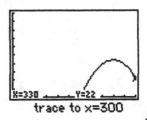

trace to x=300

73. Use m=1, v=105, ball hit from height 21ft. $f(x)=-\dfrac{32}{105^2}x^2+x+21$.

Hole is 160yds (480 ft away. Choose convenient window 0≤x≤540, 0≤y≤150. Graph below. Ball lands at a distance of 364.4ft, i.e. 115.6ft, from the pin. Radius of green is 45ft. Ball does not land on the green.

Project (a) $f(x)=|x-a|$ for a=-4,-2,0,2,4,6,8.
(b) $f(x)=2|x-2.5|+a$ for a=-6,-4,-2,0,2,4.
(c) Quadrants of circles of radius 2.5 with centers at various locations.
Circle radius 2.5,center (a,b) is $f(x)=\pm\sqrt{(x-a)^2+2.5^2)}+b$.
$f(x)=\sqrt{(-x^2+2.5^2)}$, $f(x)=\sqrt{(-(x-5)^2+2.5^2)}$,
$f(x)=-\sqrt{(-x^2+2.5^2)}+5$, $f(x)=-\sqrt{(-(x-5)^2+2.5^2)}+5$.
(d) Circles center (2.5,2.5) and radii 1 and 2, and two lines.
$f(x)=\pm\sqrt{(x-2.5)^2+1)}+2.5$, $f(x)=\pm\sqrt{(x-2.5)^2+2^2)}+2.5$,
$f(x)=x$ and $f(x)=-x+5$.

117

Section 3.5

1. $f(x) = \dfrac{2x + 1}{x}$. Domain: All real numbers except 0.

3. $f(x) = \dfrac{4x + 2}{3x + 1}$. Domain: All real numbers except $-1/3$.

5. $f(x) = \dfrac{x - 3}{(x + 2)(x - 1)}$. Domain: All real numbers except $-2, 1$.

7. $f(x) = \dfrac{(x - 8)(x + 4)}{x(x - 3)(x + 7)}$. All real numbers except $-7, 0, 3$.

9. $f(x) = \dfrac{1}{x^2 - 4x} = \dfrac{1}{x(x - 4)}$. All real numbers except $0, 4$.

11. $f(x) = \dfrac{x + 5}{x^2 + 3x - 4} = \dfrac{x + 5}{(x-1)(x+4)}$. All real numbers except $-4, 1$.

13. $f(x) = \dfrac{x(x - 2)}{x - 2} = x$, $x \neq 2$.

15. $f(x) = \dfrac{(2x + 5)(3x - 2)}{3x - 2} = 2x+5$, $x \neq 2/3$.

17. $f(x) = \dfrac{x^2 + 3x - 4}{x - 1} = \dfrac{(x-1)(x+4)}{x - 1} = x+4$, $x \neq 1$.

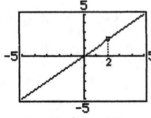

Exercise 13

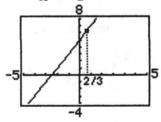

Exercise 15

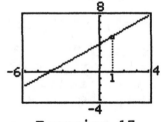

Exercise 17

19. $f(x) = \dfrac{x^2 + 3x - 4}{x + 4} = \dfrac{(x+4)(x-1)}{x + 4} = x-1$, $x \neq -4$.

21. $f(x) = \dfrac{2x^2 - 2x - 4}{2x - 4} = \dfrac{2(x-2)(x+1)}{2(x - 2)} = x+1$, $x \neq 2$.

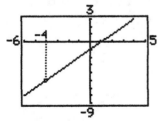

Exercise 19

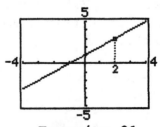

Exercise 21

Section 3.5

In exercises 23-27 R denotes right, L left, U up & D down.

23. $f(x) = \dfrac{1}{x-2}$. Vert asymptote $x=2$. Horiz $y=0$. Shift $g(x) = \dfrac{1}{x}$ R3.

25. $f(x) = \dfrac{3}{x} - 4$. Vert $x=0$. Horiz $y=-4$. Vertical stretch of $g(x) = \dfrac{1}{x}$ 3, shift D4.

27. $f(x) = -\dfrac{1}{x+3} + 2$. Vert $x=-3$. Horiz $y=2$. Shift $g(x) = \dfrac{1}{x}$ L3, reflect in x-axis, U2.

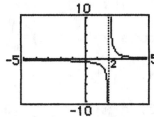
Exercise 23

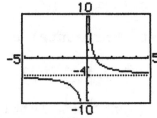
Exercise 25

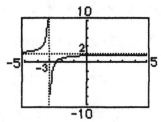
Exercise 27

29. $f(x) = \dfrac{x-1}{x^2-2x-3} = \dfrac{x-1}{(x+1)(x-3)}$. Vert $x=-1, x=3$. Horiz $y=0$. Zero $x=1$.

31. $f(x) = \dfrac{2}{x^2-5x+4} = \dfrac{2}{(x-1)(x-4)}$. Vert $x=1, x=4$. Horiz $y=0$. No zeros.

33. $f(x) = \dfrac{2x^2-5x-25}{x^2-x-12} = \dfrac{(2x+5)(x-5)}{(x+3)(x-4)}$. Vert $x=-3, x=4$. Horiz $y=2$. Zeros $x=-5/2, 5$.

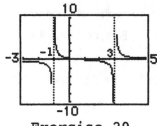
Exercise 29

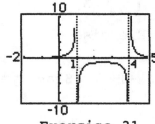
Exercise 31

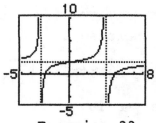
Exercise 33

35. $f(x) = \dfrac{x^2-2x-3}{x^2+2x-8} = \dfrac{(x+1)(x-3)}{(x+4)(x-2)}$. Vert $x=-4, x=2$. Horiz $y=1$. Zeros $x=-1, 3$.

37. $f(x) = \dfrac{3x+5}{x-2}$. Vert $x=2$. Horiz $y=3$. Zero $x=-5/3$.

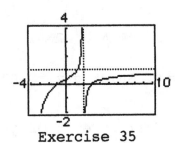

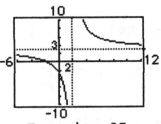

Exercise 35   Exercise 37

39. (a) Vertical asymptote x=0, Horizontal asymptote y=0. $f(x)=\frac{1}{x}$.

(b) Vert x=0. Horiz y=2. Shift $f=\frac{1}{x}$ up2. $g(x)=\frac{1}{x}+2$.

(c) Vert x=3. Horiz y=2. Shift $f=\frac{1}{x}$ up2, right3. $g(x)=\frac{1}{x-3}+2$.

41. (a) Vertical asymptote x=0, Horizontal asymptote y=0. $f(x)=\frac{1}{x^2}$.

(b) Vert x=0, Horiz y=3. Shift $f=\frac{1}{x^2}$ up3. $g(x)=\frac{1}{x^2}+3$.

(c) Vert x=-1, Horiz y=3. Shift $f=\frac{1}{x^2}$ left1, up3. $g(x)=\frac{1}{(x+1)^2}+3$.

43. $f(x)=x+3+\frac{7}{x-2}$. Slant asymptote y=x+3, vertical asymptote x=2.

45. $f(x)=\frac{x^2-4x+5}{x-3}$. $f(x)=x-1+\frac{2}{x-3}$. Slant y=x-1. Vert x=3.

47. $f(x)=\frac{x^2+x+17}{x-4}$. $f(x)=x+5+\frac{37}{x-4}$. Slant y=x+5. Vert x=4.

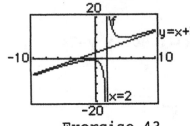

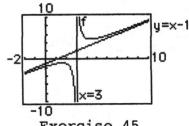

Exercise 43   Exercise 45   Exercise 47

49. $f(x)=\frac{-3x^3+x^2+4}{x^2}$. $f(x)=-3x+1+\frac{4}{x^2}$. Slant y=-3x+1. Vert x=0, the y-axis.

51. $f(x)=x^{-1/n}$, n positive even. y-axis is a vertical asymptote. x-axis is a horizontal asymptote. Domain (0,∞), range (0,∞). Graphs intersect at (1,1). See graphs below. If $f(x)=x^{-1/n}$ & $g(x)=x^{-1/m}$ with n>m, f>g on (0,1) and g>f on (1,∞).

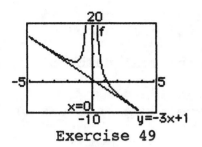
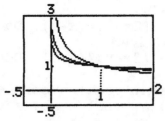

Exercise 49   Exercise 51

Section 3.6

Group Discussion:

(a)

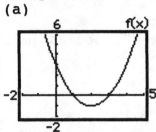
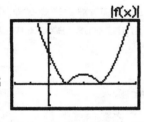

(b)

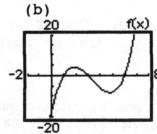
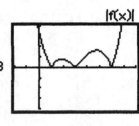

(c) In |f(x)| parts of the graph of f(x) below the x-axis are flipped about x-axis. These parts of |f(x)| are the mirror image of those of f(x) in the x-axis.

(d) If f(x)<-k, a negative number, for any x value, then |f(x)| = k for that same x value.

1. f(x) = 2x+4, g(x) = x-3.
   (f+g)(x) = (2x+4)+(x-3) = 3x+1. Domain: real numbers.
   (f-g)(x) = (2x+4)-(x-3) = x+7. Domain: real numbers.
   (fg)(x) = (2x+4)(x-3) = $2x^2-2x-12$. Domain: real numbers.
   (f/g)(x) = (2x+4)/(x-3). Domain: all real numbers except x=3,
   or $(-\infty,3)\cup(3,\infty)$.

3. f(x) = $3x^2-2x+1$, g(x) = $8x^2+2x-1$.
   (f+g)(x)= $11x^2$. Domain: real numbers.
   (f-g)(x)= $-5x^2-4x+2$. Domain: real numbers.
   (fg)(x)= $24x^4-10x^3+x^2+4x-1$. Domain: real numbers.
   (f/g)(x)= $(3x^2-2x+1)/(8x^2+2x-1)$= $(3x^2-2x+1)/((4x-1)(2x+1))$.
   Domain: real numbers except x=1/4,-1/2, or
   $(-\infty,-1/2)\cup(-1/2,1/4)\cup(1/4,\infty)$.

5. f(x) = 2|x| + 3, g(x) = |x| - 3.
   (f+g)(x)= 3|x|. Domain: real numbers.
   (f-g)(x)= |x|+6. Domain: real numbers.
   (fg)(x)= $2x^2-3$|x|-9. Domain: real numbers.
   (f/g)(x)= (2|x|+3)/(|x|-3). Domain:real numbers except x=-3,3,
   or $(-\infty,-3)\cup(-3,3)\cup(3,\infty)$.

121

Section 3.6

7. f(x) = 2, g(x) = 1/(x-4).
    (f+g)(x)= 2 + 1/(x-4)= (2x-7)/(x-4). Domain: real numbers
    except x=4, or $(-\infty, 4) \cup (4, \infty)$.
    (f-g)(x)= 2 - 1/(x-4)= (2x-9)/(x-4). Domain: real numbers
    except x=4.
    (fg)(x)= 2/(x-4). Domain: real numbers except x=4.
    (f/g)(x)= 2(x-4). Domain: real numbers except x=4, because of
    domain of g.

9. f(x) = x+3, g(x) = 1/(x+3).
    (f+g)(x)= x+3 + 1/(x+3) = $(x^2+6x+10)$/(x+3). Domain: real
    numbers except x=-3, or $(-\infty, -3) \cup (-3, \infty)$.
    (f-g)(x)= x+3 - 1/(x+3) = (x^2+6x+8)/(x+3). Domain: real numbers
    except x=-3.
    (fg)(x)= 1. Domain: Domain: real numbers except x=-3.
    (f/g)(x)= $(x+3)^2$. Domain: real numbers except x=-3, because of
    domain of g.

11. f(x) = 7, g(x) = -7.
    (f+g)(x)= 0. Domain: real numbers.
    (f-g)(x)= 14. Domain: real numbers.
    (fg)(x)= -49. Domain: real numbers.
    (f/g)(x)= -1. Domain: real numbers.

13. f(x) = $5x^2$, g(x) = $\sqrt{x-4}$.
    (f+g)(x)= $5x^2 + \sqrt{x-4}$. Domain: x≥4, or $[4, \infty)$.
    (f-g)(x)= $5x^2 - \sqrt{x-4}$. Domain: x≥4.
    (fg)(x)= $(5x^2)\sqrt{x-4}$. Domain: x≥4.
    (f/g)(x)= $5x^2/\sqrt{x-4}$. Domain: x>4, or $(4, \infty)$.

15. f(x) = $\sqrt{x-3}$ + 4, g(x) = $\sqrt{x-4}$.
    (f+g)(x)= $\sqrt{x-3} + \sqrt{x-4}$ + 4. Domain: x≥4, or $[4, \infty)$.
    (f-g)(x)= $\sqrt{x-3} - \sqrt{x-4}$ + 4. Domain: x≥4.
    (fg)(x)= $\sqrt{(x-3)(x-4)} + 4\sqrt{x-4}$. Domain: x≥4.
    (f/g)(x)= $(\sqrt{x-3} + 4)/\sqrt{x-4}$. Domain: x>4, or $(4, \infty)$.

In exercises 17-23, f(x) = 2x-1, g(x) = 5x+2, and h(x) = -x+3.

17. (f+g)(x) = 7x+1. (f+g)(3) = 22.

19. (fg)(x) = (2x-1)(5x+2) = $10x^2-x-2$. (fg)(-2) = (-5)(-8) = 40.

21. (f/g)(x) = (2x-1)/(5x+2). (f/g)(4) = 7/22.

23. (g/h)(x) = (5x+2)/(-x+3). (g/h)(3) does not exist. 3 not in
    domain of g/h.

25. $f(x)=2x+7$, $g(x)=3x+2$. $(g\circ f)(x)=g[f(x)]=g(2x+7)=3(2x+7)+2=6x+23$. $(f\circ g)(x)=f[g(x)]=f(3x+2)=2(3x+2)+7=6x+11$. Domains: real numbers.

27. $f(x)=x^2$, $g(x)=3x+1$. $(g\circ f)(x)=g[f(x)]=g(x^2)=3(x^2)+1=3x^2+1$. $(f\circ g)(x)=f[g(x)]=f(3x+1)=(3x+1)^2=9x^2+6x+1$. Domains: real numbers.

29. $f(x)=3x^2-x+5$, $g(x)=4$. $(g\circ f)(x)=g[f(x)]=g(3x^2-x+5)=4$. $(f\circ g)(x)=f[g(x)]=f(4)=3(4)^2-4+5=49$. Domains: real numbers.

31. $f(x)=x$, $g(x)=1/x$. $(g\circ f)(x)=g[f(x)]=g(x)=1/x$. $(f\circ g)(x)=f[g(x)]=f(1/x))=1/x$. Domains: reals except $x=0$, or $(-\infty,0)\cup(0,\infty)$.

33. $f(x) = |x|$, $g(x) = -3$. $(g\circ f)(x)=g[f(x)]=g(|x|)=-3$. $(f\circ g)(x)=f[g(x)]=f(-3)=3$. Domains: real numbers.

35. $f(x) = 2x^2-x+4$, $g(x) = -3x+2$. $(g\circ f)(x)=g[f(x)]=g(2x^2-x+4)=-3(2x^2-x+4)+2=-6x^2+3x-10$. $(f\circ g)(x)=f[g(x)]=f(-3x+2)=2(-3x+2)^2-(-3x+2)+4=18x^2-21x+10$. Domains: reals.

37. $f(x) = |x+1|$, $g(x) = x^2-4x+3$. $(g\circ f)(x)=g[f(x)]=g(|x+1|)=(|x+1|)^2-4|x+1|+3$. $(f\circ g)(x)=f[g(x)]=f(x^2-4x+3+1)=|x^2-4x+4|$. Domains: reals.

39. $f(x) = \sqrt{x-3}$, $g(x) = x^2+3$. $(g\circ f)(x)=g[f(x)]=g(\sqrt{x-3})= x-3+3 = x$. $(f\circ g)(x)=f[g(x)]=f(x^2+3)=\sqrt{x^2+3-3} = |x|$. Domains: reals.

41. $f(x) = \sqrt{x-3}$, $g(x) = \sqrt{x+5}$. $(g\circ f)(x)=g[f(x)]=g(\sqrt{x-3})=\sqrt{\sqrt{x-3}+5}$. Domain: $x\geq3$, or $[3,\infty)$. $(f\circ g)(x)=f[g(x)]=f(\sqrt{x+5})=\sqrt{\sqrt{x+5}-3}$. Need $\sqrt{x+5}\geq3$. $x+5\geq9$, $x\geq4$. Domain: $x\geq4$, or $[4,\infty)$.

In exercises 43–49, $f(x) = 2x+1$, $g(x) = 3x-2$ and $h(x) = x^2$.

43. $(f\circ g)(x)=f(3x-2)=2(3x-2)+1=6x-3$. $(f\circ g)(2)=9$.

45. $(f\circ h)(x)=f(x^2)=2(x^2)+1=2x^2+1$. $(f\circ h)(1)=3$.

47. $(f\circ f)(x)=f(2x+1)=2(2x+1)+1=4x+3$. $(f\circ f)(a)=4a+3$.

49. $(g\circ g)(x)=g(3x-2)=3(3x-2)-2=9x-8$. $(g\circ g)(3)=19$.

In exercises 51–55, $f(x) = 2x+3$, $g(x) = \sqrt{x+2}$, and $h(x) = x^2$.

51. $F(x)=2\sqrt{x+2} + 3 = (f\circ g)(x)$.   53. $H(x)=\sqrt{2x+5} = (g\circ f)(x)$.

55. $G(x)=x+2 = (h\circ g)(x)$.

In exercises 57-61, $f(x) = 2x+1$, $g(x) = 3x-2$, and $h(x) = x^2+1$.

57. $F(x) = 6x+1 = 3(2x+1)-2 = (g \circ f)(x)$.

59. $H(x) = 6x-3 = 2(3x-2)+1 = (f \circ g)(x)$.

61. $H(x) = 3x^2+1 = (g \circ h)(x)$.

63. $C(x) = \dfrac{x^3}{275} - 2x^2 + 658x + 2400$, $R(x) = 400x$,

$P(x) = R(x) - C(x) = -\dfrac{x^3}{275} + 2x^2 - 258x - 2400$. Current weekly production of computers is 250. P(250)=1281.82. Current profit = \$1,281.82. Max P(x) is at A(283.13871,2345.1591). A(283)=2345.14, A(284))=2344.35. Company should raise production to 283 computers/week. P(283)=\$2345.14.

65. (a) $C(x) = \dfrac{x^3}{10} - 47x^2 + 8000x + 57600$. $C_A(x) = \dfrac{C_A}{x} = \dfrac{x^2}{10} - 47x + 8000 + \dfrac{57600}{x}$.

Av cost of 150 is $C_A(150) = \dfrac{150^2}{10} - 47(150) + 8000 + \dfrac{57600}{150} = \$3,584$.

(b) Min $C_A(x)$ is at Q(240,2720). Average cost decreases to x=240, increases from x=240 on. Thus average cost of production originally decreases as production increases, but a stage is reached after which the average cost of producing a single item increases. This phenomenon is called the Law of Diminishing Returns. The initial decrease is due to the fact that it is initially more cost efficient to spread production costs out over a large number of items rather than a small number. However, this only holds true up to a point; eventually the cost per item will increase. (c) Daily production that minimizes cost is 240 units. (d) The minimum average cost is then \$2,720.

67. (a) $N(t) = 2t^3 - 156t^2 + 2530t + 66800$. t is time in years from 1945, N(t) is number of visitors in year t. $D(t) = N(t+1) - N(t) = 2(t+1)^3 - 156(t+1)^2 + 2530(t+1) + 66800 - (2t^3 - 156t^2 + 2530t + 66800) = 6t^2 - 306t + 2376$. (b) D(t) gives the increase or decrease in vistors from year t to year t+1; that is the increase during the t th year. (c) D(5)=996, D(20)=-1344, D(45)=756. There was an increase of 996 visitors during the 5th year (1950), decrease of 1344 visitors during the 20th year (1965), and an increase of 756 during the 45th year (1990). (d) Min D(t) at A(25.5, -1525.5). The largest decrease in visitors took place in 1970 when there were 1,525 fewer visitors than the year before. The number of visitors decreased annually from B (during t=9, 1954) to C (during t=41, 1986). [You might ask the students why we expect the minimum value of N to occur during the year t=41.] After that N(t+1)-N(t)>0 and the number of visitors annually started increasing.

Section 3.7

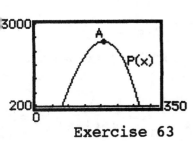
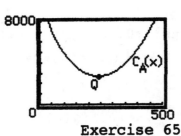
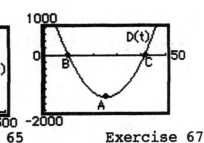

Exercise 63   Exercise 65   Exercise 67

69. $f(x)=cx(1-x)$.
    c=2.5: .5, .625, .5859375, -> .6. Steady state.
    c=3.2: .5, .8, .512, .7995392, ..., .513045, .799455, ...
    Period 2. (to 6 dec places)
    c=3.5: .5, .875, .3828125, ..., .874997, .382820, .826941, ...
    Period 3. (to 6 dec places)
    c=4: .5, 1,0,0, ->0. Steady state.

71. $f(x)=3.8397x(1-x)$.c=3.8397: .72,.77408352,.6714799169,..
    .149549, .488349, .959404, ...   Period 3.

73. $f(x)=1.2x(1-x)$. c=1.2: .08, .08832, ...->.1667 .
    If things continue in this way 16.67% of the annual budget
    will be devoted to the library.

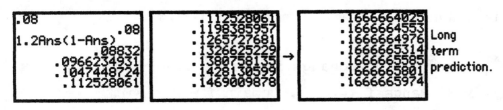

Section 3.7

1. $f(x)=2x-1, g(x)=\frac{x+1}{2}$. $g(f(x))=\frac{(2x-1)+1}{2}=x$. $f(g(x))=2(\frac{x+1}{2})-1=$
   $(x+1)-1=x$. Thus $g=f^{-1}$. Domains & ranges of both are the set
   of real numbers.

3. $f(x)=\frac{x}{4}$, $g(x)=4x$. $g(f(x))=4(\frac{x}{4})=x$. $f(g(x))=\frac{4x}{4}=x$. Thus $g=f^{-1}$.
   Domains & ranges of both are the set of real numbers.

5. $f(x)=x^3$, $g(x)=\sqrt[3]{x}$. $g(f(x)=\sqrt[3]{x^3}=x$. $f(g(x))=(\sqrt[3]{x})^3=x$.
   Thus $g=f^{-1}$. Domains & ranges of both are the set of real
   numbers.

7. $f(x)=3-2x$, $g(x)=\frac{3-x}{2}$. $g(f(x))=\frac{3-(3-2x)}{2}=x$. $f(g(x))=3-2(\frac{3-x}{2})=x$.
   Thus $g=f^{-1}$. Domains & ranges of both are the set of real
   numbers.

125

9. $f(x)=\sqrt{x-5}$, $g(x)=x^2+5$ with $x\geq0$. $g(f(x))=(\sqrt{x-5})^2+5 = x$. $f(g(x))=\sqrt{(x^2+5)-5} = x$. Domain$(g)=[0,\infty)=$Range$(f)$. Thus $g=f^{-1}$.

11. $f(x)=\dfrac{x}{x+1}$, $x\neq-1$, $g(x)=\dfrac{x}{1-x}$, $x\neq1$. $f(g(x))=\dfrac{\dfrac{x}{1-x}}{\dfrac{x}{1-x}+1} = \dfrac{x}{x+(1-x)} = x$.

$g(f(x))=\dfrac{\dfrac{x}{x+1}}{1-\dfrac{x}{x+1}} = \dfrac{x}{(x+1)-x} = x$. Range of f is $(-\infty,1)\cup(1,\infty)$

since y=1 is a horizontal asymptote of f. This is domain of g. Thus $g=f^{-1}$.

13. $f(x)=5x+2$, $y=5x+2$. Interchange x&y, $x=5y+2$. Solve for y, $y=\dfrac{x-2}{5}$. $f^{-1}(x)=\dfrac{x-2}{5}$. Range of f $(-\infty,\infty)$. Domain of f^{-1} $(-\infty,\infty)$.

15. $f(x)=2x$, $y=2x$. Interchange x&y, $x=2y$. Solve for y, $y=\dfrac{x}{2}$. $f^{-1}(x)=\dfrac{x}{2}$. Range f $(-\infty,\infty)$. Domain f^{-1} $(-\infty,\infty)$.

17. $f(x)=\dfrac{x+3}{4}$, $y=\dfrac{x+3}{4}$. Interchange x&y, $x=\dfrac{y+3}{4}$. Solve for y, $y=4x-3$. $f^{-1}(x)=4x-3$. Range f $(-\infty,\infty)$. Domain f^{-1} $(-\infty,\infty)$.

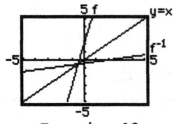

Exercise 13   Exercise 15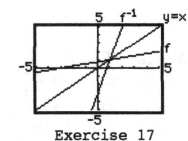
Exercise 17

19. $f(x)=x^3+3$, $y=x^3+3$. Interchange x&y, $x=y^3+3$. Solve for y, $y=\sqrt[3]{x-3}$. $f^{-1}(x)=\sqrt[3]{x-3}$. Range f $(-\infty,\infty)$. Domain f^{-1} $(-\infty,\infty)$.

21. $f(x)=\dfrac{1}{\sqrt{x}}$, $y=\dfrac{1}{\sqrt{x}}$, $x\geq0$. Interchange x&y, $x=\dfrac{1}{\sqrt{y}}$, $y\geq0$. Solve for y, $\sqrt{y}=\dfrac{1}{x}$, $y=\dfrac{1}{x^2}$. $f^{-1}(x)=\dfrac{1}{x^2}$. Range of f is $(0,\infty)$, Thus domain of f^{-1} is $(0,\infty)$.

23. $f(x)=x^2$, $y=x^2$, $x\geq0$. Interchange x&y, $x=y^2$, $y\geq0$. Solve for y, $y=\sqrt{x}$ since $y\geq0$. $f^{-1}(x)=\sqrt{x}$. Range f $[0,\infty)$. Domain f^{-1} is $[0,\infty)$.

Section 3.7

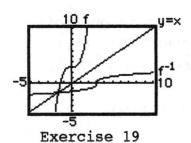
Exercise 19

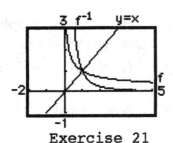
Exercise 21

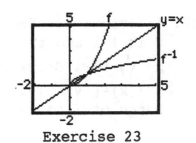
Exercise 23

25. One-to-one   27.Not one-to-one   29.Not one-to-one

<u>In the graphs of exercises 31-43 Xscl=1 and Yscl=1.</u>

31. $f(x)=2x+3$. If $a\neq b$, $2a\neq 2b$, $2a+3\neq 2b+3$, $f(a)\neq f(b)$. f is 1-1. f^{-1} exists.

33. $f(x)=4$. If $a\neq b$, $f(a)=f(b)=4$. f is not 1-1. f^{-1} not exist.

35. $f(x)=-x^2$. Let $b=-a\neq 0$. $b\neq a$. $a^2=b^2$, $-a^2=-b^2$ & $f(a)=f(b)$. f is not 1-1. f^{-1} does not exist.

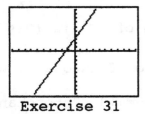
Exercise 31

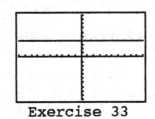
Exercise 33

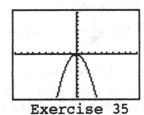
Exercise 35

37. $f(x)=x^3$. If $a\neq b$, $a^3\neq b^3$. $f(a)\neq f(b)$. f is 1-1. f^{-1} exists.

39. $f(x)=|x|$. If $b=-a\neq 0$. $b\neq a$. $|a|=|b|$ & $f(a)=f(b)$. f is not 1-1. f^{-1} does not exist.

41. $f(x)=1/x$. If $a\neq b$, $1/a\neq 1/b$. $f(a)\neq f(b)$. f is 1-1. f^{-1} exists.

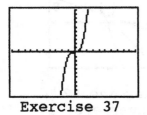
Exercise 37

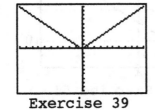
Exercise 39

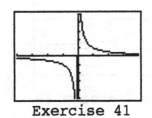
Exercise 41

43. $g(x)=(x-3)^2$. $g(1)=4, f(5)=4$. g is not 1-1. g^{-1} does not exist.

45. $f(x)=\sqrt[3]{x+4}$. Each horiz line cuts graph of f in one point (below). Thus f is 1-1 and f^{-1} exists. Let $y=\sqrt[3]{x+4}$. Interchange x & y, $x=\sqrt[3]{y+4}$. Solve for y, $y=x^3-4$. $f^{-1}(x)=x^3-4$. Range of f is $(-\infty,\infty)$. Domain of f^{-1} is $(-\infty,\infty)$.

47. $f(x)=(\sqrt{x}-5)^3$. Each horiz line cuts graph of f in one point. f is 1-1 and f^{-1} exists. Let $y=(\sqrt{x}-5)^3$. Interchange x&y: $x=(\sqrt{y}-5)^3$. Solve for y, $\sqrt[3]{x}=\sqrt{y}-5$, $\sqrt{y}=\sqrt[3]{x}+5$, $y=(\sqrt[3]{x}+5)^2$. $f^{-1}(x)=(\sqrt[3]{x}+5)^2$.
    Range of f is is $(-\infty,\infty)$. Domain of f^{-1} is $(-\infty,\infty)$.

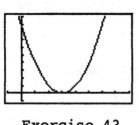

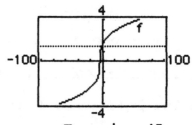

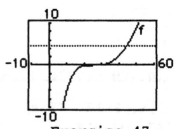

| Exercise 43 | Exercise 45 | Exercise 47 |

49. $f(x)=\dfrac{1}{x-4}$, x>4. Each horiz line cuts graph of f in one or no points (below). Thus f is 1-1 and f^{-1} exists.
    Let $y=\dfrac{1}{x-4}$. Interchange x & y, $x=\dfrac{1}{y-4}$, y>4. Solve for y, $y=\dfrac{1}{x}+4$.
    $f^{-1}(x)=\dfrac{1}{x}+4$. Range of f is $(0,\infty)$. Thus domain of f^{-1} is $(0,\infty)$.

51. $f(x)=|x+3|$, x≥-3. Each horiz line cuts graph of f in one or no points (below). Thus f is 1-1 and f^{-1} exists.
    Let $y=|x+3|$. Since x≥-3, x+3>0 & $|x+3|=x+3$. y=x+3. Interchange x & y, x=y+3. Solve for y, y=x-3. $f^{-1}(x)=x-3$.
    Range of f is $[0,\infty)$. Thus domain f^{-1} is $[0,\infty)$.

53. $f(x)=3x-2$, $g(x)=(x+2)/3$. Symmetry about y=x. (figure below, in square setting.) g is inverse of f.

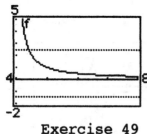

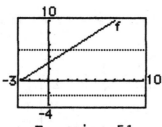

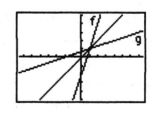

| Exercise 49 | Exercise 51 | Exercise 53 |
| | | g inverse of f |

55. $f(x)=2x-1$, $g(x)=.5x+1$. No symmetry. g not inverse f.
    Let y=2x-1. x+1=2y. $y=(x+1)/2$. $f^{-1}(x)=(x+1)/2$. Symmetry.

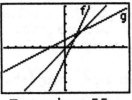

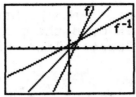

Exercise 55
g not inverse f

Exercise 55
f, f⁻¹ and y=x

57. $f(x)=x^2+3$ with domain $x \geq 0$, $g(x)=\sqrt{x}-3$ with $x \geq 0$. No symmetry. g is not inverse of f. (see fig below).
Let $y=x^2+3$, $x \geq 0$. $x=y^2+3$, $y \geq 0$. $y=\sqrt{x-3}$. $f^{-1}(x)=\sqrt{x-3}$. Symmetry. Range f is $[3,\infty)$. Domain f^{-1} is $[3,\infty)$.

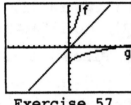

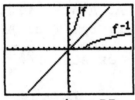

Exercise 57
g not inverse f

Exercise 57
f, f⁻¹ and y=x

59. $f(x)=\sqrt{x-5}$, $g(x)=x^2+5$, $x \geq 0$. Symmetry. g is inverse f.

61. $f(x)=x^3+kx^2+3x-4$, for various values of k. f is 1-1 and thus has inverse for $-3 \leq k \leq 3$. Use trace/zoom to see that max/min appear for other values of k. (Tangent is horiz when k=-3,3.)

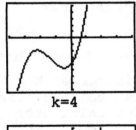

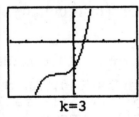

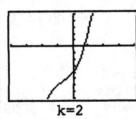

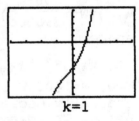

k=4   k=3   k=2   k=1

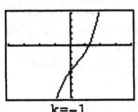

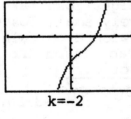

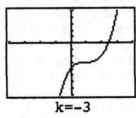

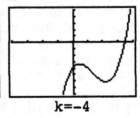

k=-1   k=-2   k=-3   k=-4

63. $f(x)=ax+b$, $a \neq 0$. Graph of f is a non-horizontal line. Thus every horizontal line cuts it at one point. Thus f has an inverse. Let $y=ax+b$. Interchange x & y, $x=ay+b$.
Solve for y, $y=(\frac{1}{a})x - \frac{b}{a}$. $f^{-1}(x)=(\frac{1}{a})x - \frac{b}{a}$, a linear function.

65. $f(x)=ax^2+bx+c$, $a\neq0$. Domain $(-\infty,\infty)$. Graph is a parabola. There is a horizontal line that cuts the parabola in two points. Thus f is not 1-1 by the horizontal line test. No inverse.

67. $f(x)=ax^3+b$, $a \neq 0$. If $p\neq q$, $p^3\neq q^3$, $ap^3\neq aq^3$, and $ap^3+b\neq aq^3+b$. Thus $f(p)\neq f(q)$. f is one-to-one. It has an inverse.
   Let $y=ax^3+b$. Interchange x&y, $x=ay^3+b$. $y=\sqrt[3]{\dfrac{x-b}{a}}$. $f^{-1}(x)=\sqrt[3]{\dfrac{x-b}{a}}$.
   Range of f is $(-\infty,\infty)$. Domain of f^{-1} is $(-\infty,\infty)$.
   Geometrically: $f(x)=x^3$ is 1-1 by horiz line test. Stretch/contract by factor "a" gives $f(x)=ax^3$ is 1-1. Slide up/down by factor b gives $f(x)=ax^3+b$ is one-to-one.

69. Circle, radius r, area A. $A=\pi r^2$, $r\geq0$. Consider two radii, r_1 and r_2, $r_1\neq r_2$. Suppose $r_1<r_2$. Thus $A(r_1)<A(r_2)$, $A(r_1)\neq A(r_2)$. A is one-to-one. Thus A^{-1} exists. $r^2=\dfrac{A}{\pi}$, $r=\pm\sqrt{\dfrac{A}{\pi}}$. Since $r>0$ take postive root. $r=\sqrt{\dfrac{A}{\pi}}$. $r(1)=\sqrt{\dfrac{1}{\pi}}$ = 0.5642 inches.

71. (a) $v(t)=u+at$. v is linear, thus one-to-one and has an inverse. $v-u=at$. $t=(v-u)/a$. This is inverse function. (b) When $u=20$, $a=25$, $v=95$, then $t=(95-20)/25=75/25=3$. Time is 3 secs.

73. (a) $C(x)=2x+700$. Graph of C is non-horizontal line. C has an inverse by horiz line test. $C=2x+700$. $C-700=2x$. $x=(C-700)/2$.
   (b)When $C=1230$, $x=(1230-700)/2=(530)/2=265$. Number items = 265.

75. $W(d)=W_E(\dfrac{3956}{3956+d})^2$. Domain W(d) is [0,1000000], translunar space.
   (a) $W(d)=140(\dfrac{3956}{3956+d})^2$.Person weighs 140 pounds on Earth. When $d=0$, $W=140$. W intercept of graph is 140. As d increases W gradually decreases. Thus graph gradually falls from 140. Graph below. The d-axis is a horizontal asymptote. The calculator graph shows that most of the weight loss takes place during the first 10,000 miles. The relatively small losses after 10,000 miles makes it difficult to draw on a calculator.
   (b) Horizontal lines cut the graph at one/no points. Thus W is 1-1 and has an inverse. Solve for d.
   $(3956+d)^2=140(3956)^2/W$. $d=\sqrt{\dfrac{140(3956)^2}{W}}$ -3956.
   $d=3956\sqrt{\dfrac{140}{W}} - 3956$. $d=3956(\sqrt{\dfrac{140}{W}}-1)$. When $W=1$, $d=42852.02324$.
   The person who weighs 140 pounds on Earth would weigh one pound at a distance of 42,852 miles from Earth.

Chapter 3 Project and Review Exercises

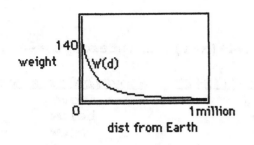

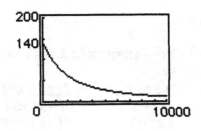

77. 13 70 82 88 67 34 88 64 67 52 16 40
    2 21 25 27 20 9 27 19 20 15 3 11
    B U Y - T I - S T O C K

79. 43 11 7 45 45 15 41
    19 3 1 20 20 5 18
    S C A T T E R

Chapter 3 Project

$f(x)=0.04x^2-x+10$, $g(x)=0.12x^2-2x+10$. (a) $f(8)=4.56$, $g(8)=1.68$.
(b) Table below gives $f(5)=6$, $g(5)=3$. Syllables retained
during sleep is twice that during the day after 5 hours.
(c) $h(x)=f(x)-g(x)=(0.04x^2-x+10)-(0.12x^2-2x+10)=-0.08x^2+x$.
Max of $h(x)$ is at A(6.25,3.125). The largest retention
difference between awake and sleep occurs after 6.25 hours. The
number of syllables remembered after 6.25 hours of sleep is
over three times the number remembered after 6.25 hours of
daytime activities.

		f	g
X		Y_1	Y_2
0		10	10
1		9.04	8.12
2		8.16	6.48
3		7.36	5.08
4		6.64	3.92
5		6	3
6		5.44	2.32

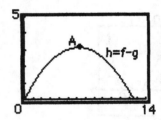

Chapter 3 Review Exercises

1. a) $f(x)=(x+1)(x-2)(x-5)$. x-intercepts=-1,2,5. y-int=f(0)=10.

Interval	Test Pt	$f(x)=(x+1)(x-2)(x-5)$	above/below x-axis
$(-\infty,-1)$	x=-2	f(-2)=-28<0	below
(-1,2)	x=0	f(0)=10>0	above
(2,5)	x=3	f(3)=-8<0	below
$(5,\infty)$	x=6	f(6)=28>0	above

Graph below.

b) $f(x)=x^2(x-3)(x+2)$. x-intercepts=-2,0,3. y-int=f(0)=0.

Interval	Test Pt	$f(x)=x^2(x-3)(x+2)$	above/below x-axis
$(-\infty,-2)$	x=-3	f(-3)=54<0	above
(-2,0)	x=-1	f(-1)=-4>0	below
(0,3)	x=1	f(1)=-6<0	below
$(3,\infty)$	x=4	f(4)=96>0	above

Graph below.

c) $f(x)=x^3+3x^2-4x=x(x^2+3x-4)=x(x+4)(x-1)$. x-intercepts=-4,0,1.
y-int=f(0)=0.

Interval	Test Pt	$f(x)=x^2(x-3)(x+2)$	above/below x-axis
$(-\infty,-2)$	x=-3	f(-3)=54<0	above
$(-2,0)$	x=-1	f(-1)=-4>0	below
$(0,3)$	x=1	f(1)=-6<0	below
$(3,\infty)$	x=4	f(4)=96>0	above

Graph below.

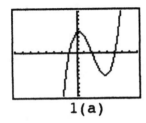 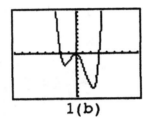 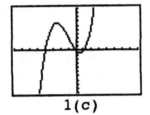

1(a)   1(b)   1(c)

2. x-int=-3,2,4. y-int=48. Try $f(x)=(x+3)(x-2)(x-4)$.

Interval	Test Pt	$f(x)=(x+3)(x-2)(x-4)$	above/below x-axis
$(-\infty,-3)$	x=-4	f(-4)=-48<0	below
$(-3,2)$	x=0	f(0)=24>0	above
$(2,4)$	x=3	f(3)=-6<0	below
$(4,\infty)$	x=5	f(5)=24>0	above

Thus $f(x)=(x+3)(x-2)(x-4)$ satisfies the x-int, (graph below).
$f(x)=k(x+3)(x-2)(x-4)$, k>0 also satisfies conditions.
f(0)=24k=48. Thus k=2. $f(x)=2(x+3)(x-2)(x-4)$.

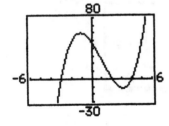

3. (a) $f(x)=2x^3+5x^2+3x+1$. Inc$(-\infty,-1.27]$,$[-.39,\infty)$.
Dec$[-1.27,-.39]$. Max(-1.27,1.16), Min(-.39,.47).
Zero x= -1.83.

(b) $f(x)=(x-3)^4+5x-17$. Dec$(-\infty,1.92)$, Inc$(1.92,\infty)$.
Min(1.92,-6.04). Zeros 1.17, 3.40.

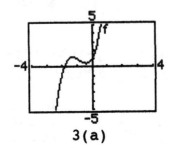 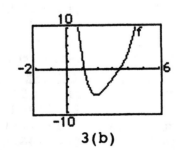

3(a)   3(b)

4. Sales after t days of advertising $s(t)=t^3-89t^2+1600t+10000$ units, t≤4. See graph below. A(11.0447,18162.112). Sales peak after 11 days.

5. (a) $s(t)=0.1t^3-16t^2+468t$. (b)A(17.4942,3825.9379),B(38.5271,0). Takes 17.49 secs to reach max height, 38.53 seconds to come back to ground level. (c) Height increasing from 0 to 17.49 secs, decreasing from 17.49 to 38.53 secs. Note max height is not half way into flight because of effect of rockets.

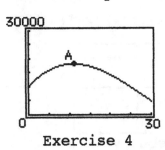

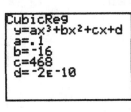

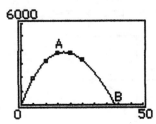

Exercise 4   Exercise 5

6. (a)$f(x)=3x-2$,domain[-3,5]. Line, y-int=-2,x-int=2/3. $f(-3)=-11,f(5)=13$.Range[-11,13].

(b)$f(x)=-2x+4$, domain [-5,5]. Line, y-int=4, x-int=2. $f(-5)=14,f(5)=-6$.Range[-6,14]

(c)$f(x)=2x^2+8x+3$,domain [-3,1]. Parabola,$x=\dfrac{-b}{2a}=\dfrac{-8}{2(2)}=-2$. $f(-2)=-5$. Vertex(-2,-5). $f(-3)=-3,f(1)=13$.Range[-5,13].

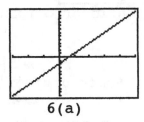

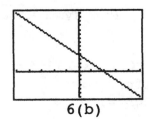

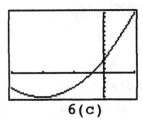

6(a)   6(b)   6(c)

7. (a) $f(x)=\begin{cases} x+3 & \text{if } x\le2 \\ -x-1 & \text{if } x>2 \end{cases}$

(b) $g(x)=\begin{cases} 2x^2+1 & \text{if } x<2 \\ -x-1 & \text{if } x\ge2 \end{cases}$

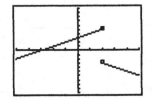

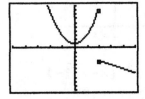

8. (a) $8 hr for 8AM to 5PM. $12 hr 5PM to midnight. Let x be # hr measured from 8AM. 8AM to 5PM corresponds to 0≤x≤9, 5PM to midnight corresponds to 9<x≤16. Let P(x) be wage function. Person works from 8AM to 7PM, x=11.

$$P(x)= \begin{cases} 8x & \text{if } 0 \le x \le 9 \\ 12(x-9)+72 & \text{if } 9 < x \le 16 \end{cases} \cdot \text{ (b) } P(11)=\$95.$$

9. $t_{inc}=9.105(M_b)^{0.167}$. 500kg is 500,000gm. M_b is in interval [2.5, 500000]. As M_b increases t_{inc} increases. Thus range is $[t_{inc}(2.5), t_{inc}(500000)]=[10.6, 81.5]$. Range of incubation times is 10.6 to 81.5 days.

10. (a) $f(x)=-\dfrac{2}{x^2}$. $f(-x)=-\dfrac{2}{(-x)^2}$.
$f(-x)=f(x), f(-x) \ne -f(x)$. Even.

(b) $f(x)=3$. $f(-x)=3$.
$f(-x)=f(x), f(-x) \ne -f(x)$. Even.

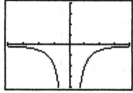

Symm about y-axis.

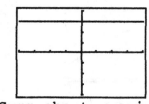

Symm about y-axis.

(c) $g(x)=-4x^3$. $g(-x)=-4(-x)^3$.
$g(-x) \ne g(x), g(-x)=-g(x)$. Odd.

(d) $h(x)=x^4+x$. $h(-x)=(-x)^4+(-x)$.
$h(-x) \ne h(x), h(-x) \ne -h(x)$. Neither.

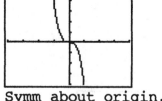

Symm about origin.

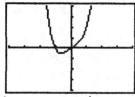

Not symm about y-axis or origin.

11. (a) $g(x)=x^2-3$
Shift $f(x)=x^2$
down 3.

(b) $g(x)=(x-2)^2+4$
Shift $f(x)=x^2$
right 2, up 4.

(c) $g(x)=|x-3|-5$
Shift $g(x)=|x|$
right 3, down 5.

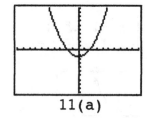

11(a)

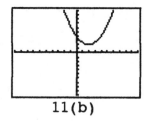

11(b)

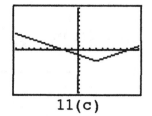

11(c)

(d) $f(x)=-(x-3)^2-4$
Shift $f(x)=x^2$
right 3, reflect
in x-axis, down 4.

(e) $f(x)=\sqrt{x-1}+3$
Shift $f(x)=\sqrt{x}$
right 1, up 3.

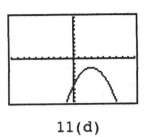

11(d)

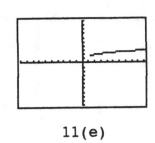

11(e)

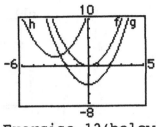

Exercise 12(below)

12. g: shift f down 4. h: shift f left 3, up 2. Graph above.

13. a) Shift f up 5. $g(x)=x^3+5$.
    b) Reflect f in y-axis, shift up 5. $g(x)=(-x)^3+5$. $g(x)=-x^3+5$.
    c) Shift f right 2, down 10. $g(x)=(x-2)^3-10$.

14. a) $f(x)=x^2$ and $g(x)=(x-1)^2$. Values of x in the table increase in steps of 1. The value of g is the same as that f one step before. Thus value of g at x is the same as that of f at $(x-1)$.
    b) $f(x)=x^2$. Value of h at each x is the same as that f plus 4. Thus $h(x)=x^2+4$.
    c) $f(x)=x^2$. Values of x increase by 1. Value of k at x is the same as that of f at $(x-1)$, plus 4. $k(x)=(x-1)^2+4$.

15.

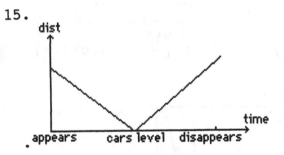

16.

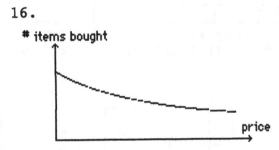

17. Let $x be parking fee per day, N(x) the number of cars using the facility when fee is $x. Graph should gradually fall as parking fee increases. N intercept will be 150. Let us use a parabola that opens down. Try $N(t)=-x^2+150$, graph (a) shown below. Suspect N would fall off more rapidly than this as price increases. Consider $N(t)=-10x^2+150$, graph (b), would probably describe it more accurately. Adjust coefficient of x^2 to control how rapidly graph falls. Domain $x \geq 0$.

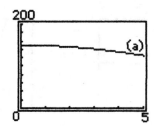 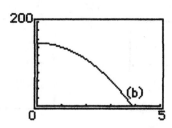

18. Tom starts from Dallas at 9AM. & travels at 60mph. Let t be time measured from 9AM. f(t)=60t gives Tom's distance from Dallas at time t. Sally starts from Dallas at 9:15AM at 75mph. g(t)=75(t-1/4) gives Sally's distance from Dallas at time t. Sally catches up with Tom at A(1.25,75) below, 75 miles from Dallas, 1.25hr after 9AM. That is at 10:15AM.

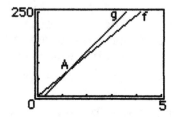

19. (a)

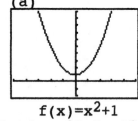

f(x)=x²+1
Dec(-∞,0],Inc[0,∞)

(b)

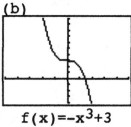

f(x)=-x³+3
Dec(-∞,∞)

(c)

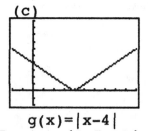

g(x)=|x-4|
Dec(-∞,4],Inc[4,∞)

(d)

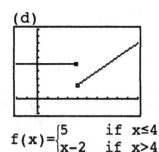

$f(x)=\begin{cases}5 & \text{if } x\le 4\\ x-2 & \text{if } x>4\end{cases}$

Constant(-∞,4],Inc(4,∞)

(e)

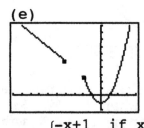

$g(x)=\begin{cases}-x+1 & \text{if } x\le -4\\ x^2+1 & \text{if } x\ge -2\end{cases}$

Dec(-∞,-4]&[-2,0],Inc[0,∞)

20. a) $f(x)=\dfrac{3x^2-4x}{x}$. Domain: All real numbers except 0.

b) $f(x)=\dfrac{x-2}{x^2+2x-8} = \dfrac{x-2}{(x+4)(x-2)} = \dfrac{1}{x+4}$, x≠2.

Domain: All real numbers except -4 and 2.

21. a) $f(x) = \dfrac{x^2+2x-3}{x-1} = \dfrac{(x+3)(x-1)}{x-1} = x+3, x \neq 1.$
    Domain: All reals except 1. No asymptotes.

  b) $f(x) = \dfrac{2}{x-4} + 3.$ Vertical asymptote x=4. Horiz y=3.
    Domain: All reals except 4.

  c) $f(x) = \dfrac{5}{x^2+x-6} = \dfrac{5}{(x+3)(x-2)}.$ Vert x=-3, x=2. Horiz y=0.
    Domain: All reals except -3,2.

  d) $g(x) = \dfrac{x-1}{(x+2)(x-4)} + 4.$ Vert x=-2, x=4. Horiz y=4.
    Domain: All reals except x=-2,4.

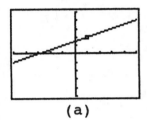
(a)

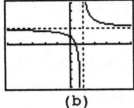
(b)

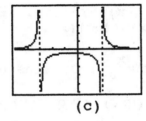
(c)

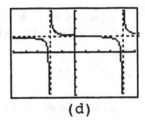
(d)

22. (a) $f(x)=3x-1$, $g(x)=-4x+5.$
    $(f+g)(x) = (3x-1) + (-4x+5) = -x+4.$ Domain: real numbers.
    $(f-g)(x) = (3x-1) - (-4x+5) = 7x-6.$ Domain: real numbers.
    $(fg)(x) = (3x-1)(-4x+5) = -12x^2+19x-5.$ Domain: real numbers.
    $(f/g)(x) = (3x-1)/(-4x+5).$ Domain: real numbers except x=5/4.

  (b) $f(x)=4x+1$, $g(x)=2\sqrt{3x+4}.$
    $(f+g)(x) = (4x+1)+(2\sqrt{3x+4})=4x+1+2\sqrt{3x+4}.$ Domain: x≥-4/3.
    $(f-g)(x) = (4x+1)-(2\sqrt{3x+4})=4x+1-2\sqrt{3x+4}.$ Domain: x≥-4/3.
    $(fg)(x) = (4x+1)(2\sqrt{3x+4}).$ Domain: x≥-4/3.
    $(f/g)(x) = (4x+1)/(2\sqrt{3x+4}).$ Domain: x>-4/3.

(c) $f(x)=x^2+3x+1$, $g(x)=1/x^2.$
    $(f+g)(x) = (x^2+3x+1)+(1/x^2)=(x^4+3x^3+x^2+1)/x^2.$ Domain: real numbers except x=0.
    $(f-g)(x) = (x^2+3x+1)-(1/x^2)=(x^4+3x^3+x^2-1)/x^2.$ Domain: real numbers except x=0.
    $(fg)(x) = (x^2+3x+1)(1/x^2)=(x^2+3x+1)/x^2.$ Domain: real numbers except x=0.
    $(f/g)(x) = (x^2+3x+1)/(1/x^2) = x^2(x^2+3x+1).$ Domain: real numbers.

23. (a) $f(x)=3x+5$, $g(x)=2x-7.$
    $(g \circ f)(x)=g[f(x)]=g(3x+5)=2(3x+5)-7=6x+3.$
    $(f \circ g)(x)=f[g(x)]=f(2x-7)=3(2x-7)+5=6x-16.$ Domains: real numbers.

  (b) $f(x)=4x+1$, $g(x)=3x^2+4x-1.$
    $(g \circ f)(x)=g[f(x)]=g(4x+1)=3(4x+1)^2+4(4x+1)-1=48x^2+40x+6.$

$(f \circ g)(x) = f[g(x)] = f(3x^2+4x-1) = 4(3x^2+4x-1)+1 = 12x^2+16x-3$.
Domains:real numbers.

(c) $f(x) = \sqrt{x+4}$, $g(x) = x^2-6x-11$.
$(g \circ f)(x) = g[f(x)] = g(\sqrt{x+4}) = (\sqrt{x+4})^2 - 6(\sqrt{x+4}) - 11 = x-7-6\sqrt{x+4}$.
 Domain: $x \geq -4$ or $[-4,\infty)$.
$(f \circ g)(x) = f[g(x)] = f(x^2-6x-11) = \sqrt{x^2-6x-11+4} = \sqrt{x^2-6x-7} = \sqrt{(x-7)(x+1)}$.
Domain: $x \geq 7$ or $x \leq -1$. $(-\infty,-1] \cup [7,\infty)$.

24. $f(x)=3x-1$ and $g(x)=-4x+3$. $F(x)=-12x+7 = -4(3x-1)+3=(g \circ f)(x)$.
$F(x)=(g \circ f)(x)$.

25. $f(x)=4x+2$ and $g(x)=5x^2+2x-1$. $H(x)=20x^2+8x-2 = 4(5x^2+2x-1)+2$
$= (f \circ g)(x)$. $H(x)=(f \circ g)(x)$.

26. $f(x)=4x+3$, $g(x)=\dfrac{x-3}{4}$. Domains & ranges of both is the set of

real numbers. $(g \circ f)(x)=g[f(x)]=g(4x+3)=\dfrac{(4x+3)-3}{4} = x$.

$(f \circ g)(x)=f[g(x)]=f(\dfrac{x-3}{4})=4(\dfrac{x-3}{4})+3 = x$. Thus g is inverse of f.

27. (a) $f(x)=5x-1$. If $a \neq b$, $5a-1 \neq 5b-1$. $f(a) \neq f(b)$. f is one-to-one.
   Every horiz line cuts graph of f at one point. f one-to-one.
   (b) $f(x)=x^3+1$. If $a \neq b$, $a^3+1 \neq b^3+1$. $f(a) \neq f(b)$. f is one-to-one.
   Every horiz line cuts graph of f at one point. f one-to-one.
   (c) $f(x)=2x^2-4x-1$. Parabola with vertex at x=1. Symmetric
   about axis x=1. $f(0)=-1$, $f(2)=-1$. f is not one-to-one.

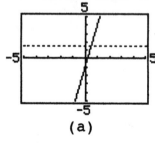

(a)

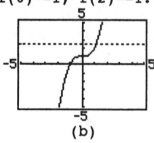

(b)

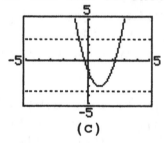

(c)

28. (a) $f(x)=7x-4$. $y=7x-4$. Interchange x & y, $x=7y-4$. Solve for y,
$y=\dfrac{x+4}{7}$. $f^{-1}(x)=\dfrac{x+4}{7}$. Graphs below, in square setting. Symmetry
about y=x. Range of f is $(-\infty,\infty)$. Domain of f^{-1} is $(-\infty,\infty)$.

(b) $f(x)=4-x^2, x \geq 0$. $y=4-x^2$. Interchange x & y, $x=4-y^2$, $y \geq 0$.
Solve for y, $y=\pm\sqrt{4-x}$. Select positive since $y \geq 0$. $f^{-1}(x)=\sqrt{4-x}$.
Graphs below, in square setting. Symmetry about y=x.
Range of f is $(-\infty,4]$. Domain of f^{-1} is $(-\infty,4]$.

Chapter 3 Test

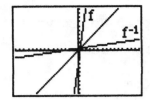
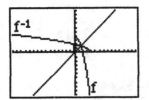

Chapter 3 Test

1. f(x)=(x+2)(x-4)(x-6). x-intercepts=-2,4,6. y-int=f(0)=48.

Interval	Test Pt	f(x)=(x+1)(x-2)(x-6)	above/below x-axis
(-∞,-2)	x=-3	f(-3)=-63<0	below
(-2,4)	x=0	f(0)=48>0	above
(4,6)	x=5	f(5)=-7<0	below
(6,∞)	x=7	f(7)=27>0	above

See graph below.

2. $f(x)=x^2+6x+1$, domain [-9,4]. Parabola, $x=\dfrac{-b}{2a}=\dfrac{-6}{2(1)}=-3$.

f(-3)=-8. Vertex(-3,-8). f(-9)=28, f(4)=41. Range[-8,41]. See graph below.

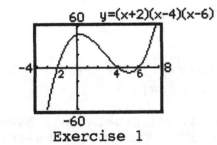

Exercise 1

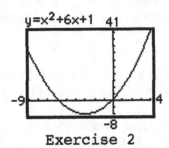

Exercise 2

3. $f(x)=x^4-3x^2$. $f(-x)=(-x)^4-3(-x)^2=x^4-3x^2$. f(-x)=f(x), f(-x)≠-f(x). Even function. See graph below. It is symmetric about y-axis.

4. $f(x) = x^2$, $g(x) = (x-4)^2 + 3$. To get the graph of g from f, shift f right 4, up 3. See graphs below.

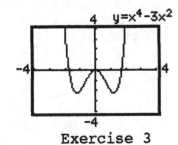

Exercise 3

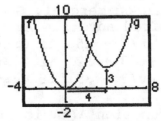

Exercise 4

5. $f(x)=x^2$ and $g(x)=(x+2)^2$: Values of x in the table increase in steps of 1. The value of g is the same as that f two steps after. Thus value of g at x is the same as that of f at (x+2).

139

6. $f(x) = \dfrac{2}{x^2-3x-4} = \dfrac{2}{(x+1)(x-4)}$. Vert x=-1,x=4. Horiz y=0. Domain: All reals except -1,4. See graph below.

7. $f(x) = 2x^3+3x^2-5x+4$. Inc$(-\infty,-1.54]$,$[0.54,\infty)$. Dec$[-1.54,0.54]$. Max(-1.54,11.51), Min(0.54,2.49). Zero x= -2.70.

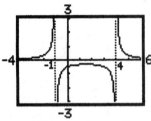

Exercise 6

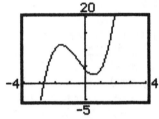

Exercise 7

8. $f(x)=3x-1$, $g(x)=\sqrt{x-2}$.

   (f+g)(x) = $(3x-1)+(\sqrt{x-2})=3x-1+\sqrt{x-2}$. Domain: x≥2.

   (f-g)(x) = $(3x-1)-(\sqrt{x-2})=3x-1-\sqrt{x-2}$. Domain: x≥2.

   (fg)(x) = $(3x-1)(\sqrt{x-2})$. Domain: x≥2.

   (f/g)(x) = $(3x-1)/(\sqrt{x-2})$. Domain: x>2.

9. $f(x)=\sqrt{x-3}$, $g(x)=x^2+5x+7$.

   (g∘f)(x)=g(f(x))=g($\sqrt{x-3}$)=$(\sqrt{x-3})^2+5(\sqrt{x-3})+7=x-3+5\sqrt{x-3}+7$.

   (g∘f)(x)=$x+4+5\sqrt{x-3}$. Domain: x≥3 or [3,∞).

10. $f(x)=3x+2$. Graph of y=3x+2 is a non-vertical line. Every horizontal line cuts it at one point. Thus f is one-to-one and has an inverse. Interchange x & y, x=3y+2. Solve for y, $y=\dfrac{x-2}{3}$. $f^{-1}(x)=\dfrac{x-2}{3}$.

11. (a)Graph below. (b)Consumer awareness, $A(t)=\begin{cases} -t/3+25 & \text{if } 0 \le t \le 75 \\ 0 & \text{if } t>75 \end{cases}$

   Value of A should not fall below 18. -t/3+25=18, -t/3=-7, t=21. Consumer awareness drops to 18 twenty one days after showing the commercial. Should show the commercial every 21 days. Desired advertising awareness function D(t) is

   $D(t) = \begin{cases} -t/3 + 25 & \text{if } 0 \le t < 21 \\ -t/3 + 32 & \text{if } 21 \le t < 42 \\ -t/3 + 39 & \text{if } 42 \le t < 63 \\ \quad \cdots \end{cases}$

   (c) The function A(t) is given a boost of 7 to bring it up from 18 to 25 every 21 days.

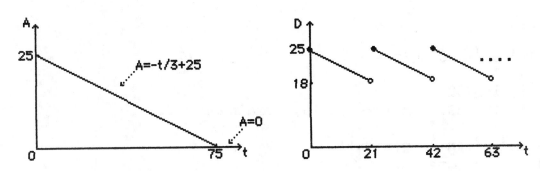

12. $N(t) = 2t^3 - 32t^2 + 83t + 1000$. A, Max(1.5108849,1059.2527).
B, Min(9.155783,612.45098). Population peaked with 1059 deer
in the middle of 1991. Was at its lowest in 1999 with 612
deer. Population is now increasing again. See graph below.

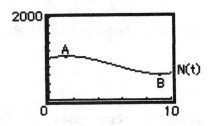

13. $C(x)=6.5x+520$. Graph of C is non-horizontal line. C has an
inverse by horiz line test. $C=6.5x+520$. $C-520=6.5x$.
$x=(C-520)/6.5$. When $C=17225$, $x=(17225-520)/6.5=2570$. Number
items = 2570.

Chapter 4

Section 4.1

1. $f(x)=2x^3+3x^2+x+3$, $g(x)=x-2$.

$$\underline{2}|\begin{array}{cccc} 2 & 3 & 1 & 3 \\ & 4 & 14 & 30 \\ \hline 2 & 7 & 15 & 33 \end{array}$$

   $q(x)=2x^2+7x+15$, remainder = 33.

3. $f(x)=-2x^3+x^2+4x-2$, $g(x)=x+1$.

$$\underline{-1}|\begin{array}{cccc} -2 & 1 & 4 & -2 \\ & 2 & -3 & -1 \\ \hline -2 & 3 & 1 & -3 \end{array}$$

   $q(x)=-2x^2+3x+1$, remainder = -3.

5. $f(x)=x^4+2x^3-3x^2+8x-7$, $g(x)=x-5$.

$$\underline{5}|\begin{array}{ccccc} 1 & 2 & -3 & 8 & -7 \\ & 5 & 35 & 160 & 840 \\ \hline 1 & 7 & 32 & 168 & 833 \end{array}$$

   $q(x)=x^3+7x^2+32x+168$, remainder = 833.

7. $f(x)=x^4+2x^3+7x+3$, $g(x)=x+6$.

$$\underline{-6}|\begin{array}{ccccc} 1 & 2 & 0 & 7 & 3 \\ & -6 & 24 & -144 & 822 \\ \hline 1 & -4 & 24 & -137 & 825 \end{array}$$

   $q(x)=x^3-4x^2+24x-137$, remainder = 825.

9. $f(x)=3x^3-4x^2+x+1$, $g(x)=x-1$.

$$\underline{1}|\begin{array}{cccc} 3 & -4 & 1 & 1 \\ & 3 & -1 & 0 \\ \hline 3 & -1 & 0 & 1 \end{array}$$

   $q(x)=3x^2-x$, remainder=1. $f(1)=1$.

11. $f(x)=4x^3+2x^2+3x+8$, $g(x)=x+2$.

$$\underline{-2}|\begin{array}{cccc} 4 & 2 & 3 & 8 \\ & -8 & 12 & -30 \\ \hline 4 & -6 & 15 & -22 \end{array}$$

   $q(x)=4x^2-6x+15$, remainder=-22. $f(-2)=-22$.

13. $f(x)=x^4+2x^3-2x^2+6x+7$, $g(x)=x+1$.

$$\underline{-1}|\begin{array}{ccccc} 1 & 2 & -2 & 6 & 7 \\ & -1 & -1 & 3 & -9 \\ \hline 1 & 1 & -3 & 9 & -2 \end{array}$$

   $q(x)=x^3+x^2-3x+9$, remainder=-2. $f(-1)=-2$.

15. $f(x)=3x^4-4x^3-x^2+2x+4$, $g(x)=x+8$.

$$\underline{-8}|\begin{array}{ccccc} 3 & -4 & -1 & 2 & 4 \\ & -24 & 224 & -1784 & 14256 \\ \hline 3 & -28 & 223 & -1782 & 14260 \end{array}$$

   $q(x)=3x^3-28x^2+223x-1782$,
   remainder=14260. $f(-8)=14260$.

17. $f(x)=4x^3-x^2+2x+4$. Want $f(1)$. Divide by $g(x)=x-1$. Remainder will then be $f(1)$.

$$\underline{1}|\begin{array}{cccc} 4 & -1 & 2 & 4 \\ & 4 & 3 & 5 \\ \hline 4 & 3 & 5 & 9 \end{array}$$

   $q(x)=4x^2+3x+5$, remainder=9. $f(1)=9$.

19. $f(x)=4x^3-x^2+6x+3$. Want $f(-2)$. Let $g(x)=x+2$.

$-2\lfloor$ 4 -1 6   3

         -8 18 -48

    4   -9 24 -45. $q(x)=4x^2-9x+24$, remainder$=-45$. $f(-2)=-45$.

21. $f(x)=-2x^3+3x^2+2x-2$. Want $f(-2)$. Let $g(x)=x+2$.

$-2\lfloor-2$ 3   2 -2

        4 -14 24

   -2 7 -12 22. $q(x)=-2x^2+7x-12$, remainder$=22$. $f(-2)=22$.

23. $f(x)=7x^3+x$. Want $f(-2)$. Let $g(x)=x+2$.

$-2\lfloor7$ 0   1   0

        -14 28 -58

   7 -14 29 -58. $q(x)=7x^2-14x+29$, remainder$=-58$. $f(-2)=-58$.

25. $f(x)=x^3-x^2-2x+8$. $r=-2$. Divide $f(x)$ by $g(x)=x+2$.

$-2\lfloor1$ -1 -2 8

        -2 6 -8

   1 -3 4   0. Remainder$=0$. $f(-2)=0$. Thus -2 is a zero.

27. $f(x)=x^3-4x^2+3x+2$. $r=2$. Divide $f(x)$ by $g(x)=x-2$.

$2\lfloor1$ -4 3   2

       2 -4 -2

   1 -2 -1 0. Remainder$=0$. $f(2)=0$. Thus 2 is a zero.

29. $f(x)=4x^3-9x^2-8x+3$. $r=3$. Divide $f(x)$ by $g(x)=x-3$.

$3\lfloor4$ -9 -8 3

       12   9   3

   4   3   1   6. Remainder$=6\neq0$. $f(3)\neq0$. Thus 3 not a zero.

31. $f(x)=3x^5-8x^3-6x^2+x-10$. $r=2$. Divide $f(x)$ by $g(x)=x-2$.

$2\lfloor3$ 0 -8 -6 1 -10

       6 12   8   4   10

   3 6   4   2   5   0. Remainder$=0$. $f(2)=0$. 2 is a zero.

33. $f(x)=x^3-4x^2+9x-10$. $x-2$. Divide $f(x)$ by $g(x)=x-2$.

$2\lfloor1$ -4 9 -10

       2 -4 10

   1 -2 5   0. Remainder$=0$. $f(2)=0$. $x-2$ is a factor.

35. $f(x)=5x^3-20x^2+5x+30$. $x-3$. Divide $f(x)$ by $g(x)=x-3$.

$3\lfloor5$ -20 5   30

       15 -15 -30

   5 -5 -10 0. Remainder$=0$. $f(3)=0$. $x-3$ is a factor.

37. $f(x)=4x^3-4x^2+7x+1$. $x+4$. Divide $f(x)$ by $g(x)=x+4$.

$-4\lfloor4$ -4 7   1

        -16 80 -348

   4 -20 87 -347. Remainder$=-347\neq0$. $f(-4)\neq0$.
   $x+4$ not a factor.

143

39. $f(x)=8x^4-2x^3+7x^2-2x-1$. $x-1/2$. Divide $f(x)$ by $g(x)=x-1/2$.

```
1/2| 8  -2   7   -2   -1
        4    1    4    1
    8    2   8    2    0.  Remainder=0. f(1/2)=0.
```
x - 1/2 is a factor.

41. Value of k for which x+1 is a factor of kx^3+2x^2-3x+4.
Divide by x+1.

```
-1| k    2    -3     4
       -k    k-2   -k+5
   k   -k+2  k-5   -k+9.  Remainder = -k+9.
```
x+1 is a factor when -k+9=0. k=9.

43. Value of k for which x-5 is a factor of $kx^4+2x^3-20kx^2+25x$.
Divide by x-5.

```
5| k    2      -20k      25        0
        5k     25k+10   25k+50   125k+375
   k   5k+2   5k+10    25k+75   125k+375.  Remainder = 125k+375.
```
x-5 is a factor when 125k+375=0. k=-3.

45. Value of k for which -2 is a zero of $kx^4+x^3-3x^2+5x-5$.
Need x+2 to be a factor. Divide by x+2.

```
-2| k    1     -3      5      -5
        -2k   4k-2   -8k+10  16k-30
    k  -2k+1  4k-5   -8k+15  16k-35.  Remainder = 16k-35.
```
x+2 is a factor if 16k-35=0,k=35/16. -2 is a zero if k=35/16.

47. $f(x)=2x^3+5x^2-2x-4$. Need f(-3).
(a) Synthetic division: Divide by g(x)=x+3.

```
-3| 2    5   -2   -4
        -6    3   -3
    2   -1    1   -7.  Remainder = -7. This f(-3)=-7.
```
(b) Substitution: $f(-3)=2(-3)^3+5(-3)^2-2(-3)-4=-7$.

Section 4.2

1. 4, real and complex.
3. 7.2, real and complex.
5. -8i, imaginary and complex.
7. -9.56, real and complex.
9. $\sqrt{-4}=\sqrt{4}i=2i$.
11. $\sqrt{-36}=\sqrt{36}i=6i$.
13. $\sqrt{121}=11$
15. $\sqrt{-49}=\sqrt{49}i=7i$.
17. i^{19}. 19=4·4+3. Thus $i^{19}=i^3=-i$.
19. i^{41}. 41=10·4+1. Thus $i^{41}=i^1=i$.
21. i^{10}. 10=2·4+2. Thus $i^{10}=i^2=-1$.
23. 2+3i. Real part = 2, imaginary part = 3.
25. 7i. Real part = 0, imaginary part = 7.
27. -8i. Real part = 0, imaginary part = -8.
29. 3+2i. Conjugate 3-2i.
31. 3-5i. Conjugate 3+5i.
33. 9i. Conjugate -9i.

Section 4.2

35. $x^2+2x+3=0.$ $a=1, b=2, c=3.$

$$x= \frac{-2\pm\sqrt{2^2-4(1)(3)}}{2(1)} = \frac{-2\pm\sqrt{-8}}{2} = \frac{-2\pm\sqrt{8}i}{2} = \frac{-2\pm2\sqrt{2}i}{2} = -1+\sqrt{2}i, \ -1-\sqrt{2}i.$$

37. $x^2-2x+1=0.$ $a=1, b=-2, c=1.$

$$x = \frac{2\pm\sqrt{(-2)^2-4(1)(1)}}{2(1)} = \frac{2\pm0}{2} = 1.$$

39. $x^2+4x+4=0.$ $a=1, b=4, c=4.$

$$x = \frac{-4\pm\sqrt{4^2-4(1)(4)}}{2(1)} = \frac{-4\pm0}{2} = -2$$

41. $x^2+4=0.$ $a=1, b=0, c=4.$

$$x = \frac{0\pm\sqrt{(0)^2-4(1)(4)}}{2(1)} = \frac{\pm\sqrt{-16}}{2} = \frac{\pm\sqrt{16}i}{2} = \frac{\pm4i}{2} = 2i, \ -2i.$$

<u>or</u>, $x^2=-4,$ $x=\pm\sqrt{-4},$ $x=\pm\sqrt{4}i,$ $x=\pm2i.$

43. $-x^2+2x-5=0.$ $a=-1, b=2, c=-5.$

$$x = \frac{-2\pm\sqrt{2^2-4(-1)(-5)}}{2(-1)} = \frac{-2\pm\sqrt{-16}}{-2} = \frac{-2\pm\sqrt{16}i}{-2} = \frac{-2\pm4i}{-2} = 1-2i, \ 1+2i.$$

45. $(2+3i) + (5+2i) = 7+5i.$

47. $(9-2i) - (7+3i) = 2-5i.$

49. $(7-3i) + (-4+2i) = 3-i.$

51. $(-3+6i) + (-3-8i) = -6-2i.$

53. $(3+i) + (9-2i) + i = 12.$

55. $(1+2i)(3+i) = 1(3+i) + 2i(3+i) = 3+i+6i+2i^2 = 3+7i-2 = 1+7i.$

57. $(6+i)(2-i) = 6(2-i)+i(2-i) = 12-6i+2i-i^2 = 12-4i+1 = 13-4i.$

59. $-4(3+2i) = -12-8i.$

61. $(2+5i)^2 = (2+5i)(2+5i) = 2(2+5i)+5i(2+5i) = 4+10i+10i+25i^2$
$= 4+20i-25 = -21+20i.$

63. $(3-4i)^2 = (3-4i)(3-4i) = 3(3-4i)-4i(3-4i) = 9-12i-12i+16i^2$
$= 9-24i-16 = -7-24i.$

65. $\dfrac{2+3i}{5+i} \cdot \dfrac{5-i}{5-i} = \dfrac{2(5-i)+3i(5-i)}{5^2 + 1^2} = \dfrac{10-2i+15i-3i^2}{26} = \dfrac{13+13i}{26} = \dfrac{1}{2} + \dfrac{1}{2}i.$

67. $\dfrac{2-3i}{1+i} \cdot \dfrac{1-i}{1-i} = \dfrac{2(1-i)-3i(1-i)}{1^2 + 1^2} = \dfrac{2-2i-3i+3i^2}{2} = \dfrac{-1-5i}{2} = -\dfrac{1}{2} - \dfrac{5}{2}i.$

69. $\dfrac{5+i}{5-i} \cdot \dfrac{5+i}{5+i} = \dfrac{5(5+i)+i(5+i)}{5^2 + 1^2} = \dfrac{25+5i+5i+i^2}{26} = \dfrac{24+10i}{26} = \dfrac{12}{13} + \dfrac{5}{13}i.$

71. $\dfrac{i}{2+i} \cdot \dfrac{2-i}{2-i} = \dfrac{i(2-i)}{2^2 + 1^2} = \dfrac{2i-i^2}{5} = \dfrac{1+2i}{5} = \dfrac{1}{5} + \dfrac{2}{5}i.$

73. $\dfrac{5-3i}{2i} \cdot \dfrac{-2i}{-2i} = \dfrac{-10i+6i^2}{(2i)(-2i)} = \dfrac{-10i-6}{-4i^2} = \dfrac{-6-10i}{4} = -\dfrac{3}{2} - \dfrac{5}{2}i.$

75. $|12+5i| = \sqrt{12^2+5^2} = \sqrt{144+25} = \sqrt{169} = 13$

77. $|3-5i| = \sqrt{3^2+(-5)^2} = \sqrt{9+25} = \sqrt{34} = 5.8310$

79. $|9| = |9+0i| = \sqrt{9^2+0^2} = \sqrt{81} = 9$

81. $|-6-3i| = \sqrt{(-6)^2+(-3)^2} = \sqrt{36+9} = \sqrt{45} = 6.7082$

83. Let the complex number be a+bi. Then (a+bi) - (a-bi)= 2bi, an imaginary number.

85. .5+i -> -.25+2i -> -3.4375+0i -> 12.3164+i ->
151.1939+25.6328i -> 22203.0431+7752.0479i -> ...
The numbers do not become increasingly large in magnitude.
.5+i does not lie in the Mandelbrot Set.

87. -1+0i -> 0+0i -> -1+0i -> 0+0i -> -1+0i -> ...
Oscillates beteen 0 and -1. Lies in Mandelbrot Set.

89. -1+.5i -> -.25-.5i -> -1.1875+.75i -> -.1523-1.2813i ->
-2.6184+.8904i -> 5.0632-4.1627i -> 7.3077-41.6535i ->
-1682.6151-608.2812i -> ... Does not lie in Mandelbrot Set.

91. .1+.5i -> -.14+.6i -> -.2404+.332i -> .0476+.3404i ->
-.0136+.5324i -> -.1832+.4855i -> ... Lies in mandelbrot Set.

93. -2+0i -> 2+0i -> 2+0i -> ... Lies in Mandelbrot Set.
-3+0i -> 6+0i -> 33+0i -> 1086+0i -> ... Does not lie in
Mandelbrot Set.
-2.000001+0i -> 2.000003+0i -> 2.000011+0i -> 2.000043+0i ->
2.000171+0i -> 2.000683+0i -> 2.002731+0i -> 2.010933+0i -> ..
The numbers are steadily incresing in magnitude. Point for
-2.000001 does not lie in the Mandelbrot Set. The boundary of
the set lies on/between points for 2 and 2.000001. Further
exploration shows that 2.000..001 always gives a point outside
the set. This evidence points to the fact that (2,0) lies in
the set and on its boundary.

Section 4.3

1. $f(x)=x^3-4ix^2+(3+i)x+3-18i$, $g(x)=x-3i$.

```
3i| 1   -4i   3+i    3-18i
         3i    3     -3+18i
    1   -i    6+i     0      
```
. Remainder = 0. Thus f(3i) = 0,
implying that x-3i is a factor of f(x).

3. $f(x)=3x^3-5ix^2+(3+4i)x+3-2i$, $g(x)=x-2i$.

$$
\begin{array}{r|rrrr}
2i & 3 & -5i & 3+4i & 3-2i \\
   &   & 6i & -2   & -8+2i \\
\hline
   & 3 & i   & 1+4i & -5
\end{array}
$$
. Remainder=$-5\neq0$. Thus $f(2i)\neq0$,
   implying that x-2i is not a factor of f(x).

5. $f(x)=2ix^3+(5-6i)x^2-(23+7i)x+12-4i$. $g(x)=x-4$.

$$
\begin{array}{r|rrrr}
4 & 2i & 5-6i & -23-7i & 12-4i \\
  &   & 8i   & 20+8i & -12+4i \\
\hline
  & 2i & 5+2i & -3+i   & 0
\end{array}
$$
. Remainder = 0. Thus f(4)=0,
   implying that x-4 is a factor of f(x).

7. $f(x)=3x^3-3ix^2+(2-4i)x+1+2i$. Want f(2). Divide by g(x)=x-2.
   Remainder will then be f(2).

$$
\begin{array}{r|rrrr}
2 & 3 & -3i   & 2-4i   & 1+2i \\
  &   & 6   & 12-6i & 28-20i \\
\hline
  & 3 & 6-3i & 14-10i & 29-18i
\end{array}
$$
. Remainder=29-18i. f(2)=29-18i.

9. $f(x)=2ix^3-4ix^2+(7+3i)x+2+5i$. Want f(-2i).
   Divide by g(x)=x+2i. Remainder will then be f(-2i).

$$
\begin{array}{r|rrrr}
2i & 2i & -4i   & 7+3i & 2+5i \\
   &   & 4   & -8-8i & -10+2i \\
\hline
   & 2i & 4-4i & -1-5i & -8+7i
\end{array}
$$
. Remainder=-8+7i. f(3i)=-8+7i.

11. $f(x)=(2-4i)x^3+(4-2i)x^2+(3+4i)x+7+2i$. Want f(2-3i).
    Divide by g(x)=x-(2-3i). Remainder will then be f(2-3i).

$$
\begin{array}{r|rrrr}
2-3i & 2-4i & 4-2i   & 3+4i   & 7+2i \\
     &   & -8-14i & -56-20i & -154+127i \\
\hline
     & 2-4i & -4-16i & -53-16i & -147+129i
\end{array}
$$
    Remainder=-147+129i. f(2-3i)=-147+129i.

13. $f(x)=(x-2)(x+3)(x+5)$. Zeros: 2,-3,-5, each multiplicity one.
15. $f(x)=(x-4)(x+2)^2$. Zeros: 4 (mult. one), -2 (mult. two).
17. $f(x)=(x-5)^3(x+7)^4$. Zeros: 5 (mult.3), -7 (mult. four).
19. $f(x)=(x-2i)(x+3-4i)^2$. Zeros:2i (mult. one), -3+4i (mult. two).

21. Zeros 1,3,5. Let $f(x)=a(x-1)(x-3)(x-5)$ where a is a complex
    number. There will be a polynomial with these zeros for each
    a. Let a=1. $f(x)=(x-1)(x^2-8x+15)=x^3-9x^2+23x-15$.

23. Zeros -2,0,6. $f(x)=(x+2)(x-0)(x-6)=(x+2)(x^2-6x)=x^3-4x^2-12x$.
25. Zeros -3,2,2. $f(x)=(x+3)(x-2)(x-2)=(x+3)(x^2-4x+4)=x^3-x^2-8x+12$.

27. Zeros 2,1-2i,3+i. $f(x)=(x-2)(x-(1-2i))(x-(3+i))$
    $=x^3-(6-i)x^2+(13-7i)x+(-10+10i)$.

29. Zeros 2,i,i. $f(x)=(x-2)(x-i)(x-i)=x^3-(2+2i)x^2+(-1+4i)x+2$.

31. $f(x)=x^3-6x^2+11x-6$. x=1 is a zero. Thus (x-1) is a factor.

$\underline{1}|1 \quad -6 \quad 11 \quad -6$
$\quad\quad \underline{1 \quad -5 \quad 6}$
$\quad 1 \quad -5 \quad 6 \quad\quad 0$. $f(x)=(x-1)(x^2-5x+6)=(x-1)(x-2)(x-3)$.
Zeros are 1,2,3.

33. $f(x)=x^3-7x+6$. x=-3 is a zero. Thus (x+3) is a factor.

$\underline{-3}|1 \quad 0 \quad -7 \quad 6$
$\quad\quad \underline{-3 \quad 9 \quad -6}$
$\quad 1 \quad -3 \quad 2 \quad\quad 0$. $f(x)=(x+3)(x^2-3x+2)=(x+3)(x-1)(x-2)$.
Zeros are -3,1,2.

35. $f(x)=x^3-6x^2+12x-8$. x=2 is a zero. Thus (x-2) is a factor.

$\underline{2}|1 \quad -6 \quad 12 \quad -8$
$\quad\quad \underline{2 \quad -8 \quad 8}$
$\quad 1 \quad -4 \quad 4 \quad\quad 0$. $f(x)=(x-2)(x^2-4x+4)=(x-2)(x-2)(x-2)$.
Single zeros are 2, of multiplicity three.

37. $f(x)=3x^3-14x^2+13x+6$. x=2 is a zero. Thus (x-2) is a factor.

$\underline{2}|3 \quad -14 \quad 13 \quad 6$
$\quad\quad \underline{6 \quad -16 \quad -6}$
$\quad 3 \quad -8 \quad -3 \quad\quad 0$. $f(x)=(x-2)(3x^2-8x-3)=(x-2)(3x+1)(x-3)$.
Zeros are 2,3,-1/3.

39. $f(x)=x^3+2x^2+x+2$. x=-i is a zero. Thus (x+i) is a factor.

$\underline{-i}|1 \quad 2 \quad 1 \quad 2$
$\quad\quad \underline{-i \quad -1-2i \quad -2}$
$\quad 1 \quad 2-i \quad -2i \quad\quad 0$. $f(x)=(x+i)(x^2+(2-i)x-2i)=(x+i)(x+2)(x-i)$.
Zeros are -i,-2,i.

41. Polynomial degree 3 with zeros -1,2,2 and f(3)=8.
Let $f(x)=a(x+1)(x-2)^2$. $f(3)=a(4)(1)^2=4a$. Thus 4a=8, a=2.
$f(x)=2(x+1)(x-2)^2=2x^3-6x^2+8$.

43. Polynomial degree 3 with zeros 2,2,2 and f(3)=-4.
Let $f(x)=a(x-2)^3$. $f(3)=a(1)(1)(1)=a$. Thus a=-4.
$f(x)=-4(x-2)^3=-4x^3+24x^2-48x+32$.

45. Polynomial degree 3 with zeros -2, -2,-4, and f(-5)=18.
Let $f(x)=a(x+2)^2(x+4)$. $f(-5)=a(-3)^2(-1)=-9a$. Thus -9a=18,
a=-2. $f(x)=-2(x+2)^2(x+4)=-2x^3-16x^2-40x-32$.

47. Polynomial degree 4 with zeros 1, 3, 3, 3, and f(5)=16.
Let $f(x)=a(x-1)(x-3)^3$. $f(5)=a(4)(2)^3=32a$. Thus 32a=16,
a=1/2. $f(x)=(1/2)(x-1)(x-3)^3=(1/2)x^4-5x^3+18x^2-27x+27/2$.

49. Polynomial degree 3 with zeros 2, 1+i, 3-2i and f(2+i)=3+i.
Let $f(x)=a(x-2)(x-(1+i))(x-(3-2i))$. $f(2+i)=a(i)(1)(-1+3i)$
$=(-3-i)a$. Thus (-3-i)a=3+i,a=-1.
$f(x)=-1(x-2)(x-(1+i))(x-(3-2i))=-x^3+(6-i)x^2+(-13+i)x+10+2i$.

148

51. Polynomial degree 3, with 4, 2-i among the zeros. By conjugate zeros theorem, 3rd zero must be 2+i.
$f(x)=a(x-4)(x-(2-i))(x-(2+i))$. Let a=1. $f(x)=x^3-8x^2+21x-20$.

53. Polynomial degree 3, with -2, 3-5i among the zeros. 3rd zero must be 3+5i. $f(x)=a(x+2)(x-(3-5i))(x-(3+5i))$. Let a=1. $f(x)=x^3-4x^2+22x+68$.

55. Polynomial degree 4, with 1, 3, 5-i among the zeros. 4th zero must be 5+i. $f(x)=a(x-1)(x-3)(x-(5-i))(x-(5+i))$. Let a=1. $f(x)=x^4-14x^3+69x^2-134x+78$.

57. Polynomial degree 4, with -3+i, 2-4i among the zeros. 3rd & 4th zero must be -3-i, 2+4i.
$f(x)=a(x-(-3+i))(x-(2-4i))(x-(-3-i))(x-(2+4i))$.
Let a=1. $f(x)=x^4+2x^3+6x^2+80x+200$.

Section 4.4

1. $f(x)=x^3-6x^2+11x-6$. Factors of p=-6 are $\pm1,\pm2,\pm3,\pm6$. Coefficient of $x^3=1$, q=1. Thus possible zeros are $\pm1,\pm2,\pm3,\pm6$. Check 1.
```
1| 1  -6  11  -6
       1  -5   6
   1  -5   6   0
```
Thus (x-1) is a factor.
$f(x)=(x-1)(x^2-5x+6)=(x-1)(x-2)(x-3)$. Zeros are 1,2,3.

3. $f(x)=x^3-7x+6$. Factors of p=6 are $\pm1,\pm2,\pm3,\pm6$. Coefficient of $x^3=1$, q=1. Thus possible zeros are $\pm1,\pm2,\pm3,\pm6$. Check 1,-1... .
```
1| 1  0  -7   6
      1   1  -6
   1  1  -6  -0
```
(x-1) is a factor.
$f(x)=(x-1)(x^2+x-6)=(x-1)(x-2)(x+3)$. Zeros are 1,2,-3.

5. $f(x)=x^3+5x^2-2x-24$. Factors -24 are $\pm1,\pm2,\pm3,\pm4,\pm6,\pm8,\pm12,\pm24$. Coefficient of $x^3=1$, q=1. Thus possible zeros are $\pm1,\pm2,\pm3,\pm4,\pm6,\pm8,\pm12,\pm24$. Check 1,-1... .
```
1| 1  5  -2  -24       -1| 1   5  -2  -24      2| 1  5  -2  -24
      1   6    4             -1  -4    6            2  14   24
   1  6   4  -20          1   4  -6  -18         1  7  12    0
```
(x-1) is not a factor. (x+1) is not a factor. (x-2) is a factor.
$f(x)=(x-2)(x^2+7x+12)=(x-2)(x+4)(x+3)$. Zeros are 2,-4,-3.

7. $f(x)=x^4-2x^3-9x^2+2x+8$. Factors of 8 are $\pm1,\pm2,\pm4,\pm8$. Coefficient of $x^3=1$, q=1. Thus possible zeros are $\pm1,\pm2,\pm4,\pm8$. Check 1,-1... .
```
1| 1  -2  -9   2   8
       1  -1 -10  -8
   1  -1 -10  -8   0
```
(x-1) is a factor. $f(x)=(x-1)(x^3-x^2-10x-8)$.

Consider $x^3-x^2-10x-8$. Possible factors are $\pm1,\pm2,\pm4,\pm8$.
Check $x=1,-1,\ldots$

```
1|1  -1  -10  -8          -1|1  -1  -10  -8
      1   0  -10                -1   2   8
   1  0  -10  -18.          1  -2  -8   0.
```
(x-1) is not a factor.   (x+1) is a factor.
$f(x)=(x-1)(x+1)(x^2-2x-8)=(x-1)(x+1)(x+2)(x-4)$. Zeros $-1,1,-2,4$.

9. $f(x)=x^4-2x^3-4x^2+2x+3$. Factors of $p=3$ are $\pm1,\pm3$.
   Coefficient of $x^3=1$, $q=1$. Thus possible zeros are $\pm1,\pm3$.
   Check $1,-1\ldots$.

```
1|1  -2  -4   2   3
      1  -1  -5  -3
   1  -1  -5  -3   0.  (x-1) is a factor.  f(x)=(x-1)(x^3-x^2-5x-3).
```
   Consider x^3-x^2-5x-3. Possible factors $\pm1,\pm3$. Check $x=1,-1,\ldots$

```
1|1  -1  -5  -3          -1|1  -1  -5  -3
      1   0  -5                -1   2   3
   1  0  -5  -8.           1  -2  -3   0.
```
   (x-1) is not a factor.   (x+1) is a factor.
   $f(x)=(x-1)(x+1)(x^2-2x-3)=(x-1)(x+1)(x+1)(x-3)$.
   Zeros are $1,3,-1$ (multiplicity two).

11. $f(x)=3x^3+x^2-8x+4$. Possible zeros are $\pm1,\pm2,\pm4,\pm1/3,\pm2/3,\pm4/3$.
    Check $1,-1,\ldots$.

```
1|3   1  -8   4
      3   4  -4
   3  4  -4   0.  (x-1) is a factor.  f(x)=(x-1)(3x^2+4x-4).
```
    $f(x)=(x-1)(3x-2)(x+2)$. Zeros are $1,2/3,-2$.

13. $f(x)=2x^3-5x^2+4x-1$. Possible zeros are $\pm1,\pm1/2$.
    Check $1,-1,\ldots$.

```
1|2  -5   4  -1
      2  -3   1
   2  -3   1   0.  (x-1) is a factor.  f(x)=(x-1)(2x^2-3x+1).
```
    $f(x)=(x-1)(2x-1)(x-1)$. Zeros are $1/2$, 1 (multiplicity two).

15. $f(x)=4x^3-11x^2+5x+2$. Possible zeros are $\pm1,\pm2,\pm1/4,\pm1/2$.
    Check $1,-1,\ldots$.

```
1|4  -11   5   2
       4  -7  -2
   4  -7  -2   0.  (x-1) is a factor.  f(x)=(x-1)(4x^2-7x-2).
```
    $f(x)=(x-1)(4x+1)(x-2)$. Zeros are $1,-1/4,2$.

17. $f(x)=6x^3+7x^2-9x+2$. Possible zeros are $\pm1,\pm2,\pm1/6,\pm1/3,\pm1/2$.
    Check $1,-1,\ldots$.

```
1|6   7  -9   2      -1|6   7  -9   2      2|6   7  -9   2
      6  13   4           -6  -1  10          12  38  58
   6  13   4   6.      6   1 -10  12.      6  19  29  60.
```
    (x-1) not a factor.   (x+1) not a factor.   (x-2) not a factor.

```
-2|6   7   -9    2
        -12  10   -2
    6   -5    1    0.  (x+2) not a factor.
```
$f(x)=(x+2)(6x^2-5x+1)=(x+2)(3x-1)(2x-1)$. Zeros are -2,1/3,1/2.

9.$f(x)=2x^4-3x^3-x^2+3x-1$. Possible zeros $\pm 1,\pm 1/2$. Check 1,-1,... .
```
1|2   -3   -1    3   -1
         2   -1   -2    1
    2   -1   -2    1    0.  (x-1) is a factor.
```
$f(x)=(x-1)(2x^3-x^2-2x+1)$.

```
1|2   -1   -2    1
         2    1   -1
    2    1   -1    0.  (x-1) is a factor of multiplicity two.
```
$f(x)=(x-1)^2(2x^2+x-1)=(x-1)^2(2x-1)(x+1)$.
Zeros are 1 (multiplicity two), 1/2,-1.

21.$f(x)=x^3 + \dfrac{5}{2}x^2 + \dfrac{1}{2}x - 1$. Multiply by 2 to get $g(x)=2x^3+5x^2+x-2$.
f and g have the same zeros. Possible zeros are $\pm 1,\pm 2,\pm 1/2$.
Check 1,-1,... .
```
1|2    5    1   -2          -1|2    5    1   -2
         2    7    8                -2   -3    2
    2    7    8    6.            2    3   -2    0
  (x-1) not a factor.          (x+1) is a factor.
```
$g(x)=(x+1)(2x^2+3x-2)=(x+1)(2x-1)(x+2)$. Zeros are -1,1/2,-2.

23.$f(x)=x^3 + \dfrac{19}{4}x^2 + \dfrac{11}{4}x - 1$. Multiply by 4, $g(x)=4x^3+19x^2+11x-4$.
f and g have the same zeros. Possible zeros are $\pm 1,\pm 2,\pm 4,\pm 1/2$.
Check 1,-1,... .
```
1|4   19   11   -4          -1|4   19   11   -4
         4   23   34                -4  -15    4
    4   23   34   30.            4   15   -4    0
  (x-1) not a factor.          (x+1) is a factor.
```
$g(x)=(x+1)(4x^2+15x-4)=(x+1)(4x-1)(x+4)$. Zeros are -1,1/4,-4.

25.$f(x)=3x^4 - \dfrac{7}{2}x^3 - \dfrac{23}{2}x^2 + 14x - 2$. Multiply by 2 to get

$g(x)=6x^4-7x^3-23x^2+28x-4$. f and g have the same zeros.
Possible zeros are $\pm 1,\pm 2,\pm 4,\pm 1/2,\pm 1/3,\pm 1/6,\pm 2/3,\pm 4/3$.
Check 1,-1,... .
```
1|6   -7   -23   28   -4
         6   -1   -24    4
    6   -1   -24    4    0.  (x-1) is a factor.
```
$g(x)=(x-1)(6x^3-x^2-24x+4)$. Must now find factors of
$6x^3-x^2-24x+4$. Find that (x+1) and (x-1) are not factors.

```
2|6   -1   -24    4
        12   22   -4
    6   11   -2    0.  (x-2) is a factor.
```

151

$f(x)=(x-1)(x-2)(6x^2+11x-2)=(x-1)(x-2)(6x-1)(x+2)$.
Zeros are $1,2,1/6,-2$.

27. $f(x)=x^4-x^3-2x-4$. Possible zeros are $\pm1,\pm2,\pm4$. Check $1,-1,\ldots$.

```
1|1  -1   0  -2  -4        1|1  -1   0  -2  -4
    1   0   0  -2              -1   2  -2   4
  1   0   0  -2  -6.          1  -2   2  -4   0.
```
(x-1) not a factor.   (x+1) is a factor.
$f(x)=(x+1)(x^3-2x^2+2x-4)$. Now find factors of x^3-2x^2+2x-4.
Find that (x+1) is not a factor. Try (x-2) next:

```
2|1  -2   2  -4
    2   0   4
  1   0   2   0.
```
(x-2) is a factor.
$f(x)=(x+1)(x-2)(x^2+2)=(x+1)(x-2)(x-\sqrt{2}i)(x+\sqrt{2}i)$.
Zeros are $-1,2,\sqrt{2}i,-\sqrt{2}i$.

29. $f(x)=3x^4+x^3+x^2+x-2$. Possible zeros $\pm1,\pm2,\pm1/3,\pm2/3$.
Check $1,-1,\ldots$. Find that -1 and $2/3$ are zeros:

```
-1|3   1   1   1  -2
     -3   2  -3   2
   3  -2   3  -2   0.
```
(x+1) is a factor.
$f(x)=(x+1)(3x^3-2x^2+3x-2)$.

```
2/3|3  -2   3  -2
       2   0   2
    3   0   3   0.
```
(x-2/3) is a factor.
$f(x)=(x+1)(x-2/3)(3x^2+3)=3(x+1)(x-2/3)(x^2+1)$
$=3(x+1)(x-2/3)(x-i)(x+i)$. Zeros are $-1,2/3,i,-i$.

31. $f(x)=x^3-2x^2+3x-6$. Possible zeros $\pm1,\pm2,\pm3,\pm6$. 2 is found to be
a zero:

```
2|1  -2   3  -6
    2   0   6
  1   0   3   0.
```
(x-2) is a factor.
$f(x)=(x-2)(x^2+3)=(x-2)(x-\sqrt{3}i)(x+\sqrt{3}i)$. Zeros are $2,\sqrt{3}i,-\sqrt{3}i$.

Chapter 4 Project

Good Source: "Chaos, Fractals, and Dynamics by Robert L. Devaney,
Addison-Wesley, 1990, chapter 6.
1. e.g. $f(x) = z^2-1$. Consider the points i and 1.
   i->-2->3->8->63->3968-> ... Not in the set.
   1->0->-1->0->-1->0-> ... In the set.
2. $f(x) = z^2-1$. Consider the points 1+.2i, .5+.5i, 1-.2i, 2, 1.5.
   1+.2i->-.4+.4i->-1.1584+0.032i->.3409+.0741i-> ... In set.
   .5+.5i->-1+.5i->-.25-i->-1.9375+.5i-> ...Not in set.
   1-.2i->.04-.4i->-1.1584+0.032i->.3409-.0741i-> ... In set.
   2->3->8->63->3968-> . . .   Not in set.
   1.5->1.25->.5625->-.6836->-.5327->-.7162->-.487-> ... In set.

3. $f(z) = z^2 + (-.1+.8i)$. Consider the points $1+i$, $.1+.1i$, $.5i$.
$1+i$->$-.1+2.8i$->$-7.93+.24i$->$62.7273-3.0064i$... Not in set.
$.1+.1i$->$-.1+.82i$->$-.7624+.636i$->$.0768-.1698i$... In set.
$.5i$->$-.35+.8i$->$.6175+.24i$->$.2237+.5036i$->$-.3036+1.0253$
->$-1.0591+.1775i$->$.9902+.424i$->$.7008+1.6398i$... Not in set.

Chapter 4 Review Exercises

1.(a) $f(x)=x^3-3x^2+4x-1$, $g(x)=x-3$.

```
3|1  -3   4   -1
      3   0   12
  1   0   4   11.  q(x)=x²+4, remainder=11.
```
q(x)=x^2+4, remainder=11.

(b) $f(x)=2x^3+5x^2+2x+6$, $g(x)=x+2$.

```
-2|2   5   2   6
      -4  -2   0
   2   1   0   6.  q(x)=2x²+x, remainder=6.
```
q(x)=$2x^2+x$, remainder=6.

(c) $f(x)=2x^4-3x^3-4x^2+6x-5$, $g(x)=x-4$.

```
4|2  -3  -4    6    -5
     8   20   64   280
  2   5   16   70   275.
```
q(x)=$2x^3+5x^2+16x+70$, remainder=275.

2.(a) $f(x)=x^3+3x^2-5x-1$. Want f(2). Divide f(x) by g(x)=x-2. Remainder will be f(2).

```
2|1   3   -5   -1
     2   10   10
  1   5    5    9 . Remainder=9.  f(2)=9.
```

(b) $f(x)=4x^3-3x^2-7x+3$. Want f(5). Divide by g(x)=x-5.

```
5|4  -3   -7    3
     20   85   390
  4   17   78   393. Remainder=393.  f(5)=393.
```

(c) $f(x)=-x^4+2x^3-5x^2+4x-6$. Want f(-3). Divide by g(x)=x+3.

```
-3|-1   2   -5    4   -6
        3  -15   60  -192
   -1   5  -20   64  -198. Remainder=-198.  f(-3)=-198.
```

3. (a) $f(x)=x^3-7x^2+14x-8$, r=4. Divide f(x) by g(x)=x-4.

```
4|-1  -7   14   -8
      4  -12    8
  -1  -3    2    0. Remainder=0.  f(4)=0.  Thus 4 is a zero.
```

(b) $f(x)=2x^3+7x^2-3x-2$, r=2. Divide f(x) by g(x)=x-2.

```
2|2   7   -3   -2
     4   26   46
  2   13   23   44. Remainder=44≠0.  f(2)≠0.  Thus 2 not a zero.
```

(c) $f(x)=3x^4+4x^3-8x^2+16x-15$, $r=-3$. Divide $f(x)$ by $g(x)=x+3$.

```
-3|3   4   -8   16  -15
       -9   15  -21   15
   3  -5    7   -5    0.
```
Remainder=0. $f(-3)=0$. Thus -3 is a zero.

4. (a) $f(x)=x^3-2x^2-2x+1$, $g(x)=x-1$.

```
1|1  -2  -2   1
      1  -1  -3
  1  -1  -3  -2.
```
Remainder=$-2\neq 0$. x-1 not a factor.

(b) $f(x)=2x^3-3x^2+4x+36$, $g(x)=x+2$.

```
-2|2  -3    4   36
      -4   14  -36
   2  -7   18    0.
```
Remainder=0. x+2 is a factor.

(c) $f(x)=-2x^4+x^3+12x^2-23x+12$, $g(x)=x+3$.

```
-3|-2   1   12  -23   12
        6  -21   27  -12
   -2   7   -9    4    0.
```
Remainder=0. x+3 is a factor.

5. (a) $\sqrt{-16} = \sqrt{16}i = 4i$.   (b) $\sqrt{-25} = \sqrt{25}i = 5i$.
   (c) $\sqrt{36} = 6$.   (d) $-\sqrt{-169} = -\sqrt{169}i = -13i$.

6. (a) i^7. $7=1\cdot 4+3$. Thus $i^7=i^3=-i$.
   (b) i^{21}. $21=5\cdot 4+1$. Thus $i^{21}=i^1=i$.
   (c) i^{35}. $35=8\cdot 4+3$. Thus $i^{35}=i^3=-i$.
   (d) i^{82}. $82=20\cdot 4+2$. Thus $i^{82}=i^2=-1$.

7. (a) $x^2+2x+5=0$. $a=1,b=2,c=5$.

$$x = \frac{-2\pm\sqrt{2^2-4(1)(5)}}{2(1)} = \frac{2\pm\sqrt{-16}}{2} = \frac{2\pm\sqrt{16}i}{2} = \frac{2\pm 4i}{2} = 1+2i, \ 1-2i.$$

(b) $2x^2-3x+4=0$. $a=2,b=-3,c=4$.

$$x = \frac{3\pm\sqrt{(-3)^2-4(2)(4)}}{2(2)} = \frac{3\pm\sqrt{-23}}{4} = \frac{3\pm\sqrt{23}i}{4}\ .$$

8. (a) $(3-4i) + (7+6i) = 10+2i$.
   (b) $(1+4i) - (8+5i) = -7-i$
   (c) $(8-2i) + (3-9i) = 11-11i$
   (d) $(1-2i)(7+5i) = 1(7+5i) - 2i(7+5i) = 7+5i-14i-10i^2 = 7-9i+10$
       $= 17-9i$.
   (e) $(6+4i)(7+i) = 6(7+i) + 4i(7+i) = 42+6i+28i+4i^2 = 42+34i-4$
       $= 38+34i$.
   (f) $(2-3i)^2 = (2-3i)(2-3i) = 2(2-3i) - 3i(2-3i) = 4-6i-6i+9i^2$
       $=4-12i-9 = -5-12i$.

9. (a) $\dfrac{1+4i}{-2-3i} \cdot \dfrac{-2+3i}{-2+3i} = \dfrac{1(-2+3i)+4i(-2+3i)}{(-2)^2+(3)^2} = \dfrac{-2+3i-8i-12}{4+9} = \dfrac{-14-5i}{13}$

$= -\dfrac{14}{13} - \dfrac{5}{13}i.$

(b) $\dfrac{6-2i}{1+i} \cdot \dfrac{1-i}{1-i} = \dfrac{6(1-i)-2i(1-i)}{(1)^2+1^2} = \dfrac{6-6i-2i+2i^2}{1+1} = \dfrac{6-8i-2}{2} = 2-4i.$

(c) $\dfrac{5-3i}{8+3i} \cdot \dfrac{8-3i}{8-3i} = \dfrac{5(8-3i)-3i(8-3i)}{(8)^2+(3)^2} = \dfrac{40-15i-24i-9}{64+9} = \dfrac{31-39i}{73}$

$= \dfrac{31}{73} - \dfrac{39}{73}i.$

10. $x^3-6x^2+kx-10$. Want $x-2$ to be a factor. Divide by $x-2$.

```
2|1   -6     k       -10
        2    -8      -16+2k
  1   -4   -8+k    -26+2k.
```
Remainder is $-26+2k$. $x-2$ is a factor if $-26+2k=0$. i.e. if $k=13$.

11. $f(x)=x^3+(-4+2i)x^2+(1-8i)x+2i$. Is $x+2i$ a factor? Divide by $x+2i$.

```
-2i|1   -4+2i   1-8i    2i
          -2i    8i    -2i
    1   -4       1      0.
```
Remainder is zero. Thus $x+2i$ a factor.

12. $f(x)=2x^3+(2-3i)x^2+(4-i)x+5i$. Want $f(-3i)$.

```
-3i|2   2-3i      4-i         5i
         -6i    -27-6i     -21+69i
    2   2-9i   -23-7i    -21+74i.
```
$f(-3i)=-21+74i.$

13. To construct a polynomial of degree 3 with zeros $-3,2,5$.
Function $f(x)=a(x+3)(x-2)(x-5)$ has zeros $-3,2,5$. Let $a=1$ say,
$f(x)=(x+3)(x-2)(x-5)=(x+3)(x^2-7x+10)=x^3-4x^2-11x+30.$

14. Given 3 is a zero of $f(x)=x^3+x^2-17x+15$ find the other zeros.
Divide $f(x)$ by $x-3$.

```
3|1    1    -17    15
       3     12   -15
  1    4     -5     0.
```
$f(x)=(x-3)(x^2+4x-5)=(x-3)(x-1)(x+5).$
Zeros are $3,1,-5$.

15. Want polynomial $f(x)$ with zeros $1,2,-3$ and $f(3)=36$.
Let $f(x)=a(x-1)(x-2)(x+3)$. $f(3)=a(2)(1)(6)$. $12a=36$, $a=3$.
$f(x)=3(x-1)(x-2)(x+3)=3(x-1)(x^2+x-6)=3(x^3+x^2-6x-x^2-x+6)$
$=3x^3+3x^2-18x-3x^2-3x+18=3x^3-21x+18.$

16. (a) $f(x)=2x^3-3x^2-11x+6$. Possible zeros $\pm1,\pm2,\pm3,\pm6,\pm1/2,\pm3/2$.
Found that $1,-1,2$ are not zeros. -2 is a zero:

```
-2|2   -3   -11    6
       -4    14   -6
   2   -7     3    0.
```
$f(x)=(x+2)(2x^2-7x+3)=(x+2)(2x-1)(x-3).$

Zeros are -2,1/2,3.

(b) $f(x)=x^3 + \frac{5}{2}x^2 + \frac{1}{2}x - 1$. Consider $g(x)=2x^3+5x^2+x-2$.
Possible zeros $\pm1,\pm2,\pm1/2$.

```
-1|2   5   1  -2
     -2  -3   2
   2   3  -2   0.
```
$f(x)=(x+1)(2x^2+3x-2)=(x+1)(2x-1)(x+2)$.
Zeros are -1,1/2,-2.

(c) $f(x)=x^3-2x^2+3x-6$. Possible zeros $\pm1,\pm2,\pm3,\pm6$.
Found that 1,-1 are not zeros. 2 is a zero:

```
2|1  -2   3  -6
      2   0   6
  1   0   3   0.
```
$f(x)=(x-2)(x^2+3)=(x-2)(x+\sqrt{3}i)(x-\sqrt{3}i)$.
Zeros are $2,\sqrt{3}i,-\sqrt{3}i$.

(d) $f(x)=x^3+5x^2+9x+5$. Possible zeros $\pm1,\pm5$.
Find 1 not a zero. -1 is a zero:

```
-1|1   5   9   5
      -1  -4  -5
   1   4   5   0.
```
$f(x)=(x+1)(x^2+4x+5)$.
Quadratic formula gives roots of $x^2+4x+5=0$ to be $-2+i$, $-2-i$.
Zeros are $-1,-2+i,-2-i$.

Chapter 4 Test

1. $f(x)=x^3-2x^2+5x-2$, $g(x)=x-3$.
```
3|1  -2   5  -2
      3   3  24
  1   1   8  22.
```
$q(x)=x^2+x+8$, remainder=22.

2. $f(x)=x^3+5x^2-4x-3$. Want $f(3)$. Divide $f(x)$ by $g(x)=x-3$.
Remainder will be $f(3)$.
```
3|1   5  -4  -3
      3  24  60
  1   8  20  57.
```
Remainder=57. $f(3)=57$.

3. $f(x)=2x^3+6x^2-4x-5$, $r=2$. Divide $f(x)$ by $g(x)=x-2$.
```
2|2   6  -4  -5
      4  20  34
  2  10  17  29.
```
Remainder=$29\neq0$. $f(2)\neq0$. Thus 2 not a zero.

4. (a) $\sqrt{-144} = \sqrt{144}i = 12i$.   (b) $-\sqrt{-81} = -\sqrt{81}i = -9i$.

5. (a) i^9. $9=2\cdot4+1$. Thus $i^9=i^1=i$.
   (b) i^{23}. $23=5\cdot4+3$. Thus $i^{23}=i^3=-i$.

6. $2x^2+3x+7=0$. $a=2, b=3, c=7$.

$$x = \frac{-3\pm\sqrt{(3)^2-4(2)(7)}}{2(2)} = \frac{-3\pm\sqrt{-47}}{4} = \frac{-3\pm\sqrt{47}i}{4}.$$

7. (a) $(2-5i) + (3+7i) = 5+2i$.
   (b) $(1+6i)(3-4i) = 1(3-4i) + 6i(3-4i) = 3-4i+18i-24i^2 =$
   $3+14i+24 = 27+14i$.

8. $\dfrac{1+3i}{2-5i} \cdot \dfrac{2+5i}{2+5i} = \dfrac{1(2+5i)+3i(2+5i)}{(2)^2+(5)^2} = \dfrac{2+5i+6i-15}{4+25} = \dfrac{-13+11i}{29}$

   $= -\dfrac{13}{29} + \dfrac{11}{29}i$.

9. x^3-3x^2+kx-8. Want x-2 to be a factor. Divide by x-2.

   ```
   2|1  -3    k      -8
        2    -2    -4+2k
     1  -1  -2+k  -12+2k.
   ```
   Remainder is $-12+2k$. x-2 is a factor if
   $-12+2k=0$. i.e. if $k=6$.

10. $f(x)=x^3+(-4+3i)x^2+(1-8i)x-12+3i$. Is x+3i a factor?
    Divide by x-3i.

    ```
    -3i|1  -4+3i   1-8i   -12+3i
             -3i    12i    12-3i
         1   -4    1+4i      0.
    ```
    Remainder is zero. Thus x-3i a
    factor.

11. Given 2 is a zero of $f(x)=x^3-4x^2-11x+30$ find the other zeros.
    Divide f(x) by x-2.

    ```
    2|1  -4  -11   30
         2   -4   -30
      1  -2  -15    0.
    ```
    $f(x)=(x-2)(x^2-2x-15)=(x-2)(x+3)(x-5)$.
    Zeros are 2,-3,5.

12. $f(x)=2x^3+3x^2-9x-10$. Possible zeros $\pm1, \pm2, \pm5, \pm10, \pm1/2, \pm5/2$.
    Found that 1 is not a zero. -1 is a zero:

    ```
    -1|2   3   -9   -10
          -2   -1    10
       2   1  -10     0.
    ```
    $f(x)=(x+1)(2x^2+x-10)=(x+1)(2x+5)(x-2)$.
    Zeros are -1,-5/2,2.

Chapter 5

Section 5.1

Group Discussion Assume that there are 4 months per semester. There will be a total of 32 months during the four years. If you make good grades during the whole time the $1 will grow to 2^{32} =$4,294,967,296. Things grow quickly in exponential growth.

1.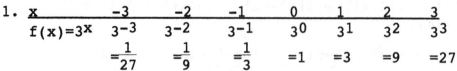

x	-3	-2	-1	0	1	2	3
$f(x)=3^x$	3^{-3}	3^{-2}	3^{-1}	3^0	3^1	3^2	3^3
	$=\frac{1}{27}$	$=\frac{1}{9}$	$=\frac{1}{3}$	$=1$	$=3$	$=9$	$=27$

Points $(-3,\frac{1}{27}),(-2,\frac{1}{9}),(-1,\frac{1}{3}),(0,1),(1,3),(2,9),(3,27)$ Graph below.

3.

x	-3	-2	-1	0	1	2	3
$f(x)=2^{3x}$	2^{-9}	2^{-6}	2^{-3}	2^0	2^3	2^6	2^9
	$=\frac{1}{512}$	$=\frac{1}{64}$	$=\frac{1}{8}$	$=1$	$=8$	$=64$	$=512$

Points $(-3,\frac{1}{512}),(-2,\frac{1}{64}),(-1,\frac{1}{8}),(0,1),(1,8),(2,64),(3,512)$

   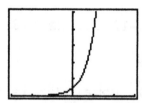

Exercise1,$f(x)=3^x$ Exercise3,$f(x)=2^{3x}$

5. $f(x)=2^x$. (a) $f(1.7)=3.2490$, (b) $f(3.42)=10.7034$, (c) $f(-2.1)=0.2333$. (to 4 decimal places)

7. $g(x)=4^{-x}$. (a) $g(1.6)=0.1088$, (b) $g(3.75)=0.0055$, (c) $g(-5.2)=1351.1761$.

In exercise 9-19 all windows have Xscl=1, Yscl=1. D stands for domain, R for range.

9. $g(x)=2^x+3$.
   Shift $f(x)=2^x$
   up 3.
   $D(-\infty,\infty)$, $R(3,\infty)$

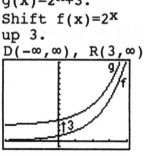

11. $g(x)=3^{-x}+5$.
    Shift $f(x)=3^{-x}$
    up 5.
    $D(-\infty,\infty)$, $R(5,\infty)$

13. $g(x)=3^{x+1}$.
    Shift $f(x)=3^x$
    left 1.
    $D(-\infty,\infty)$, $R(0,\infty)$

15. $g(x)=5^{-x-3}$
    $=5^{-(x+3)}$.
    Shift $f(x)=5^{-x}$
    left 3.
    $D(-\infty,\infty)$, $R(0,\infty)$

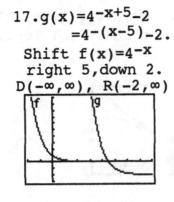

17. $g(x)=4^{-x+5}-2$
    $=4^{-(x-5)}-2$.
    Shift $f(x)=4^{-x}$
    right 5,down 2.
    $D(-\infty,\infty)$, $R(-2,\infty)$

19. $g(x)=3(2^x)$.
    $f(x)=2^x$. $g=3f$.
    Mult every value
    of f by 3.
    $D(-\infty,\infty)$, $R(0,\infty)$

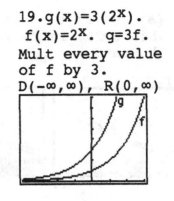

21. $f(x)=3^x$, $g(x)=4^x$. f>g when x<0. f=g when x=0. f<g when x>0.

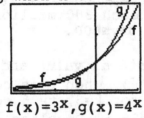

$f(x)=3^x$, $g(x)=4^x$

23. $g(x)=2^x+1$. Shift f up 1.
25. $g(x)=-2^x+1$. Reflect f in x-axis, shift up 1.
27. $g(x)=2^{-x}-3$. Reflect f in y-axis, shift down 3.
29. g is exponential growth. Thus let $g(x)=2^{x+a}+b$. g(2)=4. Thus $4=2^{2+a}+b$. Satisfied if a=1,b=-4. But graph $g(x)=2^{x+1}-4$ not correct. Satisfied if a=-1,b=2. Graph $g(x)=2^{x-1}+2$ correct.

31. Graph of f is reflected in x-axis and then vertical shift. Thus let $g(x)=-2^{x+a}+b$. g(2)=4. Thus $4=-2^{2+a}+b$. Satisfied if a=-2,b=5. Graph $g(x)=-2^{x-2}+5$ correct.

33. $f(x)=k2^{x/c}$. f(0)=8. Thus initial value is 8; k=8. Graph gives, f(0)=8, f(12)=16, f(24)=32. As x increases by 12, values of f double. Doubling time is 12; c=8. $f(x)=8(2^{x/12})$.

35. Initial value is 6. f(0)=6, f(9)=12, f(18)=24. Function doubles every 9 units. $f(x)=6(2^{x/9})$.

37. (a) World population given by $P(t)=5.94(2^{t/40})$ with t=0 in 1998.In year 2020, t=22. P(25)=8.70 billion.
    (b) In 1970, t=-28. P(-28)=3.66 billion. See www.census.gov for access to World and U.S. popclocks and data. World population was 3,706,601,448 in 1970 according to this source.

39. Doubling time for transistors on a chip is 18 months. In 1995 there were 5.5 million transistors on a Pentium chip. Let t=0 correspond to 1995. $N(t)=5.5(2^{t/1.5})$. In the year 2002, t=7. N(7)=139.6912926 million. 139.7 million transistors.

41. (a) Let $P_1(t)=150(2^{t/20})$, $P_2(t)=200(2^{t/30})$. Graphs intersect when t=25, in 25 years time. (b) Populations will then be 356 thousand.

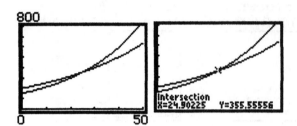

43. $P=268.5(2^{t/65})$. t=0 corresponds to 1997. P=300 when t=10.4. Population of U.S.will reach 300 million in the year 2007. Suitable window 0≤x≤50, 0≤y≤600.

45. $P=k(2^{t/c})$, where k is initial value and c is doubling time. Can interpret $P(c)=k(2^{t/c})$, when t is constant and c varies. Want to find value of c for k=8, t=15 corresp to P=12. Graph $P=8(2^{15/c})$. P=12 when c=25.6. Doubling time = 25.6 yrs. Suitable window 0≤x≤50, 0≤y≤20.

47. Pseudomonas natriegens, doubling time 10 minutes, and initial count 100. Times 10, 20, 30, 40, 50 minutes. $N(t)=100(2^{t/10})$. Intervals increase by doubling time. Thus population doubles.

t	0	10	20	30	40	50	60
N	100	200	400	800	1600	3200	6400

49. Bacillus coagulans, doubling time 13 minutes, initial count 600. Times 26, 52, 78, 104, 130 minutes. $N(t)=600(2^{t/13})$. Intervals increase by twice doubling time. Thus population quadruples.

t	0	26	52	78	104	130
N	600	2400	9600	38400	153600	6144000

51. $N=1000(2^{t/25})$. Takes 58 minutes (to nearest minute) to grow to 5000. Suitable window 0≤x≤100, 0≤y≤10000.

53. Bacillus subtilis. Initial count of 1000 grew to 1500 in 15 minutes. $N=1000(2^{t/c})$. When t=15,N=1500. Thus $1500=1000(2^{15/c})$. Graph $y=1000(2^{15/x})$ to see which value of x gives y=1500. Get c=26 minutes (to nearest minute). Doubling time is 26 minutes. Suitable window 0≤x≤50, 0≤y≤3000.

Project: (a) Source: "The World Almanac", 1997, page 838.
1996 populations (thousands) China:1,210,005. India:952,108.
2020 predicted populations   China:1,413,251. India:1,289,473.
$P=k(2^{t/c})$. Solve for c, $P/k=2^{t/c}$, $t/c=\log_2(P/k)$, $c=t/\log_2(P/k)$, $c=(t\ln2)/\ln(P/k)$.
China: $c=(t\ln2)/\ln(P/k)=(24\ln2)/\ln(1413251/1210005)=107.14$.
India: $c=(t\ln2)/\ln(P/k)=(24\ln2)/\ln(1289473/952108)=54.85$.

The predicted doubling time for China is 107.14 years while the predicted doubling time for India is shorter, 54.85 years. China predicted growth from 1996: $P=1210005(2^{t/107.14})$. India predicted growth from 1996: $P=1413251(2^{t/54.85})$. We plot these functions, below. The population of India is predicted to pass that of China when t=38.7, in the year 2034. The Chinese government has passed a law restricting the number of children allowed. There is no restriction in India, and little use of birth control.

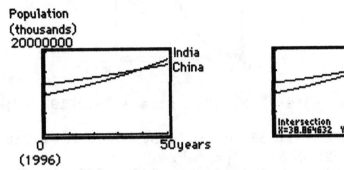

(b) $r \approx \dfrac{70}{c}$. Chinese rate = 70/107.14=.65% per annum.

Indian rate = 70/54.85 = 1.28% per annum.

(c) $f(t)=k2^{t/c}$. To show $r=100(2^{1/c} - 1)$.

r = % growth in 1 year = $\dfrac{f(1) - f(0)}{f(0)}$ x 100 =

$\dfrac{k2^{1/c} - k2^{0/c}f(0)}{k2^{0/c}}$ x 100 = $(2^{1/c} - 1)100$.

We graph r(c)=70/c and $r(c)=100(2^{1/c} - 1)$ below in a window that allows doubling time c of up to 200 years and growth rates of up to 10% per annum. Observe that the graphs appear to coincide indicating the validity of the rule of thumb. The tables for Y1=70/c and $Y2=100(2^{1/c} - 1)$ reveal how close the values are.

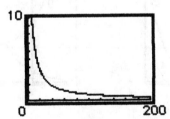

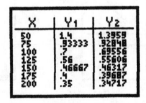

X	Y1	Y2
50	1.4	1.3959
75	.93333	.92848
100	.7	.69556
125	.56	.55606
150	.46667	.46317
175	.4	.39687
200	.35	.34717

Graphs and tables of r(c)=70/c and $r(c)=100(2^{1/c} - 1)$

Section 5.2

Group Discussion $f(x)=\dfrac{100}{1+100e^{-kx}}$, learning function. Select window

$0 \le x \le 1$ (1hour), $0 \le y \le 100$ (%), and k=10. This curve might describe the learning pattern in a university class where students initially take time to settle down (0 to A), there is a solid time of teaching in the middle (A to B), but concentration slips toward the end of the class (B to C). See figure below. The professor wishes he/she had more time (C to D). The professor should be guided by this graph in the presentation of material? The presentation should at first be slow, picking up the pace during the middle of the class and then slow down toward the end of the class. The presentation would then match the typical learning curve of the students. Ask the students for estimates of A and B!

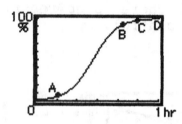

1. (a)$e^{.25}$=1.2840 (b)$e^{.85}$=2.3396 (c)$e^{-1.6}$=.2019 (d)$e^{-5.4}$=.0045.

3. (a) e^{-3}=.0498. e^{-2}=.1353. e^{-1}=.3679. e^0=1. e^1=2.7183. e^2=7.3891. e^3=20.0855. f(x)=e^x: f(-3)=.0498. f(-2)=.1353. f(-1)=.3679. f(0)=1. f(1)=2.7183. f(2)=7.3891. f(3)=20.0855. (b) See graph below.

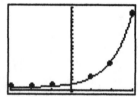

Exercise 3. f(x)=e^x

5. g(x)=$-e^x$.
   Reflect f(x)=e^x
   in x-axis.

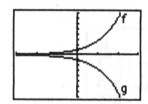

7. g(x)=e^{x-3}
   Shift f(x)=e^x
   right 3.

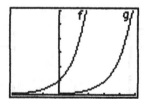

9. g(x)=e^{1-x}+3
   Reflect f(x)=e^x
   in y-axis, shift
   right 1, up 3.

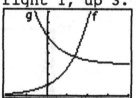

11. $3^{x+1}=3^7$. x+1 = 7. x=6.

13. $7^{5x+2}=7^{3x-5}$. 5x+2 = 3x-5. 2x = -7. x=-7/2.

15. $3^{1-2x}=9^{x-1}$. $3^{1-2x}=3^{2(x-1)}$. 1-2x = 2x-2. -4x=-3. x=3/4.

17. $9^{2x+3}=27$. $3^{2(2x+3)}=3^3$. 2(2x+3)=3. 4x+6=3. 4x=-3. x=-3/4.

19. $8^{2x+1}=4^{3x+1}$. $2^{3(2x+1)}=2^{2(3x+1)}$. 3(2x+1)=2(3x+1). 6x+1=6x+2. No values of x that satisfy this equation. No solution.

162

21. $e^{5-x}=e^{x-1}$. $5-x=x-1$. $-2x=-6$. $x=3$.

23. $2^n = 32$. $2^n = 2^5$. $n=5$. $n^3=5^3=125$.

25. $n^4 = 16$. $n^4 = 2^4$. $n=2$. $n^6=2^6=64$.

27. $3^{(x+2)}=12$. A(0.26185951,12). Soln is $x=0.2619$ to 4 dec places.

29. $5^{(4x+9)}=15$. A(-1.829348,15). Soln is $x=-1.8293$ to 4 dec places.

31. $2^{(3x^2+2x-5)}=11$. A(-2.045327,11), B(1.3786601,11). Two solutions, $x=-2.0453$ and $x=1.3787$, to 4 dec places.

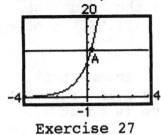

Exercise 27   Exercise 29   Exercise 31

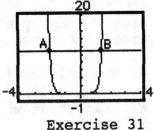

33. $5^{x^3}=13$. A(1.1680682,13). Soln is $x=1.1681$ to 4 dec places.

35. $e^{x^2}-e^{2x}-31=0$. A(-1.853316,0), B(2.1598681,0). Two solutions, $x=-1.8533$ and $x=2.1599$, to 4 dec places.

37. $e^x+e^{-x}=12$. A(-2.477889,12), B(2.4778887,12). Two solutions, $x=-2.4779$ and $x=2.4779$, to 4 dec places.

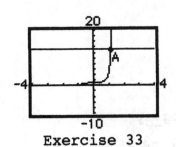

Exercise 33

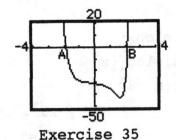

Exercise 35

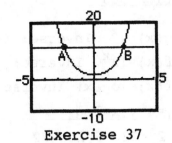

Exercise 37

39. \$8,000 at 6%, for 4 years, (a) compounded annually.
$A_1=P(1 + r/n)^{nt}$. $A_2= Pe^{it}$. A=8000. r=.06. t=4. n=1.
$A_1=P(1 + r/n)^{nt} = 8000(1+.06)^4 = \$10,099.82$.
(b) Continuously, $A_2= Pe^{rt} = 8000e^{(.06)(4)} = \$10,169.99$.

41. \$12,000 at 9%, for 3 years, (a) compounded semiannually.
$A_1=P(1 + r/n)^{nt} = 12000(1+.045)^6 = \$15,627.12$.
(b) Continuously, $A_2= Pe^{rt} = 12000e^{(.09)(3)} = \$15,719.57$.

43. $5,325 at 10%, for 7 years 3 months, (a)compounded quarterly.
$A_1 = P(1 + r/n)^{nt} = 5325(1+.025)^{29} = \$10,897.12.$
(b) Continuously, $A_2 = Pe^{rt} = 5325e^{(.1)(7.25)} = \$10994.69.$

45. $50,250 at 7% for 3 years, (a) compounded quarterly.
$A_1 = P(1 + r/n)^{nt} = 50250(1+.07/4)^{12} = \$61,879.83.$
(b)$A_2 = Pe^{rt} = 50250e^{(.07)(3)} = \$61,992.32.$

47. Graphs $A=5000e^{0.05t}$ & A=6000 intersect at t=3.65; 3yrs 8months.

49. Graphs $A=4000e^{0.06t}$ & A=8000 intersect t=11.6; 11yrs 7months.
Let P be the initial principle. When A=2P, $2P=Pe^{0.06t}$,
$2=e^{0.06t}$. The solution t to this equation is the 'doubling
time'. Graph $y=e^{0.06t}$ and y=2. Intersect at t=11.55. Doubling
time 11yrs 7months.

51. Graphs $A=1000e^{7r}$ & A=1522 intersect at r=.06, interest=6%.
53. Graphs $A=8000e^{10r}$ & A=12560 intersect at r=.045, interest=4.5%.
55. Wavelength at which the eye is most sensitive is about
561.26nm.

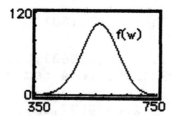
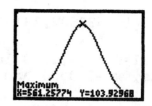

Section 5.3

1. $g(x)=4^x$. Inverse: $f(x)=\log_4 x$
3. $f(x)=12^x$. Inverse: $g(x)=\log_{12} x$
5. $f(x)=\log_8 x$. Inverse: $g(x)=8^x$

7. $y=4^x$. $x=\log_4 y$
9. $y=13^x$. $x=\log_{13} y$

11. $2^5=32$. $5=\log_2 32$
13. $2^{-3}=0.125$. $-3=\log_2 0.125$

15. $y=\log_5 x$. $x=5^y$
17. $y=\log_{12} x$. $x=12^y$

19. $\log_2 8=3$. $8=2^3$
21. $\log_7 1=0$. $7^0=1$

23. $\log_3 9=y$, $9=3^y$, $3^2=3^y$, $y=2$
25. $\log_6 1=y$, $1=6^y$, $6^0=6^y$, $y=0$
27. $\log_7 7=y$, $7=7^y$, $7^1=7^y$, $y=1$
29. $\log_{10} 1000=y$, $1000=10^y$, $10^3=10^y$, $y=3$
31. $\log_6(1/216)=y$, $1/216 = 6^y$, $6^{-3}=6^y$, $y=-3$
33. $\log_{16} 2=y$, $2=16^y$, $16^{1/4}=16^y$, $y=1/4$

35. (a) ln 2.13 = 0.7561   (b) ln 6.78 = 1.9140
    (c) ln 0.41 = -0.8916   (d) ln 0.07823 = -2.5481

37. (a) ln 0.52 = -0.6539   (b) ln 5.92 = 1.7783
    (c) ln 16.93 = 2.8291   (d) ln 478.29 = 6.1702

39. (a) log 0.361 = -0.4425   (b) log 4.982 = 0.6974
    (c) log 16.032 = 1.2050   (d) log 543.21 = 2.7350

41. g(x)=lnx+3.
    Shift f(x)=lnx
    up 3.

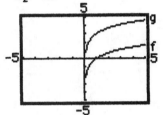

43. g(x)=log(x-4).
    Shift f(x)=logx
    right 4.

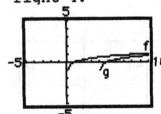

45. g(x)=log(x+1)-3.
    Shift f(x)=logx
    left 1, down 3.

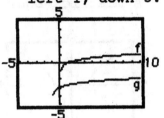

47. g(x)=ln(-x).
    Reflect f(x)=lnx
    in y-axis.

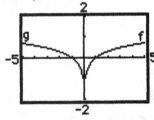

49. $f(x)=2^x-3$. $y=2^x-3$. Interchange x & y: $x=2^y-3$. Solve for y:
    $x+3=2^y$. $y=\log_2(x+3)$. $f^{-1}(x)=\log_2(x+3)$.

51. $f(x)=7^{(2x+5)}+3$. $y=7^{(2x+5)}+3$. Interchange x & x: $x=7^{(2y+5)}+3$.
    Solve for y: $x-3=7^{(2y+5)}$. $2y+5=\log_7(x-3)$, $y=(\log_7(x-3) - 5)/2$.
    $f^{-1}(x)=(\log_7(x-3) - 5)/2$.

53. $f(x)=\log_6(4x+7)$. $y=\log_6(4x+7)$. Interchange x & y: $x=\log_6(4y+7)$.
    Solve for y: $4y+7=6^x$, $y=(6^y-7)/4$. $f^{-1}(x)=(6^x-7)/4$.

55. $f(x)=e^x+1$. $y=e^x+1$. Interchange x & y: $x=e^y+1$. Solve for y:
    $x-1=e^y$, $y=\ln(x-1)$. $f^{-1}(x)=\ln(x-1)$.

57. $f(x)=e^{(2x+4)}+1$. $y=e^{(2x+4)}+1$. Interchange x & y: $x=e^{(2y+4)}+1$.
    Solve for y: $x-1=e^{(2y+4)}$, $\ln(x-1)=2y+4$, $y=\frac{1}{2}\ln(x-1) - 2$.
    $f^{-1}(x)=(\ln(x-1))/2 - 2$.

59. $f(x)=\ln(3x+6)$. $y=\ln(3x+6)$. Interchange x&y: $x=\ln(3y+6)$. Solve for y: $e^x=3y+6$, $y=(e^x-6)/3$. $f^{-1}(x)=\frac{1}{3}e^x - 2$.

61. $g(x) = \ln(x+1)$. Shift left 1.
63. $g(x) = \ln(-x)$. Reflect in y-axis.
65. $g(x) = \ln(-(x+1))$. Reflect in y-axis, shift left 1.

67. $f(x) = \ln x^2$. Vertical asymptote y-axis. $f(x)=f(-x)$. f is symmetric about y-axis. See graph below.

69. $f(x)=\ln|x|$. Vertical asymptote y-axis. $f(x)=f(-x)$. f is symmetric about y-axis.

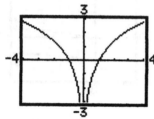

Exercise 67   Exercise 69

Section 5.4

1. $\log_3(x-4)=2$. $x-4=3^2$. $x=4+9=13$.
3. $\log_2(2x-3)=4$. $2x-3=2^4$. $2x=16+3=19$. $x=19/2$.
5. $\log_5(3x-2)=3$. $3x-2=5^3$. $3x=125+2=127$. $x=127/3$.
7. $\log_x(3-2x)=2$. $3-2x=x^2,x>0$. $x^2+2x-3=0$. $(x+3)(x-1)=0$. $x=1$.
9. $\log_x81=4$. $81=x^4$, $x>0$. $x=3$.
11. $\log_216=x$. $16=2^x$. $x=4$.
13. $2y = 3^{(x+1)}$. $x+1=\log_3(2y)$. $x=\log_3(2y)-1$.
15. $4y = 2^{(x-5)}$. $x-5=\log_2(4y)$. $x=\log_2(4y)+5$.
17. $2y = \log_3(4x+1)$. $4x+1=3^{2y}$. $x=(3^{2y}-1)/4$.
19. $\log_218y = \log_2(3x+6)$. $2^{\log_218y} = 2^{\log_2(3x+6)}$. $18y=3x+6$. $x=6y-2$.
21. $4(\log_33^2)=4(2)=8$.   23. $5(\log_44^2)=5(2)=10$.
25. $-2(5^{\log_58})=-2(8)=-16$.

27. $\log_a6 = 1.7918$, $\log_a2 = 0.6931$.
    a) $\log_a12 = \log_a(6\cdot2) = \log_a6 + \log_a2 = 1.7918+0.6931 = 2.4849$
    b) $\log_a3 = \log_a(6/2) = \log_a6 - \log_a2 = 1.7918-0.6931 = 1.0987$
    c) $\log_a36 = \log_a(6\cdot6) = \log_a6 + \log_a6 = 1.7918+1.7918 = 3.5836$
    d) $\log_a(1/3) = \log_a(2/6) =\log_a2 - \log_a6= 0.6931-1.7918= -1.0987$

29. $\log_a 7 = 1.9459$, $\log_a 4 = 1.3863$.
   a) $\log_a 1.75 = \log_a(7/4) = \log_a 7 - \log_a 4 = 1.9459 - 1.3863 = 0.5596$
   b) $\log_a 2401 = \log_a(7^4) = 4\log_a 7 = 4(1.9459) = 7.7836$
   c) $\log_a(4/7) = \log_a 4 - \log_a 7 = 1.3863 - 1.9459 = -0.5596$
   d) $\log_a 16 = \log_a 4^2 = 2\log_a 4 = 2(1.3863) = 2.7726$

31. $\log_a(xy/z) = \log_a x + \log_a y - \log_a z$.

33. $\log_b(x^3 y^2/z^{-1}) = \log_b x^3 + \log_b y^2 - \log_b z^{-1} = 3\log_b x + 2\log_b y + \log_b z$.

35. $\log_3(25yz/x) - \log_3(5x^3 y/z^{-2})$
   $= \log_3 25 + \log_3 y + \log_3 z - \log_3 x - \log_3 5 - \log_3 x^3 - \log_3 y - \log_3 z^2$
   $= \log_3 5^2 + \log_3 y + \log_3 z - \log_3 x - \log_3 5 - 3\log_3 x - \log_3 y - 2\log_3 z$
   $= -4\log_3 x - \log_3 z + \log_3 5$

37. $3\log_a 2x + \log_a x - 2\log_a 4x = \log_a 2^3 x^3 + \log_a x + \log_a 4^{-2} x^{-2}$
   $= \log_a(2^3 x^3 x 2^{-4} x^{-2}) = \log_a(2^{-1} x^2) = \log_a(x^2/2)$

39. $7\log_c x + 3\log_c x(5x+2)^{-1} - \log_c x^2(5x+2)^{-2}$
   $= \log_c x^7 + \log_c x^3(5x+2)^{-3} + \log_c x^{-2}(5x+2)^2$
   $= \log_c(x^7 x^3(5x+2)^{-3} x^{-2}(5x+2)^2) = \log_c(x^8/(5x+2))$

41. $\log_{10} x(x-2) + 2\log_{10} y(x-2) - 3\log_{10} xy(x-2) + 4\log_{10}(x-2)$
   $= \log_{10} x(x-2) + \log_{10} y^2(x-2)^2 + \log_{10} x^{-3} y^{-3}(x-2)^{-3} + \log_{10}(x-2)^4$
   $= \log_{10}(x(x-2)y^2(x-2)^2 x^{-3} y^{-3}(x-2)^{-3}(x-2)^4) = \log_{10}(x^{-2} y^{-1}(x-2)^4)$

45. $\log_2(x-4) + \log_2 6 = 3$. $\log_2 6(x-4) = 3$. $6(x-4) = 2^3$. $6x = 8 + 24$.
   $6x = 32$. $x = 16/3$.

47. $\log_3(3x+6) - \log_3(x-2) = 2$. $\log_3(3x+6)/(x-2) = 2$.
   $(3x+6)/(x-2) = 3^2$. $3x+6 = 9(x-2)$. $-6x = -24$. $x = 4$.

49. $\log_8(5x+2) = \log_8 6 - \log_8 3$. $\log_8(5x+2) - \log_8 6 + \log_8 3 = 0$.
   $\log_8(3(5x+2)/6) = 0$. $(5x+2)/2 = 8^0 = 1$. $5x+2 = 2$. $x = 0$.

51. $2\log_7 3x - \log_7 2x = 1$. $\log_7(3x)^2 - \log_7 2x = 1$. $\log_7(3x)^2/2x = 1$.
   $\log_7 9x/2 = 1$. $9x/2 = 7^1$. $x = 14/9$.

53. $2\log_9 3x - \log_9 x = 1$. $\log_9(3x)^2/x = 1$. $\log_9 9x = 1$. $9x = 9^1$. $x = 1$.

55. $3\log_6 x = \log_6 125$. $\log_6 x^3 = \log_6 125$. $x^3 = 125$. $x = 5$.

57. $\log_2(x^2-7x+10)=2$. $x^2-7x+10=2^2$. $x^2-7x+6=0$. $(x-1)(x-6)=0$. $x=1,6$.

59. $\ln(2x-5) = -2x + 13$. $x=5.5894$ to 4 decimal places.
61. $\ln(x^2 + 2x + 7) = 3x - 7$. $x=3.4125$ to 4 decimal places.
63. $\ln(x^3 + 2x) = 2x - 8$. $x=6.9226$ to 4 decimal places.

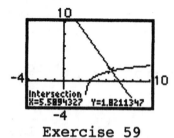
Exercise 59

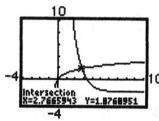
Exercise 61

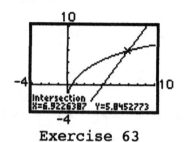
Exercise 63

65. $\ln(2x + 1) = 3^{4-x} - 2$. $x=2.7666$ to 4 decimal places.

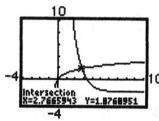
Exercise 65

67. $\log_3 5 = \ln5/\ln3 = 1.4650$
69. $\log_7 2.4 = \ln2.4/\ln7 = 0.4499$
71. $\log_6 75.43 = \ln75.43/\ln6 = 2.4128$
73. $\log_3 5 + \log_3 7 = \log_3(5\cdot7) = \log_3 35 = \ln35/\ln3 = 3.2362$
75. $3\log2 - 2\log10 = \log(2^3/10^2) = \log(0.08) = -1.0969$
77. $\log_8 820 + 3\log_8 13 - \log_8 5 = \log_8(820\cdot13^3/5)$
    $= \ln(820\cdot13^3/5)/\ln8 = 6.1530$

79. $5=e^{3x}$. $3x=\ln5$. $x = \ln5/3 = 0.5365$
81. $72=10^{5x}$. $5x=\log72$. $x = \log72/5 = 0.3715$
83. $4=2^x$. $x=\log_2 4$. $x = \ln4/\ln2 = 2.0000$
85. $5.1=7^{(2x-1)}$. $2x-1=\log_7 5.1$. $x = (1/2)\ln5.1/\ln7 + 1/2 = 0.9186$
87. (a) $\log_a(x/y) = \log_a x-\log_a y$: Let $p=\log_a x$, $q=\log_a y$.
    Then $x=a^p$ and $y=a^q$. $x/y = a^p/a^q = a^{p-q}$.
    Take logs of both sides. $\log_a(x/y)=\log_a a^{p-q} =p-q = \log_a x-\log_a y$.
    (b) $\log_a(x^r) = r\log_a x$: Let $p=\log_a x$. Then $x=a^p$. $x^r=a^{rp}$.
    Take logs of both sides. $\log_a(x^r) = \log_a(a^{rp}) = rp = r\log_a x$.

Section 5.5

In exercises 89-93: Solve $R=(15.3)(2^{-t/5568})$ for t.

$R/15.3=2^{-t/5568}$, $\log_2(R/15.3)=-t/5568$, $t=-5568\ln(R/15.3)/\ln2$.

This equation can be used to find t when given R.

89. $t=-(5568\ln(R/15.3))/\ln2$. R=11.17. t=2,530 years old.

91. $t=-(5568\ln(R/15.3))/\ln2$. R=12.40. t=1,690 years old.

93. $t=-(5568\ln(R/15.3))/\ln2$. R=3.96. t=10,860 years old.

95. (a) Growth of bacteria: $N=k(2^{t/c})$. $N/k=2^{t/c}$. $t=c\log_2(N/k)$.
For computing use natural logarithm, $t(N)=c\ln(N/k)/\ln2$.
E-coli has doubling time 17mins, c=17. $t(N)=17\ln(N/k)/\ln2$.
(b) 1 bacterium (k=1) grows to 1000 bacteria (N=1000) when
$t(1000)=17\ln(1000)/\ln2=169.4183328$. Time is 169.4 minutes.

Section 5.5

1. (a) $N(t)=12.23655495(1.034783205)^t$. (b)t=0 corresponds to 1940.
In year 2000, t=60. Number of people with 4yrs high school
education is N(60)=95.1967, to 3 decimal places.

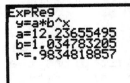 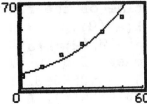

3. (a) $E(t)=892.5321814(1.057320469)^t$. t=0 corresponds to 1940.
(b) E(59)=$23,923.00, E(60)=$25,294.28. Annual earnings
predicted to reach $25,000 when t=60, in year 2000.

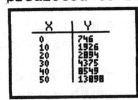 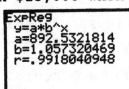 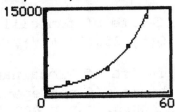

5. (a) $R(t)=603.7404884(1.102255883)^t$. t=0 corresponds to 1960.
(b) In year 2005, t=45. Predicted expenditure on medical
research is R(45)=$48,257 million.

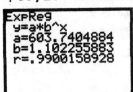 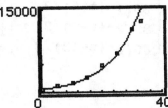

7. (a) The exponential curve does not fit the latest data well,
see windows below. (b) Use data from 1977 on. The curve fits
the data much better. (c) Prediction for 2007 (t=55) using this
function is $520 billion.

169

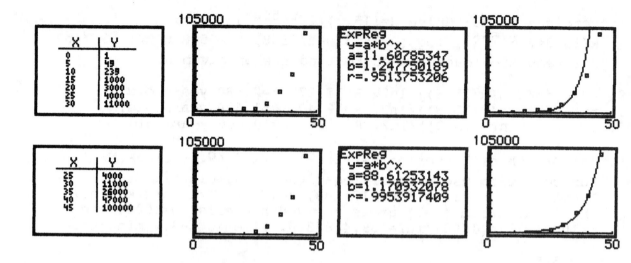

9. (a) Growth of a population given by $f(t)=k(3^{t/p})$. Discussion follows that of doubling time in this section.
$f(t+p)=k(3^{(t+p)/p})=k(3^{(t/p+1)})=k(3^{t/p})(3)=3k(3^{t/p})=3f(t)$.
Thus increase of p in time triples the value of f. p is the tripling time of the function.
   (b) $f(t)=k(a^{t/q})$.
$f(t+q)=k(a^{(t+q)/q})=k(a^{(t/q+1)})=k(a^{t/q})(a)=ak(a^{t/q})=af(t)$.
Thus increase of q in time multiplies the value of f by a.
   (c) $N(t+1) = 3.256628189(1.047281015)^{t+1} = 1.047281015N(t)=N(t)+.047281015N(t)$. Annual growth is 4.7281015%.
In general, if a calculator gives $f(t)=k(a^t)$ as exponential function of best fit then an increase of 1 in t multiplies f by a. If we write a = 1+b then the growth in 1 unit of time is 100b%.

11. (a) Half-life of penicillin is 30 minutes. Patient given 250mg. $Q(t)=250(2^{-t/30})$. (b) After 90 minutes, Q(90)=31.25mg.

13. (a) Half-life of Procainamide is 3 hours. 80mg every 3 hours for 12 hours. If no repetitions $Q(t)=80(2^{-t/3})$. With doses every 3 hours, use translations of 3 as in example 7 to get Q(t), below. (b) $Q(20)=80(2^{-(20-12)/3})=12.60$mg.

$$Q(t)= \begin{cases} 80(2^{-t/3}) & \text{if } 0 \le t < 3 \\ 80(2^{-(t-3)/3}) & \text{if } 3 \le t < 6 \\ 80(2^{-(t-6)/3}) & \text{if } 6 \le t < 9 \\ 80(2^{-(t-9)/3}) & \text{if } 9 \le t < 12 \\ 80(2^{-(t-12)/3}) & \text{if } 12 \le t \end{cases}$$

hapter 5 Review Exercises

.a) $f(x)=6^x$.

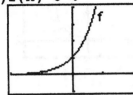

b) $f(x)=3^{-x}$.

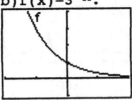

2. a) $g(x)=3^x+2$.
Shift $f(x)=3^x$
up 2.

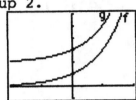

b) $g(x)=2^{x+1}$.
Shift $f(x)=2^x$
left 1.

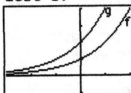

c) $g(x)=6^{x-2}-3$.
Shift $f(x)=6^x$
right 2, down 3.

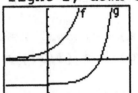

d) $g(x)=-3^{x+2}$.
Shift $f(x)=3^x$ left 2,
reflect in x-axis.

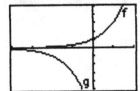

3. a) $g(x)=e^x-3$.
Shift $f(x)=e^x$ down 3.

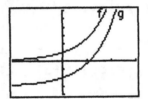

3. b) $g(x)=e^{-x}+1$.
Reflect $f(x)=e^x$
in y-axis, shift up 1.

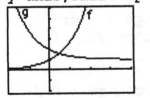

c) $g(x)=e^{x+1}+2$.
Shift $f(x)=e^x$
left 1, up 2.

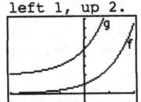

d) $g(x)=-e^x+6$.
Reflect $f(x)=e^x$
in x-axis, shift up 6.

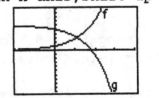

4. (a) $4^{x-3} = 4^5$. x-3 = 5. x=8.
(b) $2^{4x+1}=2^{7x-3}$. 4x+1 = 7x-3. -3x=-4. x=4/3.
(c) $4^{7x-2} = 8^3$. $2^{2(7x-2)} = 2^{3(3)}$. 2(7x-2)=9. 14x=13. x=13/14.
(d) $9^{4x-3} = 27$. $3^{2(4x-3)} = 3^3$. 2(4x-3)=3. 8x=9. x=9/8.
(e) $e^{3x+2} = e^{5x-4}$. 3x+2=5x-4. -2x=-6. x=3.

5. (a) $3^{(x-4)}=5$. A(5.4649735,5). Soln is x=5.46 to 2 dec places.
(b) $2^{(x^2+2x-4)}=9$. A(-3.858308,9), B(1.8583081,9). x=-3.86,1.86.
(c) $2^{(x^3+3x-2)}=7$. A(1.1262537,7). x=1.13.

171

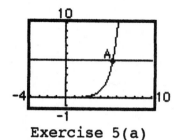

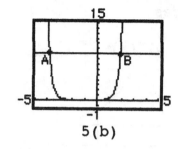

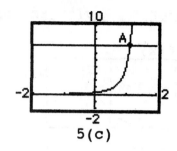

Exercise 5(a)   5(b)   5(c)

6. (a) $y=3^x$. $x=\log_3 y$.   (b) $y=2^{4x}$. $4x=\log_2 y$.
    (c) $3^{-2} = 1/9$. $-2=\log_3(1/9)$. We can write this in a more convenient way, $2=-\log_3(1/9)$, $2=\log_3(1/9)^{-1}$, $2=\log_3(9)$.

7. (a) $y=\log_3 x$. $x=3^y$.   (b) $y=\log_7 5x$. $5x = 7^y$.
    (c) $\log_3(1/27) = -3$. $1/27 = 3^{-3}$.

8. (a) $\log_2 4=y$. $4=2^y$. $2^2=2^y$. $y=2$.
    (b) $\log_3 27=y$. $27=3^y$. $3^3=3^y$. $y=3$.
    (c) $\log_5(1/125)=y$. $1/125=5^y$. $1/5^3=5^y$. $5^{-3}=5^y$. $y=-3$.
    (d) $\log_{27}(1/3)=y$. $1/3=27^y$. $3^{-1}=(3^3)^y$. $3^{-1}=3^{3y}$. $3y=-1$. $y=-1/3$.

9. (a) $f(x)=3^x - 5$, $y=3^x-5$. Interchange x & y: $x=3^y-5$. Solve for y: $x+5=3^y$. $y=\log_3(x+5)$. $f^{-1}(x)=\log_3(x+5)$.

    (b) $f(x)=5^{(3x+5)}-7$, $y=5^{(3x+5)}-7$. Interchange x & y: $x=5^{(3y+5)}-7$. Solve for y: $x+7=5^{(3y+5)}$, $\log_5(x+7)=3y+5$, $y=(\log_5(x+7)-5)/3$. $f^{-1}(x)=(\log_5(x+7)-5)/3$.

    (c) $f(x)=\log_4 5x + 8$. $y=\log_4 5x + 8$. Interchange x & y: $x=\log_4 5y + 8$. Solve for y: $x-8=\log_4 5y$. $y=(4^{x-8})/5$. $f^{-1}(x)=(4^{x-8})/5$.

    (d) $f(x)=\log_3(2x-5)+9$. $y=\log_3(2x-5)+9$. Interchange x & y: $x=\log_3(2y-5)+9$. Solve for y: $x-9=\log_3(2y-5)$, $y=(3^{x-9}+5)/2$. $f^{-1}(x)=(3^{x-9}+5)/2$.

10. (a) $\log_2(2x+4)=3$. $2x+4=2^3$. $2x+4=8$. $x=2$.
    (b) $\ln(3x-2)=5$. $3x-2=e^5$. $3x=e^5+2$. $x=(e^5+2)/3=50.1377197$.
    (c) $\log_x 8=3$. $8=x^3$. $2^3=x^3$. $x=2$.
    (d) $3y=4^{(7x+1)}$. $7x+1=\log_4 3y$. $x=(\log_4 3y - 1)/7$.

11. (a) $4\log_a x+2\log_a(x-3)-\log_a 5x = \log_a x^4+\log_a(x-3)^2-\log_a 5x$
        $= \log_a(x^4(x-3)^2/5x) = \log_a(x^3(x-3)^2/5)$
    (b) $4\log_7(x+1) - 3\log_7 x(x+1)^2 + \log_7 5x^3$

172

$$= \log_7(x+1)^4 + \log_7 x^{-3}(x+1)^{-6} + \log_7 5x^3$$
$$= \log_7((x+1)^4(x^{-3})(x+1)^{-6}(5x^3)) = \log_7(5(x+1)^{-2})$$
$$= \log_7(5/(x+1)^2)$$

12. (a) $\log_3 x + \log_3 2 = 4$. $\log_3(2x) = 4$. $2x=3^4$. $3x=81$. $x=81/2=40.5$.

(b) $\log_2(7x-4) - \log_2 2 = 3$. $\log_2((7x-4)/2)=3$. $7x-4=2(2^3)$.
7x=16+4. x=20/7.

(c) $\log_2(x^2+4x-1) = 2$. $x^2+4x-1 = 2^2$. $x^2+4x-5 = 0$. $(x+5)(x-1)=0$.
x=-5, 1.

(d) $\log_3(x-1)^2 - 2\log_3(x+4) = 4$. $\log_3(x-1)^2 - \log_3(x+4)^2 = 4$.
$\log_3(x-1)^2/(x+4)^2 = 4$. $(x-1)^2/(x+4)^2 = 3^4$.
$(x-1)^2 = 3^4(x+4)^2$. $x-1=\pm 3(x+4)$. Leads to x=-3.5 or -37/8.
x=-37/8 not possible since leads to log of negative number
in original equation. Thus x=-3.5.

13. a) $\ln(3x+1)=-2x+5$. A(1.6167562,1.7664876). x=1.62.
b) $\ln(x^2+3x+5)=3x^2-7$. A(-1.634849,1.0181916),
B(1.789566,2.607968). x=-1.63, 1.79.

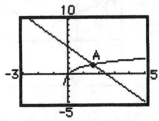

Exercise 13(a)

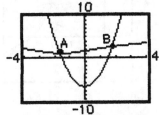

Exercise 13(b)

14.

x	-3.2	-1.8	0	4.3	5.7
e^x	0.0408	0.1653	1	73.6998	298.8674
e^{-x}	24.5325	6.0496	1	0.0136	0.0033

15. (a)

x	0.00038	0.75	1	3.45	12.78
$\ln x$	-7.8753	-0.2877	0	1.2384	2.5479
$\log x$	-3.4202	-0.1249	0	0.5378	1.1065

(b) When 0<x<1, lnx<logx. At x=1, lnx=logx, When x>1, lnx>logx.
Suggests that the graph of f(x)=lnx lies below the graph of
g(x)=logx for 0<x<1 and above it for x>1. That the graphs cross
at x=1. This is true, see fig below.

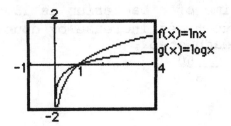

16. (a) $\log_2 6 = \ln 6/\ln 2 = 2.5850$

   (b) $\log_5 3.5 = \ln 3.5/\ln/5 = .7784$

   (c) $\log_4 5.8 - \log_4 2 = \log_4(5.8/2) = \ln(5.8/2)/\ln 4 = .7680$

   (d) $\log_5 6.7 + \log_5 2.3 = \log_5(6.7 \times 2.3) = \ln(6.7 \times 2.3)/\ln 5$
      $= 1.6994$

17. (a) $g(x)=\ln x+3$
   Shift $f(x)=\ln x$
   up 3.

   (b) $g(x)=\ln x-1$
   Shift $f(x)=\ln x$
   down 1.

   (c) $g(x)=\ln(x-1)$
   Shift $f(x)=\ln x$
   right 1.

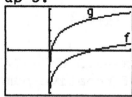 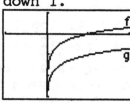 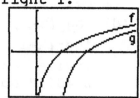

   (d) $g(x)=\ln(x+1)-2$
   Shift $f(x)=\ln x$
   left 1, down 2.

   (e) $g(x)=-\ln(x-2)$
   Shift $f(x)=\ln x$
   right 2, refl in x-axis

   (f) $g(x)=\ln(3-x)$
   Reflect $f(x)=\ln x$
   in y-axis, right 3.

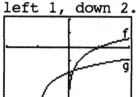 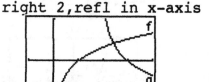 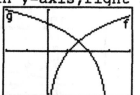

18. (a) $P(t)=35(2^{t/25})$. A(12.864329,50) in fig below, or use trace/zoom. Population reaches 50 thousand in 12th year after 1990, i.e. in 2002.
   (b) $P=35(2^{t/25})$. $2^{t/25}=P/35$. $t/25=\log_2(P/35)$. $t=25\log_2(P/35)$ $=25\ln(P/35)/\ln 2$ with P=50. t=12.86432932. In 12th year.

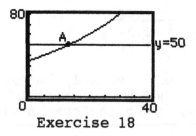

Exercise 18

19. (a) Initial count=500. Doubling time of a bacterium is 15 minutes. $N(t)=500(2^{t/15})$. (b) Note times increase by doubling time. Thus population doubles at each step.

t(min)	0	15	30	45	60
N	500	1000	2000	4000	8000

0. (a) Doubling time of vibro marinus is 80 minutes. Initial count=200. $N(t)=200(2^{t/80})$. (b) There are 4hr (240min) between each measurement. Count doubles 3 times in 240min. Thus there are three doublings at every step.

t(hrs)	0	4	8	12	16
N	200	1600	12800	102400	819200

21. Doubling time of a rhizobium melitoli is 75 minutes. $N(t)=N(0)(2^{t/75})$. $N(t)=10N(0)$. $10N(0)=N(0)2^{t/75}$. $10=2^{t/75}$. $t/75=\log_2 10$. $t=75\log_2 10=75\ln10/\ln2=249.1446071$. Count will multiply by 10 in 4hr 9.14min.

22. $t= -(5568\ln(R/15.3))/\ln2$. $R=1.99$. $t=16384.90525$. The till is approximately 16,380 years old.

23. $f(x)=k(2^{x/c})$. Graph tells us that $f(0)=8$, $f(9)=16$, $f(18)=32$, $f(27)=64$. Thus the initial value is 8, k=8. Doubling time is 9, c=9. Thus the function is $f(x)=8(2^{x/9})$. Get the correct graph.

Chapter 5   Test

1. $f(x)=2^x$. $g(x)=2^{x+4}-5$. Shift f left 4 and down 5.

2. (a) $2^{x-4} = 4^3$. $2^{x-4} = (2^2)^3$. $2^{x-4} = 2^6$. x-4 = 6. x=10.
   (b) $9^{2x-3} = 81$. $3^{2(2x-3)} = 3^4$. 2(2x-3)=4. 4x=10. x=5/2.
   (c) $e^{-x+4} = e^{3x-5}$. -x+4=3x-5. -4x=-9. x=9/4.

3. (a) $2^{4x/3} = 16$. $2^{4x/3} = 2^4$. 4x/3 = 4. x=3.
   (b) $3^{-2x/5} = 9$. $3^{-2x/5} = 3^2$. -2x/5 = 2. x=-5.

4. $4^{(2x-5)}=3$. A(2.8962406,3). Soln is x=2.90 to 2 dec places. See figure below.

5. (a) $y=5^x$. $x=\log_5 y$.   (b) $y=3^{7x}$. $7x=\log_3 y$.

6. (a) $y=\log_6 x$. $x=6^y$.   (b) $y=\log_3 4x$. $4x = 3^y$.

7. $\log_3(1/81)=y$. $1/81=3^y$. $1/(3^4)=3^y$. $3^{-4}=3^y$. y=-4.

8. $f(x)=\log_2(x-3)+7$. $y=\log_2(x-3)+7$. Interchange x & y: $x=\log_2(y-3)+7$. Solve for y: $x-7=\log_2(y-3)$, $y=2^{x-7}+3$. $f^{-1}(x)=2^{x-7}+3$.

9. (a) $\ln(4x-3)=8$. $4x-3=e^8$. $4x=e^8+3$. $x=(e^8+3)/4=745.9894968$.
   (b) $2y=3^{(5x-1)}$. $5x-1=\log_3 2y$. $x=(\log_3 2y + 1)/5$.

10. (a) $\log_4 x + \log_4 2 = 1$. $\log_4(2x) = 1$. $2x = 4^1$. $2x = 4$. $x = 2$.
    (b) $\log_2(x+1)^2 - 2\log_2(x-3) = 6$. $2\log_2(x+1) - 2\log_2(x-3) = 6$.
        $\log_2(x+1) - \log_2(x-3) = 3$. $\log_2((x+1)/(x-3)) = 3$.
        $(x+1)/(x-3) = 2^3$. $(x+1) = 8(x-3)$. $-7x = -25$. $x = 25/7$.

11. $\ln(2x-3) = -2x+9$. A(3.7483321, 1.5033358). $x = 3.75$. (graph below.)

12. (a) $\log_3 4.7 = \ln 4.7 / \ln 3 = 1.4087$.
    (b) $\log_2 8.3 + \log_2 5.8 = \log_2(8.3 \times 5.8) = \ln(8.3 \times 5.8)/\ln 2$
        $= 5.5892$

13. (a) $P(t) = 20(2^{t/53})$. (b) A(31.003013, 30) in fig below, or use trace/zoom. Population reaches 30 thousand in 31st year after 1995, i.e. in 2026.
    (c) $P = 20(2^{t/53})$. $2^{t/53} = P/20$. $t/53 = \log_2(P/20)$.
    $t = 53\log_2(P/20) = 53\ln(P/20)/\ln 2 = 53\ln(30/20)/\ln 2 = 31.003013$.

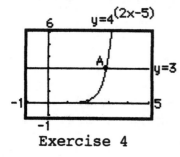
Exercise 4

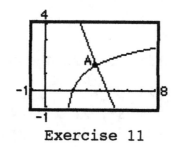
Exercise 11

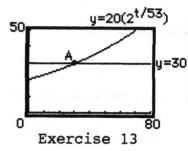
Exercise 13

14. Solve $R = (15.3)(2^{-t/5568})$ for t.
    $R/15.3 = 2^{-t/5568}$, $\log_2(R/15.3) = -t/5568$, $t = -5568\ln(R/15.3)/\ln 2$.
    This equation can be used to find t when given R.
    When $R = 10.26$, $t = -(5568\ln(10.26/15.3))/\ln 2 = 3209.957134$.
    The wall is approximately 3,210 years old.

Cumulative Test Chapters 3,4,5

1. $f(x) = x^2 - 4x + 2$, domain [-3,8]. Parabola, $x = \dfrac{-b}{2a} = \dfrac{4}{2(1)} = 2$.
   $f(2) = -2$. Vertex(2,-2). $f(-3) = 23$, $f(8) = 34$. Range[-2,34]. See graph below.

2. $f(x) = x^2$, $g(x) = (x-3)^2 - 5$. To get the graph of g from f, shift f right 3, down 5. See graphs below.

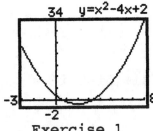

Exercise 1

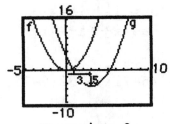

Exercise 2

3. $f(x) = \dfrac{2}{x^2-x-2} = \dfrac{2}{(x+1)(x-2)}$. Vert $x=-1, x=2$. Horiz $y=0$.
   Domain: All reals except $-1, 2$. See graph below.

4. $f(x) = 2x^3+4x^2-6x+3$. Inc$(-\infty, -1.87]$, $[0.54, \infty)$.
   Dec$[-1.87, 0.54]$. Max $A(-1.87, 15.13)$, Min $B(0.54, 1.24)$.
   Zero C $x = -3.12$. See graph below.

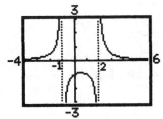

Exercise 3

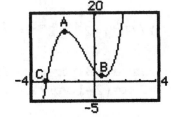

Exercise 4

5. $f(x)=\sqrt{x-4}$, $g(x)=x^2+3x+5$.
   $(g \circ f)(x) = g(f(x)) = g(\sqrt{x-4}) = (\sqrt{x-4})^2 + 3(\sqrt{x-4}) + 5 = x+1+3\sqrt{x-4}$.
   Domain: $x \geq 4$ or $[4, \infty)$.

6. $f(x)=-2x+6$. Graph of $y=-2x+6$ is a non-vertical line. Every
   horizontal line cuts it at one point. Thus f is one-to-one and
   has an inverse. Interchange x & y, $x=-2y+6$. Solve for y,
   $y=\dfrac{x-6}{-2}$. $f^{-1}(x)=\dfrac{6-x}{2}$.

7. $f(x)=x^3+3x^2-4x+1$, $g(x)=x-2$.

   $$\underline{2}|\begin{array}{cccc} 1 & 3 & -4 & 1 \\ & 2 & 10 & 12 \\ \hline 1 & 5 & 6 & 13 \end{array}$$

   $q(x)=x^2+5x+6$, remainder$=13$.

8. $f(x)=x^3+2x^2-8x-5$. Want $f(4)$. Divide $f(x)$ by $g(x)=x-4$.
   Remainder will be $f(3)$.

   $$\underline{4}|\begin{array}{cccc} 1 & 2 & -8 & -5 \\ & 4 & 24 & 64 \\ \hline 1 & 6 & 16 & 59 \end{array}$$

   Remainder$=59$. $f(4)=59$.

9. (a) $\sqrt{-81} = \sqrt{81}i = 9i$.   (b) $-\sqrt{-36} = -\sqrt{36}i = -6i$.

10. (a) i^{14}. 14=3·4+2. Thus $i^{14}=i^2=-1$.
    (b) i^{35}. 35=8·4+3. Thus $i^{35}=i^3=-i$.

11. (a) (3-i) + (4+9i) = 7+8i.
    (b) (2+3i)(5-7i) = 2(5-7i) + 3i(5-7i) = 10-14i+15i-21i^2 = 10+i+21 = 31+i.

12. $\dfrac{2+7i}{1-i} \cdot \dfrac{1+i}{1+i} = \dfrac{2(1+i)+7i(1+i)}{(1)^2+(1)^2} = \dfrac{2+2i+7i-7}{2} = \dfrac{-5+9i}{2}$
    $= -\dfrac{5}{2} + \dfrac{9i}{2}$.

13. $f(x)=2^x$. $g(x)=2^{x-3}+7$. Shift f right 3 and up 7.

14. $3^{(-2x+7)}=4$. A(2.8690702,4). Soln is x=2.87 to 2 dec places. See figure below.

15. $\log_2(1/32)=y$. $1/32=2^y$. $1/(2^5)=2^y$. $2^{-5}=2^y$. y=-5.

16. (a) ln(5x+3)=7. $5x+3=e^7$. $5x=e^7-3$. $x=(e^7-3)/5=218.7266$.
    (b) $y=2^{(2-3x)}$. $2-3x=\log_2 y$. $x=(2-\log_2 y)/3$.

17. $N(t) = 1.8t^3 - 28t^2 + 90t$. (a) A, Max(1.988392,82.402323). Number peaks with 82 cases after 2 years. (b) Zero (4.5383601). If present trend continues should be no more new cases after about four and one half years. See graph below.

18. (a) $P(t)=27(2^{t/34})$. A(12.729448,35) in fig below, or use trace/zoom. Population reaches 35 thousand in 12.7 years.
    (b) $P=27(2^{t/34})$. $2^{t/34}=P/27$. $t/34=\log_2(P/27)$.
    $t=34\log_2(P/27)=34\ln(P/27)/\ln2=34\ln(35/27)/\ln2=12.729448$.

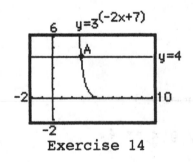

Exercise 14

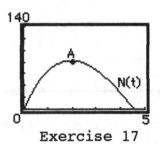

Exercise 17

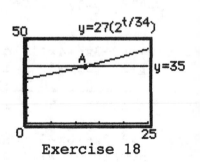

Exercise 18

Chapter 6

1. x+2y=5. (1)   x=5-2y.(3) In (2) -(5-2y)+y=1. 3y=6. <u>y=2</u>.
    -x+ y=1. (2)   Substitute for y in (3), x=5-2(2). <u>x=1</u>.

3. 2x+3y=9 (1)   x=(9-3y)/2. In (2) -(9-3y)+2y=6. 5y=15. <u>y=3</u>.
    -x+2y=6 (2)   Substitute for y, x=(9-3(3))/2. <u>x=0</u>.

5. 3x+2y=1 (1) x=(1-2y)/3. In(2) 4(1-2y)/3+y=2. -5y/3=2/3. <u>y=-2/5</u>.
    4x+ y=2 (2)   Substitute for y, x=(1-2(-2/5))/3. <u>x=3/5</u>.

7. 2x+ y= 8 (1) x=(8-y)/2. In (2) 3(8-y)/2-2y=12. -7y/2=0. <u>y=0</u>.
    3x-2y=12 (2)   Substitute for y, x=(8-0)/2. <u>x=4</u>.

9. x+2y= 6 (1) x=6-2y. In(2) -2(6-2y)+y=-2. 5y=10.<u>y=2</u>.
    -2x+ y=-2 (2)   Substitute for y, x=6-2(2). <u>x=2</u>.

11. x+2y= 5 (1) x=5-2y. In(2) -(5-2y)+2y=-1.4y=4. <u>y=1</u>.
    -x+2y=-1 (2)   Substitute for y, x=5-2(1). <u>x=3</u>.

13. x+2y=3 (1) Subtract (2) from (1). 3y=3. <u>y=1</u>.
    x- y=0 (2) Substitute for y in (2). <u>x=1</u>.

15. x+4y=-1 (1)   Mult (1) by -3. -3x-12y= 3
    3x-2y=11 (2)   <u>3x- 2y=11</u>
                        Add   -14y=14. <u>y=-1</u>.
    Substitute for y into (1). x+4(-1)=-1. <u>x=3</u>.

17. 3x+ y=-4 (1)   Mult (1) by -2. -6x-2y= 8
    6x+2y=-8 (2)   <u>6x+2y=-8</u>
                        Add   0=0.
    Many solutions. (1) gives y=-3x-4. Solutions are (x,-3x-4).

19. 2x+ y= 8 (1)   Mult (1) by 3.   6x+3y=24
    3x-2y=12 (2)   Mult (2) by -2. <u>-6x+4y=-24</u>
                        Add   7y=0. <u>y=0</u>.
    Substitute for y into (1). 2x+0=8. 2x=2. <u>x=4</u>.

21. 4x+2y=16 (1)   4x+2y=16
    2x-3y= 0 (2)   Mult (2) by -2. <u>-4x+6y= 0</u>
                        Add   8y=16. <u>y=2</u>.
    Substitute for y into (1). 4x+2(2)=16. 4x=12. <u>x=3</u>.

23. 3x+6y=3 (1)   Mult (1) by 2.   6x+12y= 6
    2x+4y=2 (2)   Mult (2) by -3. <u>-6x-12y=-6</u>
                        Add   0=0.
    Many solutions. (1) gives y=(1-x)/2. Solutions are (x,(1-x)/2).

25. 3x =12 (1) x=4. Mult (1) by 2. 6x = 24.
    2x-y= 7 (2)   Mult (2) by -3. -6x+3y=-21.
                          Add   3y=3. y=1.

27. 2x+y=4 (1)   2x+y=4
   -2x-y=3 (2) -2x-y=3
            Add   0=7. No solution.

29. x+2y=3   31. x+4y=-2   33. x+3y=-4
     x- y=0   2x-3y=10   2x+6y=-8
   x=1.0000,y=1.0000 x=3.0909, y=-1.2727 One line, many solns
                                              (x,-(x+4)/3)

35. 2.1x+ y=8.7   37. 0.8x-5.2y=1.8
    5.3x-2y = 13.2   4.2x-1.1y=9.3
   x=3.2210,y=1.9358   x=2.2128,y=-.0057

39. x=-1, y=2. e.g. x+2y=3, 2x+y=0.
41. x=5, y=-3. e.g. x-2y=11, 2x+y=7.
43. x=6, y=0. e.g. 3x-2y=18. 2x-y=12.
45. x=3, y=-7. e.g. x+2y=-11. x-y=10.
47. Many solutions (x,x-3). y=x-3. e.g. system x-y=3. 2x-2y=6.
49. x+y=4(1) slope=-1,y-int=4. cx+2y=2(2) slope=-c/2,y-int=1.
    (a) Unique solution if slopes differ. -c/2≠-1. c≠2.
    (b) Slopes same if -c/2=-1. c=2. Distinct lines since y-int
    differ. Thus no solutions if c=2.
    (c) Many solutions only if same line - lines are distinct since
    y-int differ.

51. (2,3),(1,1). y=mx+b. 2m+b=3, m+b=1. m=2. b=-1. y=2x-1.

53. (0,1),(2,-2). y=mx+b. b=1, 2m+b=-2. 2m=-3. m=-3/2. y=-3x/2+1.

55. (-2,-8),(0,-2). y=mx+b. -2m+b=-8, b=-2. m=3. y=3x-2.

57. (1,-1),(-1,-7). y=mx+b. m+b=-1,-m+b=-7. 2m=6. m=3.b=-4.y=3x-4.

59. Let algebra books cost $x each and trig books $y each.
    400x+200y=23800. 4x+2y=238. Mult by -13. -52x-26y=-3094
     52x+ 11y= 2569.   52x+11y= 2569
                                    Add   -15y=-525. y=35.
    Substitute, 4x+2(35)=238. 4x=168. x=42. Algebra $42. Trig $35.

61. Let Roma cost $x each and Seabreeze $y each.
    4x+2y=256.   4x+2y= 256
     x+ y= 96. Mult by -4 -4x-4y=-384
                     Add   -2y=-128. y=64.
    Substitute, x+64=96. x=32. Roma $32. Seabreeze $64.

63. $2.30 made up of quarters and dimes. Total of 14 coins.
Let x quarters, y dimes.
.25x+.10y=2.30 Mult by 4.   x+.40y= 9.20
   x+ y= 14   Mult by -1   x+   y=14.00
                     Subtract   -.60y=-4.8. y=8.
Substitute, x+6=14. x=6. Man has 6 quarters, 8 dimes.

65. Let ages be x & y, with x>y. Sum of ages 30, difference is 2.
        x+y=30
        x-y= 2
Subtract 2y=28. y=14. x=16. Ages are 16 and 14 years old.

67. $5000 in two types of investments. X pays 8%, Y pays 10%.
Total return is $420. Let $x at 8%, $y at 10%.
    x+ y=5000   Mult by 8   8x+ 8y= 40000
.08x+.10y= 420   Mult by -100 -8x-10y=-42000
                         Add   -2y=-2000. y=1000.
Substitute, x+1000=5000. x=4000. $4000 at 8%, $1000 at 10%.

69. Let numbers be x and y.
3x+4y=70 Mult by 2   6x+8y=140
2x+2y=16 Mult by -3 -6x-6y=-48
                    Add   2y= 92. y=46.
Substitute, 2x+92=16. 2x=-76. x=-38.Two numbers are -38 and 46.

71. Perimeter of rectangle = 200. Length 20 greater than width.
2L+2W=200, L+W=100.
 L- W= 20, L-W= 20. Add, 2L=120, L=60
Substitute, 60+W=100. W=40.

73. Nuts at $2.50, raisins at $1.75. 25 lb of mix at $2.20 lb.
Let x lb of nuts, y lb of raisins. 2.5x+1.75y=(25)(2.20).
2.5x+1.75y=55   2.5x+1.75y= 55
   x+ y=25 Mult by -2.5 -2.5x- 2.5y=-62.5
                  Add   -.75y= -7.5. y=10.
Substitute, x-10=25. x=15. 15 lb nuts & 10 lb of raisins.

75. $25 to transport from Kent, $30 from Denton. 50 transported
for $1340. Let x be transported from Kent and y from Sanford.
25x+30y=1340   25x+30y= 1340
   x+ y= 50 Mult by -25 -25x-25y=-1250
                 Add   5y=   90. y=18.
Substitute,x+18=50,x=32. Take 32 from Kent,18 from Sanford.

77. v(t) = v₀ + at. v(3)=26 and v(7)=58.
v₀ + 3a = 29, v₀ + 7a = 61. subtract from 1st equ.-4a=-32. a=8.
Sustitute, v₀ + 7(8) = 61. v₀=5.

79. c=mx+b. 500 units cost $1700, 1000 cost $2,820. Cost of 1200?
 500m+b=1700
1000m+b=2820 Subtract, -500m=-1120, m=2.24.
Substitute, 500(2.24)+b=1700. b=1700-1120=580.
Formula is c=2.24m+580. c(1200)=$3,268.

1. $\begin{bmatrix} 1 & 3 \\ 2 & -5 \end{bmatrix}$, $\begin{bmatrix} 1 & 3 & 7 \\ 2 & -5 & -3 \end{bmatrix}$

3. $\begin{bmatrix} -1 & 3 & -5 \\ 2 & -2 & 4 \\ 1 & 3 & 0 \end{bmatrix}$, $\begin{bmatrix} -1 & 3 & -5 & -3 \\ 2 & -2 & 4 & 8 \\ 1 & 3 & 0 & 6 \end{bmatrix}$

5. $\begin{bmatrix} 1 & 0 & 0 \\ 0 & 1 & 0 \\ 0 & 0 & 1 \end{bmatrix}$, $\begin{bmatrix} 1 & 0 & 0 & 8 \\ 0 & 1 & 0 & 2 \\ 0 & 0 & 1 & -7 \end{bmatrix}$

7. $\begin{aligned} x + 2y &= 3 \\ 4x + 5y &= 6 \end{aligned}$

9. $\begin{aligned} x + 9y &= -3 \\ 5x \quad &= 2 \end{aligned}$

11. $\begin{aligned} 2x - 3y + 6z &= 4 \\ 7x - 5y - 2z &= 3 \\ 2y + 4z &= 0 \end{aligned}$

13. $\begin{aligned} x \quad\quad &= 3 \\ y \quad &= 8 \\ z &= 4 \end{aligned}$

15. $\begin{bmatrix} 2 & 6 & -4 & 0 \\ 1 & 2 & -3 & 6 \\ 8 & 3 & 2 & 5 \end{bmatrix}$ $(1/2)R1$ $\begin{bmatrix} 1 & 3 & -2 & 0 \\ 1 & 2 & -3 & 6 \\ 8 & 3 & 2 & 5 \end{bmatrix}$

17. $\begin{bmatrix} 1 & 2 & 3 & -1 \\ -1 & 1 & 7 & 1 \\ 2 & -4 & 5 & -3 \end{bmatrix}$ Add R1 to R2, Add (-2)R1 to R3 $\begin{bmatrix} 1 & 2 & 3 & -1 \\ 0 & 3 & 10 & 0 \\ 0 & -8 & -1 & -1 \end{bmatrix}$

19. $\begin{bmatrix} 1 & 0 & 4 & -3 \\ 0 & 1 & -3 & 2 \\ 0 & 0 & 1 & 5 \end{bmatrix}$ Add (-4)R3 to R1, Add (3)R3 to R2 $\begin{bmatrix} 1 & 0 & 0 & -23 \\ 0 & 1 & 0 & 17 \\ 0 & 0 & 1 & 5 \end{bmatrix}$

21. To create zeros below the leading 1 in the first column.

23. To get a leading one in the right location for the 2nd row.

25. To create zeros above the leading one in the third column.

27. To create a leading one in the third row.

29. $\begin{bmatrix} 1 & 0 & 1 & 3 \\ 0 & 2 & -2 & -4 \\ 0 & 1 & -2 & 5 \end{bmatrix}$ $(-1/2)R2$ $\begin{bmatrix} 1 & 0 & 1 & 3 \\ 0 & 1 & -1 & -2 \\ 0 & 1 & -2 & 5 \end{bmatrix}$ Add (-1)R2 to R3

$\begin{bmatrix} 1 & 0 & 1 & 3 \\ 0 & 1 & -1 & -2 \\ 0 & 0 & -1 & 7 \end{bmatrix}$ $(-1)R3$ $\begin{bmatrix} 1 & 0 & 1 & 3 \\ 0 & 1 & -1 & -2 \\ 0 & 0 & 1 & -7 \end{bmatrix}$ Add (-1)R3 to R1, Add R3 to R2

$$\begin{bmatrix} 1 & 0 & 0 & 10 \\ 0 & 1 & 0 & -9 \\ 0 & 0 & 1 & -7 \end{bmatrix} \quad \begin{matrix} x = 10 \\ y = -9 \\ z = -7 \end{matrix}$$

31. $\begin{bmatrix} 1 & -1 & 3 & 3 \\ 2 & -1 & 2 & 2 \\ 3 & 1 & -2 & 3 \end{bmatrix}$ Add $(-2)\tilde{R}1$ to R2 $\begin{bmatrix} 1 & -1 & 3 & 3 \\ 0 & 1 & -4 & -4 \\ 0 & 4 & -11 & -6 \end{bmatrix}$ Add $R\tilde{2}$ to R1
Add $(-3)R1$ to R3 Add $(-4)R2$ to R3

$\begin{bmatrix} 1 & 0 & -1 & -1 \\ 0 & 1 & -4 & -4 \\ 0 & 0 & 5 & 10 \end{bmatrix}(1/\tilde{5})R3 \begin{bmatrix} 1 & 0 & -1 & -1 \\ 0 & 1 & -4 & -4 \\ 0 & 0 & 1 & 2 \end{bmatrix}$ Add $R\tilde{3}$ to R1 $\begin{bmatrix} 1 & 0 & 0 & 1 \\ 0 & 1 & 0 & 4 \\ 0 & 0 & 1 & 2 \end{bmatrix} \begin{matrix} x = 1 \\ y = 4 \\ z = 2 \end{matrix}$
Add $(4)R3$ to R2

33. $\begin{bmatrix} 0 & 2 & 4 & 8 \\ 2 & 2 & 0 & 6 \\ 1 & 1 & 1 & 5 \end{bmatrix}$ R1<$\tilde{-}$>R2 $\begin{bmatrix} 2 & 2 & 0 & 6 \\ 0 & 2 & 4 & 8 \\ 1 & 1 & 1 & 5 \end{bmatrix}(1/\tilde{2})R1 \begin{bmatrix} 1 & 1 & 0 & 3 \\ 0 & 2 & 4 & 8 \\ 1 & 1 & 1 & 5 \end{bmatrix}$

Add $(-1)\tilde{R}1$ to R3 $\begin{bmatrix} 1 & 1 & 0 & 3 \\ 0 & 2 & 4 & 8 \\ 0 & 0 & 1 & 2 \end{bmatrix}(1/\tilde{2})R2 \begin{bmatrix} 1 & 1 & 0 & 3 \\ 0 & 1 & 2 & 4 \\ 0 & 0 & 1 & 2 \end{bmatrix}$

Add $(-1)\tilde{R}2$ to R1 $\begin{bmatrix} 1 & 0 & -2 & -1 \\ 0 & 1 & 2 & 4 \\ 0 & 0 & 1 & 2 \end{bmatrix}$ Add $(2)\tilde{R}3$ to R1
Add $(-2)R3$ to R2

$\begin{bmatrix} 1 & 0 & 0 & 3 \\ 0 & 1 & 0 & 0 \\ 0 & 0 & 1 & 2 \end{bmatrix} \begin{matrix} x = 3 \\ y = 0 \\ z = 2 \end{matrix}$

35. $\begin{bmatrix} 1 & 2 & 3 & 14 \\ 2 & 5 & 8 & 36 \\ 1 & -1 & 0 & -4 \end{bmatrix}$ Add $(-2)\tilde{R}1$ to R2 $\begin{bmatrix} 1 & 2 & 3 & 14 \\ 0 & 1 & 2 & 8 \\ 0 & -3 & -3 & -18 \end{bmatrix}$ Add $(-2)\tilde{R}2$ to R1
Add $(-1)R1$ to R3 Add $(3)R2$ to R3

$\begin{bmatrix} 1 & 0 & -1 & -2 \\ 0 & 1 & 2 & 8 \\ 0 & 0 & 3 & 6 \end{bmatrix}(1/\tilde{3})R3 \begin{bmatrix} 1 & 0 & -1 & -2 \\ 0 & 1 & 2 & 8 \\ 0 & 0 & 1 & 2 \end{bmatrix}$ Add $R\tilde{3}$ to R1
Add $(-2)R3$ to R2

$\begin{bmatrix} 1 & 0 & 0 & 0 \\ 0 & 1 & 0 & 4 \\ 0 & 0 & 1 & 2 \end{bmatrix} \begin{matrix} x = 0 \\ y = 4 \\ z = 2 \end{matrix}$

37. $\begin{bmatrix} 2 & 2 & -4 & 14 \\ 3 & 1 & 1 & 8 \\ 2 & -1 & 2 & -1 \end{bmatrix}(-1/\tilde{2})R1 \begin{bmatrix} 1 & 1 & -2 & 7 \\ 3 & 1 & 1 & 8 \\ 2 & -1 & 2 & -1 \end{bmatrix}$ Add $(-3)\tilde{R}1$ to R2
Add $(-2)R1$ to R3

$$\begin{bmatrix} 1 & 1 & -2 & 7 \\ 0 & -2 & 7 & -13 \\ 0 & -3 & 6 & -15 \end{bmatrix} \underset{(-1/2)R2}{\sim} \begin{bmatrix} 1 & 1 & -2 & 7 \\ 0 & 1 & -3.5 & 6.5 \\ 0 & -3 & 6 & -15 \end{bmatrix}$$ Add (-1)R2 to R1
Add (3)R2 to R3

$$\begin{bmatrix} 1 & 0 & 1.5 & 0.5 \\ 0 & 1 & -3.5 & 6.5 \\ 0 & 0 & -4.5 & 4.5 \end{bmatrix} \underset{(-1/4.5)R3}{\sim} \begin{bmatrix} 1 & 0 & 1.5 & 0.5 \\ 0 & 1 & -3.5 & 6.5 \\ 0 & 0 & 1 & -1 \end{bmatrix}$$ Add (-1.5)R3 to R1
Add (3.5)R3 to R2

$$\begin{bmatrix} 1 & 0 & 0 & 2 \\ 0 & 1 & 0 & 3 \\ 0 & 0 & 1 & -1 \end{bmatrix} \quad \begin{matrix} x = 2 \\ y = 3 \\ z = -1 \end{matrix}$$

39. $$\begin{bmatrix} 1.5 & 0 & 3 & 15 \\ -1 & 7 & -9 & -45 \\ 2 & 0 & 5 & 22 \end{bmatrix} \underset{(1/1.5)R1}{\sim} \begin{bmatrix} 1 & 0 & 2 & 10 \\ -1 & 7 & -9 & -45 \\ 2 & 0 & 5 & 22 \end{bmatrix}$$ Add R1 to R2
Add (-2)R1 to R3

$$\begin{bmatrix} 1 & 0 & 2 & 10 \\ 0 & 7 & -7 & -35 \\ 0 & 0 & 1 & 2 \end{bmatrix} \underset{(1/7)R2}{\sim} \begin{bmatrix} 1 & 0 & 2 & 10 \\ 0 & 1 & -1 & -5 \\ 0 & 0 & 1 & 2 \end{bmatrix}$$ Add (-2)R3 to R1
Add R3 to R2

$$\begin{bmatrix} 1 & 0 & 0 & 6 \\ 0 & 1 & 0 & -3 \\ 0 & 0 & 1 & 2 \end{bmatrix} \quad \begin{matrix} x = 6 \\ y = -3 \\ z = 2 \end{matrix}$$

41. Points $(1,4)$, $(2,11)$, $(3,22)$. $y = ax^2 + bx + c$.

$$\begin{matrix} a + b + c = 4 \\ 4a + 2b + c = 11 \\ 9a + 3b + c = 22 \end{matrix}, \begin{bmatrix} 1 & 1 & 1 & 4 \\ 4 & 2 & 1 & 11 \\ 9 & 3 & 1 & 22 \end{bmatrix}$$ Add (-4)R1 to R2
Add (-9)R1 to R3 $$\begin{bmatrix} 1 & 1 & 1 & 4 \\ 0 & -2 & -3 & -5 \\ 0 & -6 & -8 & -14 \end{bmatrix}$$

$$(-1/2)R2 \begin{bmatrix} 1 & 1 & 1 & 4 \\ 0 & 1 & 3/2 & 5/2 \\ 0 & -6 & -8 & -14 \end{bmatrix}$$ Add (-1)R2 to R1
Add (6)R2 to R3 $$\begin{bmatrix} 1 & 0 & -1/2 & 3/2 \\ 0 & 1 & 3/2 & 5/2 \\ 0 & 0 & 1 & 1 \end{bmatrix}$$

Add $(-1/2)$R3 to R1
Add $(-3/2)$R3 to R2 $$\begin{bmatrix} 1 & 0 & 0 & 2 \\ 0 & 1 & 0 & 1 \\ 0 & 0 & 1 & 1 \end{bmatrix} \begin{matrix} a = 2 \\ b = 1 \\ c = 1 \end{matrix}, \quad y = 2x^2 + x + 1.$$

43. Points $(-1,0)$, $(1,6)$, $(2,12)$. $y = ax^2 + bx + c$.

$$\begin{matrix} a - b + c = 0 \\ a + b + c = 6 \\ 4a + 2b + c = 12 \end{matrix}, \begin{bmatrix} 1 & -1 & 1 & 0 \\ 1 & 1 & 1 & 6 \\ 4 & 2 & 1 & 12 \end{bmatrix}$$ Add (-1)R1 to R2
Add (-4)R1 to R3 $$\begin{bmatrix} 1 & -1 & 1 & 0 \\ 0 & 2 & 0 & 6 \\ 0 & 6 & -3 & 12 \end{bmatrix}$$

184

$(1/\tilde{2})R2$
$\begin{bmatrix} 1 & -1 & 1 & 0 \\ 0 & 1 & 0 & 3 \\ 0 & 6 & -3 & 12 \end{bmatrix}$
Add $(1)\tilde{R}2$ to R1
Add $(-6)R2$ to R3
$\begin{bmatrix} 1 & 0 & 1 & 3 \\ 0 & 1 & 0 & 3 \\ 0 & 0 & -3 & -6 \end{bmatrix}$

$(-1/\tilde{3})R3$
$\begin{bmatrix} 1 & 0 & 1 & 3 \\ 0 & 1 & 0 & 3 \\ 0 & 0 & 1 & 2 \end{bmatrix}$
Add $(-1)\tilde{R}3$ to R1
$\begin{bmatrix} 1 & 0 & 0 & 1 \\ 0 & 1 & 0 & 3 \\ 0 & 0 & 1 & 2 \end{bmatrix}$,

$a=1$, $b=3$, $c=2$. $y = x^2 + 3x + 2$.

45. Points $(-1,1)$, $(1,5)$, $(2,19)$. $y = ax^2 + bx + c$.

$\begin{aligned} a - b + c &= 1 \\ a + b + c &= 5 \\ 4a + 2b + c &= 19 \end{aligned}$,
$\begin{bmatrix} 1 & -1 & 1 & 1 \\ 1 & 1 & 1 & 5 \\ 4 & 2 & 1 & 19 \end{bmatrix}$
Add $(-1)\tilde{R}1$ to R2
Add $(-4)R1$ to R3
$\begin{bmatrix} 1 & -1 & 1 & 1 \\ 0 & 2 & 0 & 4 \\ 0 & 6 & -3 & 15 \end{bmatrix}$

$(1/\tilde{2})R2$
$\begin{bmatrix} 1 & -1 & 1 & 1 \\ 0 & 1 & 0 & 2 \\ 0 & 6 & -3 & 15 \end{bmatrix}$
Add $(1)\tilde{R}2$ to R1
Add $(-6)R2$ to R3
$\begin{bmatrix} 1 & 0 & 1 & 3 \\ 0 & 1 & 0 & 2 \\ 0 & 0 & -3 & 3 \end{bmatrix}$

$(-1/\tilde{3})R3$
$\begin{bmatrix} 1 & 0 & 1 & 3 \\ 0 & 1 & 0 & 2 \\ 0 & 0 & 1 & -1 \end{bmatrix}$
Add $(-1)\tilde{R}3$ to R1
$\begin{bmatrix} 1 & 0 & 0 & 4 \\ 0 & 1 & 0 & 2 \\ 0 & 0 & 1 & -1 \end{bmatrix}$,

$a=4$, $b=2$, $c=-1$. $y = 4x^2 + 2x - 1$.

47. Points $(2,8)$, $(4,6)$, $(6,-4)$. $y = ax^2 + bx + c$.

$\begin{aligned} 4a + 2b + c &= 8 \\ 16a + 4b + c &= 6 \\ 36a + 6b + c &= -4 \end{aligned}$,
$\begin{bmatrix} 4 & 2 & 1 & 8 \\ 16 & 4 & 1 & 6 \\ 36 & 6 & 1 & -4 \end{bmatrix}$
$(1/\tilde{4})R1$
$\begin{bmatrix} 1 & 1/2 & 1/4 & 2 \\ 16 & 4 & 1 & 6 \\ 36 & 6 & 1 & -4 \end{bmatrix}$

Add $(-16)\tilde{R}1$ to R2
Add $(-36)R1$ to R3
$\begin{bmatrix} 1 & 1/2 & 1/4 & 2 \\ 0 & -4 & -3 & -26 \\ 0 & -12 & -8 & -76 \end{bmatrix}$
$(-1/\tilde{4})R2$

$\begin{bmatrix} 1 & 1/2 & 1/4 & 2 \\ 0 & 1 & 3/4 & 26/4 \\ 0 & -12 & -8 & -76 \end{bmatrix}$
Add $(-1/2)\tilde{R}2$ to R1
Add $(12)R2$ to R3
$\begin{bmatrix} 1 & 0 & -1/8 & -5/4 \\ 0 & 1 & 3/4 & 26/4 \\ 0 & 0 & 1 & 2 \end{bmatrix}$

Add $(1/8)\tilde{R}3$ to R1
Add $(-3/4)R3$ to R2
$\begin{bmatrix} 1 & 0 & 0 & -1 \\ 0 & 1 & 0 & 5 \\ 0 & 0 & 1 & 2 \end{bmatrix}$
$\begin{aligned} a &= -1 \\ b &= 5 \\ c &= 2 \end{aligned}$, $y = -x^2 + 5x + 2$.

49. Let the student have x one dollar bills, y fives, and z tens. 10 bills, total value $44, 2 more fives than tens.

$$x + y + z = 10 \qquad x = 4 \qquad \text{4 ones}$$
$$x + 5y + 10z = 44 \qquad y = 4 \qquad \text{4 fives}$$
$$y - z = 2 \qquad z = 2 \qquad \text{2 tens}$$

51. Let the person have x one dollar bills, y fives, and z tens.
    13 bills, 4 more ones than fives, 3 more fives than tens.

$$x + y + z = 13 \qquad x = 8 \qquad \text{8 ones}$$
$$x - y = 4 \qquad y = 4 \qquad \text{4 fives} \qquad \text{\$38 total.}$$
$$y - z = 3 \qquad z = 1 \qquad \text{1 tens}$$

53. Let there be x nickels, y dimes, and z quarters.
    780 coins, value \$137, 80 more nickels than dimes.

$$x + y + z = 780 \qquad x = 200 \qquad \text{200 nickels}$$
$$.05x + .10y + .25z = 137 \qquad y = 120 \qquad \text{120 dimes}$$
$$x - y = 80 \qquad z = 460 \qquad \text{460 quarters}$$

55. Let x bonds at 7%, y at 8% and z at 10%. Total inv. = \$8,000.
    Total return = \$730. \$3,000 more at 10% than 8%.

$$x + y + z = 8000 \qquad x = 1000 \qquad \text{\$1000 @ 7\%}$$
$$.07x + .08y + .1z = 730 \qquad y = 2000 \qquad \text{\$2000 @ 8\%}$$
$$-y + z = 3000 \qquad z = 5000 \qquad \text{\$5000 @ 10\%}$$

57. Let x tables, y chairs, z cupboards be manufactured.

$$\text{Metal:} \quad 2x + y + 3z = 150 \qquad x = 20 \qquad \text{20 tables}$$
$$\text{Wood:} \quad 6x + 2y + 5z = 330 \qquad y = 80 \qquad \text{80 chairs}$$
$$\text{Plastic:} \quad x + y + z = 120 \qquad z = 10 \qquad \text{10 cupboards}$$

59. (a) $s(t) = (1/2)at^2 + v_o t + s_o$. a) $s(1)=30, s(2)=56, s(3)=86$.

$$.5a + v_o + s_o = 30 \qquad x = 4 \qquad a = 4$$
$$2a + 2v_o + s_o = 56 \qquad y = 20 \qquad v_o = 20$$
$$4.5a + 3v_o + s_o = 86 \qquad z = 8 \qquad s_o = 8$$

b) $s(2)=49, s(4)=73, s(6)=93$.
$$2a + 2v_o + s_o = 49 \qquad x = -1 \qquad a = -1 \text{ (deceleration)}$$
$$8a + 4v_o + s_o = 73 \qquad y = 15 \qquad v_o = 15$$
$$18a + 6v_o + s_o = 93 \qquad z = 21 \qquad s_o = 21$$

Section 6.3

Group Discussion: No. First system has a unique solution. Thus

reduced form must be of the type $\begin{bmatrix} 1 & 0 & 0 & a \\ 0 & 1 & 0 & b \\ 0 & 0 & 1 & c \end{bmatrix}$. Unique solution

is x=a, y=b, z=c. Let the matrix of coefficients of both
systems be A. During the transformations A is transformed into

$\begin{bmatrix} 1 & 0 & 0 \\ 0 & 1 & 0 \\ 0 & 0 & 1 \end{bmatrix}$. This would also apply for the second system,

implying that it also would have to have a unique solution.

1. In reduced form.

3. Not in reduced form. Nonzero element above leading 1 in row 2.

5. In reduced form.   7. In reduced form.

9. In reduced form.   11. In reduced form.

13. Not in reduced form. Nonzero elements above leading 1 in row 3.

15. In reduced form.

17. Not in reduced form. Leading 1 in row 3 not to right of
    leading 1 in row 2.

19. x=2, y=4, z=-3.   21. x=-3r+6, y=r, z=-2.

23. x=-2r+3, y=5, z=r.

25. $\begin{bmatrix} 1 & 4 & 3 & 1 \\ 2 & 8 & 11 & 7 \\ 1 & 6 & 7 & 3 \end{bmatrix}$ Add (-2)R1 to R2 $\overset{\sim}{}$ $\begin{bmatrix} 1 & 4 & 3 & 1 \\ 0 & 0 & 5 & 5 \\ 0 & 2 & 4 & 2 \end{bmatrix}$ R2 $\overset{\sim}{<->}$ R3 $\begin{bmatrix} 1 & 4 & 3 & 1 \\ 0 & 2 & 4 & 2 \\ 0 & 0 & 5 & 5 \end{bmatrix}$
Add (-1)R1 to R3

$(1/\overset{\sim}{2})$R2 $\begin{bmatrix} 1 & 4 & 3 & 1 \\ 0 & 1 & 2 & 1 \\ 0 & 0 & 5 & 5 \end{bmatrix}$ Add (-4)$\overset{\sim}{}$R2 to R1 $\begin{bmatrix} 1 & 0 & -5 & -3 \\ 0 & 1 & 2 & 1 \\ 0 & 0 & 5 & 5 \end{bmatrix}$

$(1/\overset{\sim}{5})$R3 $\begin{bmatrix} 1 & 0 & -5 & -3 \\ 0 & 1 & 2 & 1 \\ 0 & 0 & 1 & 1 \end{bmatrix}$ Add (5)$\overset{\sim}{}$R3 to R1 $\begin{bmatrix} 1 & 0 & 0 & 2 \\ 0 & 1 & 0 & -1 \\ 0 & 0 & 1 & 1 \end{bmatrix}$ x = 2
Add (-2)R3 to R2   y = -1
                                                   z = 1

27. $\begin{bmatrix} 1 & 1 & 1 & 7 \\ 2 & 3 & 1 & 18 \\ -1 & 1 & -3 & 1 \end{bmatrix}$ Add (-2)$\overset{\sim}{}$R1 to R2 $\begin{bmatrix} 1 & 1 & 1 & 7 \\ 0 & 1 & -1 & 4 \\ 0 & 2 & -2 & 8 \end{bmatrix}$
Add R1 to R3

Add (-1)$\overset{\sim}{}$R2 to R1 $\begin{bmatrix} 1 & 0 & 2 & 3 \\ 0 & 1 & -1 & 4 \\ 0 & 0 & 0 & 0 \end{bmatrix}$ x = -2z+3   x = -2r+3
Add (-2)R2 to R3   y = z+4   y = r+4
                                                            z = r

29. $\begin{bmatrix} 1 & -1 & 1 & 3 \\ 2 & -1 & 4 & 7 \\ 3 & -5 & -1 & 7 \end{bmatrix}$ Add (-2)$\overset{\sim}{}$R1 to R2 $\begin{bmatrix} 1 & -1 & 1 & 3 \\ 0 & 1 & 2 & 1 \\ 0 & -2 & -4 & -2 \end{bmatrix}$
Add (-3)R1 to R3

187

$$\begin{array}{l}\text{Add R2 to R1}\\\text{Add (2)R2 to R3}\end{array}\begin{bmatrix}1&0&3&4\\0&1&2&1\\0&0&0&0\end{bmatrix}\quad\begin{array}{l}x=-3z+4\\y=-2z+1\end{array}\quad\begin{array}{l}x=-3r+4\\y=-2r+1\\z=r\end{array}$$

31. $\begin{bmatrix}3&6&-3&6\\-2&-4&-3&-1\\3&6&-2&10\end{bmatrix}(1/3)R1\begin{bmatrix}1&2&-1&2\\-2&-4&-3&-1\\3&6&-2&10\end{bmatrix}\begin{array}{l}\text{Add (2)R1 to R2}\\\text{Add (-3)R1 to R3}\end{array}$

$\begin{bmatrix}1&2&-1&2\\0&0&-5&3\\0&0&1&4\end{bmatrix}(-1/5)R1\begin{bmatrix}1&2&-1&2\\0&0&1&-.6\\0&0&1&4\end{bmatrix}\begin{array}{l}\text{Add R2 to R1}\\\text{Add (-1)R2 to R3}\end{array}$

$\begin{bmatrix}1&2&0&1.4\\0&0&1&-.6\\0&0&0&4.6\end{bmatrix}(1/4.6)R3\begin{bmatrix}1&2&0&1.4\\0&0&1&-.6\\0&0&0&1\end{bmatrix}\begin{array}{l}\text{Add (-1.4)R3 to R1}\\\text{Add (.6)R3 to R2}\end{array}$

$\begin{bmatrix}1&2&0&0\\0&0&1&0\\0&0&0&1\end{bmatrix}$. No solution.

33. $\begin{bmatrix}1&2&-1&3\\2&4&-2&6\\3&6&2&-1\end{bmatrix}\begin{array}{l}\text{Add (-2)R1 to R2}\\\text{Add (-3)R1 to R3}\end{array}\begin{bmatrix}1&2&-1&3\\0&0&0&0\\0&0&5&-10\end{bmatrix}R2<->R3$

$\begin{bmatrix}1&2&-1&3\\0&0&5&-10\\0&0&0&0\end{bmatrix}\text{Add R2 to R1}\begin{bmatrix}1&2&0&1\\0&0&1&-2\\0&0&0&0\end{bmatrix}.\begin{array}{l}x=-2y+1\\z=-2\end{array}$

$x=-2r=1,\ y=r,\ z=-2.$

35. $\begin{bmatrix}0&1&2&5\\1&2&5&13\\1&0&2&4\end{bmatrix}R1<->R2\begin{bmatrix}1&2&5&13\\0&1&2&5\\1&0&2&4\end{bmatrix}\text{Add (-1)R1 to R3}$

$\begin{bmatrix}1&2&5&13\\0&1&2&5\\0&-2&-3&-9\end{bmatrix}\begin{array}{l}\text{Add (-2)R2 to R1}\\\text{Add (2)R2 to R3}\end{array}\begin{bmatrix}1&0&1&3\\0&1&2&5\\0&0&1&1\end{bmatrix}\begin{array}{l}\text{Add (-3)R3 to R1}\\\text{Add (-5)R3 to R2}\end{array}$

$\begin{bmatrix}1&0&0&2\\0&1&0&3\\0&0&1&1\end{bmatrix}.\begin{array}{l}x=2\\y=3\\z=1\end{array}$

37. $\begin{bmatrix}1&1&-3&10\\-3&-2&4&-24\end{bmatrix}\text{Add (3)R1 to R2}\begin{bmatrix}1&1&-3&-10\\0&1&-5&6\end{bmatrix}$

Add $(-1)\tilde{R}2$ to R1 $\begin{bmatrix} 1 & 0 & 2 & 44 \\ 0 & 1 & -5 & 6 \end{bmatrix}$ $\quad x = -2z+44$ $\quad x = -2r+4$
$\qquad\qquad\qquad\qquad\qquad\qquad\qquad\qquad\quad y = 5z-54$ $\quad y = 5r+6$
$\qquad\qquad\qquad\qquad\qquad\qquad\qquad\qquad\qquad\qquad\qquad\qquad\quad z = r$

39. (a) System with more variables than equations having no solution.

e.g. Let reduced form be $\begin{bmatrix} 1 & 0 & 3 & 1 & 0 \\ 0 & 1 & 2 & -1 & 0 \\ 0 & 0 & 0 & 0 & 1 \end{bmatrix}$ and work backwards.

$\begin{bmatrix} 1 & 0 & 3 & 1 & 0 \\ 0 & 1 & 2 & -1 & 0 \\ 0 & 0 & 0 & 0 & 1 \end{bmatrix}$ $\begin{array}{c} \text{Add } (-2)\tilde{R}3 \text{ to R1} \\ \text{Add R3 to R2} \end{array}$ $\begin{bmatrix} 1 & 0 & 3 & 1 & -2 \\ 0 & 1 & 2 & -1 & 1 \\ 0 & 0 & 0 & 0 & 1 \end{bmatrix}$

$\begin{array}{c} \text{Add } \tilde{R}2 \text{ to R1} \\ \text{Add } (2)R2 \text{ to R3} \end{array}$ $\begin{bmatrix} 1 & 1 & 5 & 0 & -1 \\ 0 & 1 & 2 & -1 & 1 \\ 0 & 2 & 4 & -2 & 3 \end{bmatrix}$ $\begin{array}{c} \text{Add } R\tilde{1} \text{ to R2} \\ \text{Add } (-1)R1 \text{ to R3} \end{array}$

$\begin{bmatrix} 1 & 1 & 5 & 0 & -1 \\ 1 & 2 & 7 & -1 & 0 \\ -1 & 1 & -1 & -2 & 4 \end{bmatrix}$. $\quad \begin{aligned} x + y + 5z &= -1 \\ x + 2y + 7z - w &= 0 \\ -x + y - z - 2w &= 4 \end{aligned}$ has no solution.

(b) System with more equations than variables, with unique solution. Let unique solution be say $x=1$, $y=2$, $z=3$. Construct four equations satisfying these conditions. e.g.
$x+y+z=6$, $x-y+z=2$, $2x+y-z=1$, $x+2y-z=2$.

41. $I_1+I_2-I_3=0$, $2I_1+6I_3=38$, $3I_2+6I_3=33$. $I_1=4, I_2=1, I_3=5$.
43. $I_1+I_2-I_3=0$, $2I_1+2I_3=20$, $3I_2+2I_3=26$. $I_1=3, I_2=4, I_3=7$.
45. $I_1+I_2-I_3=0$, $3I_1+6I_3=24$, $1I_2+6I_3=19$. $I_1=2, I_2=1, I_3=3$.

Section 6.4

1. $x^2+y^2=26$, $x=1$. $1+y^2=26$, $y^2=25$, $y=\pm5$. Solutions $(1,-5),(1,5)$.

3. $x^2+9y^2=97$, $x=4$. $16+9y^2=97$, $9y^2=81$, $y=\pm3$. Solns $(4,-3),(4,3)$.

5. $5x^2+2y^2=28$, $y=3$. $5x^2+18=28$, $x^2=2$, $x=\pm\sqrt{2}$. Solns $(\sqrt{2},3),(-\sqrt{2},3)$.

7. $2x^2+3y^2=5$, $x+y=0$. $x=-y$. $2y^2+3y^2=5$, $y^2=1$, $y=\pm1$. $(-1, 1),(1, -1)$.

9. $-3x^2-4x+y^2=26$, $2x-y=-1$. $y=2x+1$. $-3x^2-4x+(2x+1)^2=26$, $x^2+1=26$, $x^2+1=26$. $x=\pm5$. Solutions $(-5, -9),(5, 11)$.

11. $x^2+12x+2y^2=30$, $x+y=3$. $y=3-x$. $x^2+12x+2(3-x)^2=30$, $3x^2+18=30$, $3x^2=12$, $x^2=4$. $x=\pm2$. Solutions $(-2,5), (2,1)$.

13. $x^2+y^2=5$ (1) Add, $6x^2=6$, $x^2=1$. $x=\pm1$.

$5x^2-y^2=1$ (2) Substitute in (1). $y^2=5-x^2$, $y^2=5-(\pm1)^2$, $y^2=4$, $y=\pm2$. Solns, $(-1,-2),(-1,2),(1,-2),(1,2)$.

15. $2x^2+3y^2=12$ (1) $2x^2+3y^2=12$
    $3x^2-\ y^2=-4$ (2) Mult by 3, $\underline{9x^2-3y^2=-12}$
                    Add $11x^2\ \ \ \ =0$, $x=0$.
    Subst into (2). $y^2=4$, $y=\pm2$. Solns $(0,-2)$, $(0,2)$.

17. $2x^2+\ y^2=33$ (1) Mult by 3, $6x^2+3y^2=\ 99$
    $3x^2+4y^2=52$ (2) Mult by -2, $\underline{-6x^2-8y^2=-104}$
                    Add $-5y^2=-5$. $y=\pm1$.
    Subst into (1). $2x^2+(\pm1)^2=33$, $2x^2=32$. $x=\pm4$. Solns $(\pm4,\pm1)$.

19. $2x^2+3y^2=7$ (1) Mult by 2, $4x^2+6y^2=14$
    $-4x^2+2y^2=2$ (2) $\underline{-4x^2+2y^2=\ 2}$
                    Add $8y^2=16$. $y=\pm\sqrt{2}$.
    Subst into (1). $2x^2+3(\pm\sqrt{2})^2=7$, $2x^2=1$. $x=\pm1/\sqrt{2}$. Solns
    $(\pm1/\sqrt{2},\pm\sqrt{2})$.

21. $2x^2+3y^2=32$ (1) Mult by 3, $6x^2+\ 9y^2=\ 96$
    $3x^2+7y^2=48$ (2) Mult by -2, $\underline{-6x^2-14y^2=-96}$
                    Add $-5y^2=0$. $y=0$.
    Subst into (1). $2x^2=32$. $x=\pm4$. Solns $(\pm4,0)$.

23. $2x^2+\ y^2=27$ (1) $2x^2+\ y^2=\ 27$
    $-x^2+3y^2=74$ (2) Mult by 2, $\underline{-2x^2+6y^2=148}$
                    Add $7y^2=175$. $y^2=25$. $y=\pm5$.
    Subst into (1). $2x^2+(\pm5)^2=27$. $2x^2+25=27$. $x=\pm1$. Solns $(\pm1,\pm5)$.

25. $x^2-y^2=7$, $2x-y=5$. $y=2x-5$. $x^2-(2x-5)^2=7$, $-3x^2+20x-25=7$, $3x^2-20x+32=0$
    $(3x-8)(x-4)=0$. $x=8/3,4$. When $x=8/3$, $y=2(8/3)-5=1/3$.
    When $x=4$, $y=2(4)-5=3$. Solns $(8/3,1/3)$, $(4,3)$.

27. $2x^2+4x-y^2=0$, $-x+y=2$. $y=x+2$. $2x^2+4x-(x+2)^2=0$, $x^2-4=0$. $x=\pm2$
    When $x=2,y=4$. When $x=-2,y=0$. Solns $(2,4),(-2,0)$.

29. $4x^2+2y^2=22$ (1) $4x^2+2y^2=22$
    $2x^2-\ y^2=-7$ (2) Mult by -2, $\underline{-4x^2+2y^2=14}$
                    Add $4y^2=36$. $y^2=9$. $y=\pm3$.
    Subst into (2). $2x^2-(\pm3)^2=-7$. $2x^2-9=-7$. $x=\pm1$. Solns $(\pm1,\pm3)$.

31. $3xy-2x=-2$, $x-3y=1$. $x=3y+1$. $3(3y+1)y-2(3y+1)=-2$.
    $9y^2-3y=0$. $3y(3y-1)=0$. $y=0,1/3$. Solns $(1,0)$, $(2,1/3)$.

33. $xy/4-x=3$, $4x-y=-12$. $y=4x+12$. $x(4x+12)/4-x=3$.
    $x(4x+12)-4x=12$. $4x^2+8x-12=0$. $x^2+2x-3=0$. $(x+3)(x-1)=0$.
    $x=1,-3$. Solns $(1,16)$, $(-3,0)$.

35. $y^2+3xy=18$, $y^2=9$. $y=\pm3$. When $y=3$, $9+9x=18$, $x=1$.
When $y=-3$, $9-9x=18$, $x=-1$. Solns $(1,3),(-1,-3)$.

37. $x-y=-1$, $xy=2$. $y=x+1$. $x(x+1)=2$. $x^2+x-2=0$. $(x+2)(x-1)=0$.
$x=-2,1$. When $x=-2,y=-1$. When $x=1,y=2$. Solns $(-2,-1),(1,2)$.

39. $x^2-2xy+y^2=1$, $x^2-3xy+y^2=-1$. Subtract 2nd from 1st. $xy=2$. $y=2/x$
if $x\neq0$. Subst for y. $x^2-2x(2/x)+(2/x)^2=1$, $x^2-4+4/x^2=1$.
$x^4-4x^2+4=x^2$. $x^4-5x^2+4=0$. Let $z=x^2$. $z^2-5z+4=0$. $(z-4)(z-1)=0$.
$z=1,4$. $x=\pm1,\pm2$. $y=2/x$. Solns $(1,2),(-1,-2),(2,1),(-2,-1)$.
Special case, $x=0$: $y^2=1$, $y^2=-1$, not a solution.

41. $x^2+3xy+(4/9)y^2=4$, $xy=3$. $y=3/x, x\neq0$.
Substit, $x^2+3x(3/x)+(4/9)(3/x)^2=4$, $x^2+9+4/x^2=4$, $x^4+9x^2+4=4x^2$.
$x^4+5x^2+4=0$. Let $z=x^2$. $z^2+5z+4=0$. $(z+1)(z+4)=0$. $z=-1,-4$.
$x^2=-1,-4$. No solutions.

43. $2/x-3/y = -1$, $5/x+2/y = 26$. $2y-3x=-xy$ (1), $5y+2x=26xy$ (2)
Mult (1) by 5, (2) by -2. $10y-15x=-5xy$, $-10y-4x=-52xy$.
Add, $-19x=-57xy$, $x=3xy$, $x-3xy=0$, $x(1-3y)=0$. $x\neq0$, thus $y=1/3$.
Subst, $2/x - 3/(1/3) = -1$, $2/x=8$, $x=1/4$. Soln $(1/4, 1/3)$.

45. $y=x^2-3$, $y=2x+3$.
$(-1.6458,-0.2915)$,
$(3.6458,10.2915)$.

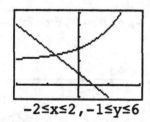

$-5\leq x\leq5, -5\leq y\leq15$

47. $y=-3x^2+x+7$,
$y=x+3$.
$(-1.1547,1.8453)$,
$(1.1547,4.1547)$.

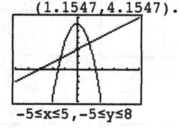

$-5\leq x\leq5, -5\leq y\leq8$

49. $y=-x^2-x+1$,
$y=x^2-6x+8$.
No solution.

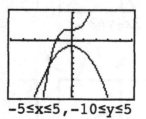

$-5\leq x\leq5, -10\leq y\leq10$

51. $y=-2x+1$,
$y=e^x+2$.
$(-0.7388,2.4777)$.

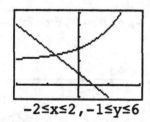

$-2\leq x\leq2, -1\leq y\leq6$

53. $y=2e^{x-2}-3$,
$y=\ln(2x-3)$.
$(1.5944,-1.6668)$,
$(2.6506,0.8334)$.

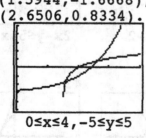

$0\leq x\leq4, -5\leq y\leq5$

55. $y=x^3+2$,
$y=-x^2-1$.
$(-1.8637,-4.4734)$

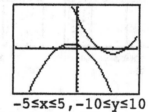

$-5\leq x\leq5, -10\leq y\leq5$

Section 6.5

1. $(1,2)$, $2x+y\leq6$. $2(1)+2=4\leq6$. On graph.

3. (-1,3), 2x+5y≤17. 2(-1)+5(3)=13≤17. On graph.

5. (7,1), 3x+y≤22. 3(7)+1=22≤22. On graph.

7. (4,1),2x+y<12. 2(4)+1=9<12. On graph.

9. (2,4), 2x+3y≤10. 2(2)+3(4)=16≠10. Not on graph.

11. (2,4), 2x+y≥6, 2(2)+4=8≥6. On graph.

13. (-2,3), 4x-6y≤6. 4(-2)-6(3)=-26≤6. On graph.

In the windows of exercises 15 - 63 Xscl=1, Yscl=1.

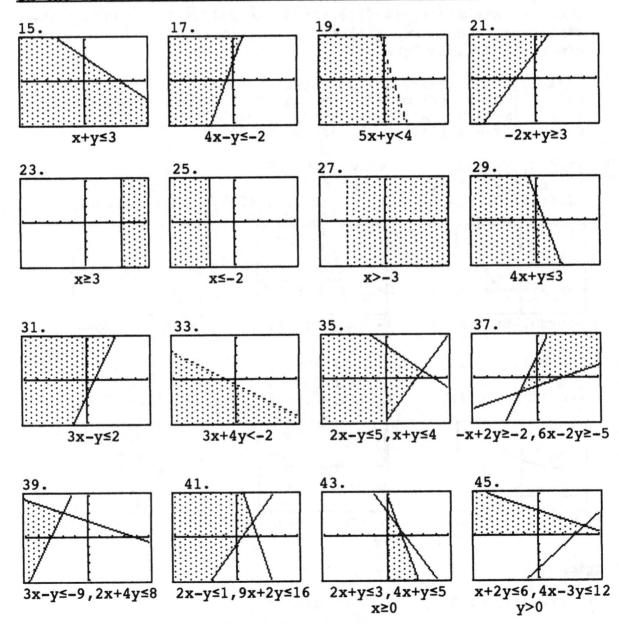

15.

x+y≤3

17.

4x-y≤-2

19.

5x+y<4

21.

-2x+y≥3

23.

x≥3

25.

x≤-2

27.

x>-3

29.

4x+y≤3

31.

3x-y≤2

33.

3x+4y<-2

35.

2x-y≤5,x+y≤4

37.

-x+2y≥-2,6x-2y≥-5

39.

3x-y≤-9,2x+4y≤8

41.

2x-y≤1,9x+2y≤16

43.

2x+y≤3,4x+y≤5
x≥0

45.

x+2y≤6,4x-3y≤12
y>0

47.

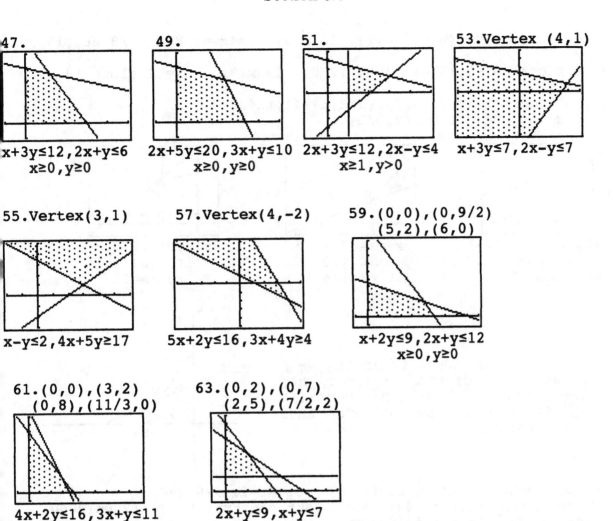

x+3y≤12,2x+y≤6
x≥0,y≥0

49.

2x+5y≤20,3x+y≤10
x≥0,y≥0

51.

2x+3y≤12,2x-y≤4
x≥1,y>0

53.Vertex (4,1)

x+3y≤7,2x-y≤7

55.Vertex(3,1)

x-y≤2,4x+5y≥17

57.Vertex(4,-2)

5x+2y≤16,3x+4y≥4

59.(0,0),(0,9/2)
(5,2),(6,0)

x+2y≤9,2x+y≤12
x≥0,y≥0

61.(0,0),(3,2)
(0,8),(11/3,0)

4x+2y≤16,3x+y≤11
x≥0,y≥0

63.(0,2),(0,7)
(2,5),(7/2,2)

2x+y≤9,x+y≤7
x≥0,y≥2

Section 6.6

1.Vertices:(0,0),(0,11),(4,3),(5,0). f=5x+2y. Max=26 at (4,3).

3.Vertices:(0,0),(0,4),(1,2),(4/3,0). f=4x+y. Max=6 at (1,2).

5.Vertices:(0,0),(0,5),(3,4),(5,0). f=4x+3y. Max=24 at (3,4).

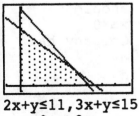

2x+y≤11,3x+y≤15
x≥0,y≥0
Exercise 1

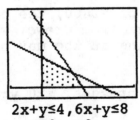

2x+y≤4,6x+y≤8
x≥0,y≥0
Exercise 3

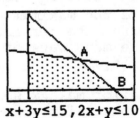

x+3y≤15,2x+y≤10
x≥0,y≥0
Exercise 5

7.Vertices:(0,0),(0,4),(4,2),(16/3,0). f=4x-y. Max=64/3 at (16/3,0)

9.Vertices:(0,0),(0,4),(1,3),(2,0). f=3x+3y. Max=12 along AB.

11.Vertices:(2,1),(2,9/2),(5/2,17/4),(14/3,0).
  f=2x+6y. Max=31 at (2,9/2).

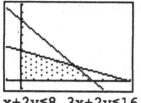

x+2y≤8,3x+2y≤16
x≥0,y≥0
Exercise 7

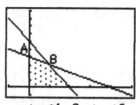

x+y≤4,3x+y≤6
x≥0,y≥0
Exercise 9

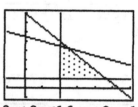

3x+2y≤16,x+2y<11
x≥2,y≥1
Exercise 11

For exercises 13-17 following we
refer to the drawing on the right.
The points A,B,C will be identified,
and the function evaluated at these
three points.

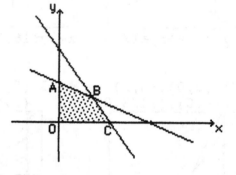

13. Let company make x of KC-1, y of KC-2. x≥0,y≥0.
    Time constraint: x+ 4y≤ 1600. Cost constraint: 30x+20y≤18000,
    3x+2y≤1800. Profit f=10x+8y.
    Vertices:A(0,400),B(400,300),C(600,0).
    f_A=3200,f_B=6400,f_C=6000. Max f=6400 at B(400,300).
    To ensure the maximum profit of $6400, the company should
    manufacture 400 of model KC-1 and 300 of model KC-2.

15. Let company take x refrigerators from Chester, y from Crewe.
    x≥0,y≥0. Time constraint: 20x+10y≤1200=> 2x+y≤120.
    Cost constraint: 60x+10y≤2400=> 6x+y≤240. Profit f=40x+20y.
    Vertices:A(0,120),B(30,60),C(40,0).
    f_A=2400,f_B=2400,f_C=1600. Max f=2400 along AC.
    To ensure the maximum profit of $2400, the company should
    ship x from Chester and 120-2x from Crewe, where 0≤x≤30.
    There are many ways of arriving at the max profit. Company has
    flexibility.

17. Let tailor make x suits, y dresses. x≥0,y≥0.
    Cotton constraint: 2x+ y≤ 80.
    Wool constraint: x+3y≤120. Profit f=90x+90y.
    Vertices:A(0,40),B(24,32),C(40,0).
    f_A=3600,f_B=5040,f_C=3600. Max f=4900 at B(24,32).
    To ensure the maximum income of $5040, the tailor should
    make 24 suits and 32 dresses.

19. Let district buy x Toros and y Sprites. x≥0,y≥17.
    Cost constraint: 18000x+ 22000y≤ 572000=>9x+11y≤286.
    Driver constraint: x+y≤30. Number of students is f=25x+30y.
    Vertices:A(0,17),B(0,26),C(11,17). (see fig below).
    f_A=510,f_B=780,f_C=785. Max f=785 at C(11,17).
    To maximize number of students the school district should purchase
    11 Torro and 17 Sprite buses. Would then carry 785 students.

21. Let hospital serve x of item M and y of item N. x≥0,y≥0.
    Vitamin A constraint: x+y≥70. Vitamin B constraint: 2x+y≥10.
    Cost f=8x+12y. Vertices:A(0,10),B(3,4),C(7,0). See fig below.
    (fig below). f_A=120,f_B=72,f_C=56. Min f=56 at C(7,0).
    The hospital can achieve minimum cost of 56 cents by serving 7 oz.
    of item M and none of item N.

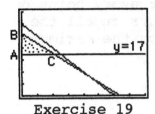

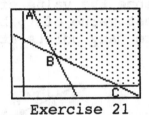

Exercise 19   Exercise 21

23. (a) Let shipper carry x packages for Pringle, y for Williams.
    Weight constraint: 5x+6y≤12000. Volume constraint: 5x+3y≤9000.
    x≥0, y≥0. Profit f=3x+4y.
    Vertices:A(0,2000),B(1200,100),C(1800,0). Fig below. f_A=8000,
    f_B=5400,f_C=7600. Max f=8000 at A(0,2000). Max profit is $8000 if
    shipper carries nothing for Pringle and 2000 for Williams.

    (b) The condition that the shiper must carry at least 240
    packages for Pringle gives constraint x≥240. Thus the point A
    (where the maximum value occurred) is no longer in the feasible
    region. The maximum value of f now occurs at the point
    A'(240,1800) and is now $7920. The clause in the fine print cost
    the shipper $80.

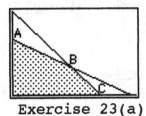

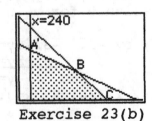

Exercise 23(a)   Exercise 23(b)

Chapter 6 Project

Let the objective function be f=ax+by. f has a value at each
point in the feasible region ABCO below. Let P(xp,yp) be an
arbitrary point in ABCO. Let f have the value fp at P. Thus,
fp=axp+byp. This implies that P lies on the line axp+byp=fp.

The y-intercept of this line is fp/b (see Figures below). In the Figures we see that the value of fp/b is largest when P is actually at B. Thus the value of fp is largest when P is at the vertex B.

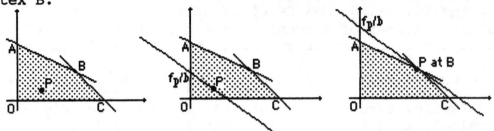

If the line through P is parallel to a side of the feasible region, such as AB below, there will be many optimal solutions. The value of fp/b, and hence the value of f at every point on AB is the same. The largest value of fp/b occurs at all the points along AB. There are many points at which the maximum occurs.

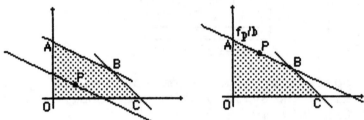

Chapter 6 Review Exercises

1.(a) x-2y= 3. (1)   x=3+2y.(3) In (2) 3(3+2y)+5y=20. 11y=11. y=1.
     3x+5y=20. (2)   Substitute for y in (3), x=3+2(1). x=5.

(b) 3x-2y=12.(1) x=4+2y/3.(3) In (2) 2(4+2y/3)+y=1. 7y/3=-7. y=-3.
     2x+ y= 1.(2)   Substitute for y in (3), x=4+2(-3)/3. x=2.

(c) -x+3y= 1.(1) x=3y-1.(3) In (2) 4(3y-1)-5y=10. 7y=14. y=2.
     4x-5y=10.(2)   Substitute for y in (3), x=3(2)-1. x=5.

2.(a) x+2y= 8 (1)   Mult (1) by -3. -3x-6y=-24
       3x+5y=20 (2)   3x+5y= 20
                                Add   -y= -4. y=4.
     Substitute for y into (1). x+2(4)=8. x=0.

(b) 3x-4y=-13 (1)   Mult (1) by 2.   6x-8y=-26
     2x+ y= -5 (2)   Mult (2) by -3. -6x-3y= 15
                                Add   -11y=-11. y=1.
     Substitute for y into (1). 3x-4(1)=-13. 3x=-9. x=-3.

(c) 5x+ y= 14 (1)   Mult (1) by 3.   15x+ 3y=42
     3x-4y=-10 (2)   Mult (2) by -5. -15x+20y=50
                                Add   23y=92. y=4.
     Substitute for y into (1). 5x+4=14. 5x=10. x=2.

3. x=2, y=-5. e.g. x+y=-3, x-y=7.

4. Many solutions (x,3x-7). y=3x-7. e.g. system -3x+y=-7,6x-2y=14.

5.(a)(1,3),(3,-1). y=mx+b. m+b=3,3m+b=-1. -2m=4,m=-2. b=5. y=-2x+5.

  (b)(4,2),(3,0). y=mx+b. 4m+b=2,3m+b=0. m=2. b=-6. y=2x-6.

  (c)(-3,6),(2,-5). y=mx+b. -3m+b=6,2m+b=-5. -5m=11, m=-11/5.
       b=6+3m=6-33/5=-3/5. y=-11x/5 - 3/5.

6. x quarters, y dimes. Total 8 coins.
     .25x+.10y=1.55   (1)   .25x+.10y= 1.55
       x+   y=8   (2)   Mult by -.25   -.25x-.25y=-2
                             Add   -.15y=-0.45. y=3.
   Substitite, x+3=8, x=5. Has 5 quarters, 3 dimes.

7. 400 math, 300 history for $25300. 20 math,25 history for $1615.
   Let math books cost $x each and history books $y each.
   400x+300y=25300.   400x+300y= 25300
   20x+ 25y= 1615. Mult by -20   -400x-500y=-32300
                             Add   -200y=-7000. y=35.
   Substitute, 20x+25(35)=1615. 20x=740. x=37. Math $37. Hist $35.

8. (a) $\begin{bmatrix} 1 & 1 & 3 & 4 \\ 1 & 2 & 4 & 7 \\ 2 & 1 & 4 & 8 \end{bmatrix}$ Add (-1)R1 to R2 $\begin{bmatrix} 1 & 1 & 3 & 4 \\ 0 & 1 & 1 & 3 \\ 0 & -1 & -2 & 0 \end{bmatrix}$
   Add (-2)R1 to R3

   Add (-1)R2 to R1 $\begin{bmatrix} 1 & 0 & 2 & 1 \\ 0 & 1 & 1 & 3 \\ 0 & 0 & -1 & 3 \end{bmatrix}$ (-1)R3 $\begin{bmatrix} 1 & 0 & 2 & 1 \\ 0 & 1 & 1 & 3 \\ 0 & 0 & 1 & -3 \end{bmatrix}$
   Add R2 to R3

   Add (-2)R3 to R1 $\begin{bmatrix} 1 & 0 & 0 & 7 \\ 0 & 1 & 0 & 6 \\ 0 & 0 & 1 & -3 \end{bmatrix}$ x = 7
   Add (-1)R3 to R2   y = 6
                                                            z =-3

(b) $\begin{bmatrix} 2 & 4 & 14 & -4 \\ 2 & 5 & 17 & -5 \\ 1 & -1 & 0 & -1 \end{bmatrix}$ (-1/2)R1 $\begin{bmatrix} 1 & 2 & 7 & -2 \\ 2 & 5 & 17 & -5 \\ 1 & -1 & 0 & -1 \end{bmatrix}$ Add (-2)R1 to R2
                                                                                 Add (-1)R1 to R3

   $\begin{bmatrix} 1 & 2 & 7 & -2 \\ 0 & 1 & 3 & -1 \\ 0 & -3 & -7 & 1 \end{bmatrix}$ Add (-2)R2 to R1 $\begin{bmatrix} 1 & 0 & 1 & 0 \\ 0 & 1 & 3 & -1 \\ 0 & 0 & 1 & 1 \end{bmatrix}$
                                          Add (3)R2 to R3

   Add (-1)R3 to R1 $\begin{bmatrix} 1 & 0 & 0 & 1 \\ 0 & 1 & 0 & 2 \\ 0 & 0 & 1 & -1 \end{bmatrix}$. x = 1, y = 2, z = -1.
   Add (-3)R3 to R2

(c) $\begin{bmatrix} 1 & 3 & -1 & 4 \\ 2 & 5 & 4 & 3 \\ 3 & 8 & 3 & 5 \end{bmatrix}$ Add $(-2)\tilde{R}1$ to R2 $\begin{bmatrix} 1 & 3 & -1 & 4 \\ 0 & -1 & 6 & -5 \\ 0 & -1 & 6 & -7 \end{bmatrix}$ $(-\overset{\sim}{1})R2$
Add $(-3)R1$ to R3

$\begin{bmatrix} 1 & 3 & -1 & 4 \\ 0 & 1 & -6 & 5 \\ 0 & -1 & 6 & -7 \end{bmatrix}$ Add $(-3)\tilde{R}2$ to R1 $\begin{bmatrix} 1 & 0 & 17 & -11 \\ 0 & 1 & -6 & 5 \\ 0 & 0 & 0 & -2 \end{bmatrix}$ $(1/\overset{\sim}{-2})R3$
Add R2 to R3

$\begin{bmatrix} 1 & 0 & 17 & -11 \\ 0 & 1 & -6 & 5 \\ 0 & 0 & 0 & 1 \end{bmatrix}$ No solution. Note this is really known before

the multiplication by $(1/-2)$ since $0x+0y+0z=-2$ cannot be
satisfied for any values of x,y,z.

9. (a) Points $(1,0)$, $(2,7)$, $(-1,-2)$. $y = ax^2 + bx + c$.

$\begin{aligned} a + b + c &= 0 \\ 4a + 2b + c &= 7 \\ a - b + c &= -2 \end{aligned}$ $\begin{aligned} a &= 2 \\ b &= 1 \\ c &= -3 \end{aligned}$ $y = 2x^2 + x - 3.$

(b) Points $(-2,5)$, $(0,1)$, $(3,40)$. $y = ax^2 + bx + c$.

$\begin{aligned} 4a - 2b + c &= 5 \\ c &= 1 \\ 9a + 3b + c &= 40 \end{aligned}$ $\begin{aligned} a &= 3 \\ b &= 4 \\ c &= 1 \end{aligned}$ $y = 3x^2 + 4x + 1.$

10. (a) $\begin{bmatrix} 1 & 1 & 3 & 4 \\ 1 & 2 & 4 & 7 \\ 2 & 1 & 4 & 8 \end{bmatrix}$ Add $(-1)\tilde{R}1$ to R2 $\begin{bmatrix} 1 & 1 & 3 & 4 \\ 0 & 1 & 1 & 3 \\ 0 & -1 & -2 & 0 \end{bmatrix}$
Add $(-2)R1$ to R3

Add $(-1)\tilde{R}2$ to R1 $\begin{bmatrix} 1 & 0 & 2 & 1 \\ 0 & 1 & 1 & 3 \\ 0 & 0 & -1 & 3 \end{bmatrix}$ $(-\overset{\sim}{1})R3$ $\begin{bmatrix} 1 & 0 & 2 & 1 \\ 0 & 1 & 1 & 3 \\ 0 & 0 & 1 & -3 \end{bmatrix}$
Add R2 to R3

Add $(-2)\tilde{R}3$ to R1 $\begin{bmatrix} 1 & 0 & 0 & 7 \\ 0 & 1 & 0 & 6 \\ 0 & 0 & 1 & -3 \end{bmatrix}$ $\begin{aligned} x &= 7 \\ y &= 6 \\ z &= -3 \end{aligned}$
Add $(-1)R3$ to R2

(b) $\begin{bmatrix} 2 & 4 & 14 & -4 \\ 2 & 5 & 17 & -5 \\ 1 & -1 & 0 & -1 \end{bmatrix}$ $(1/\overset{\sim}{2})R1$ $\begin{bmatrix} 1 & 2 & 7 & -2 \\ 2 & 5 & 17 & -5 \\ 1 & -1 & 0 & -1 \end{bmatrix}$ Add $(-2)\tilde{R}1$ to R2
Add $(-1)R1$ to R3

$\begin{bmatrix} 1 & 2 & 7 & -2 \\ 0 & 1 & 3 & -1 \\ 0 & -3 & -7 & 1 \end{bmatrix}$ Add $(-2)\tilde{R}2$ to R1 $\begin{bmatrix} 1 & 0 & 1 & 0 \\ 0 & 1 & 3 & -1 \\ 0 & 0 & 1 & -1 \end{bmatrix}$ Add $(-1)\tilde{R}3$ to R1
Add $(3)R2$ to R3 Add $(-3)R3$ to R2

$\begin{bmatrix} 1 & 0 & 0 & 1 \\ 0 & 1 & 0 & 2 \\ 0 & 0 & 1 & -1 \end{bmatrix}$. $x = 1$, $y = 2$, $z = -1$.

c) $\begin{bmatrix} 1 & -1 & 1 & -4 \\ -1 & 2 & 1 & 7 \\ 1 & 1 & 5 & 2 \end{bmatrix}$ $\overset{\sim}{\underset{\text{Add }(-1)R1 \text{ to R3}}{\text{Add R1 to R2}}}$ $\begin{bmatrix} 1 & -1 & 1 & -4 \\ 0 & 1 & 2 & 3 \\ 0 & 2 & 4 & 6 \end{bmatrix}$

$\underset{\text{Add }(-2)R2 \text{ to R3}}{\text{Add R2 to R1}}\overset{\sim}{}$ $\begin{bmatrix} 1 & 0 & 3 & -1 \\ 0 & 1 & 2 & 3 \\ 0 & 0 & 0 & 0 \end{bmatrix}$ $\begin{array}{l} x = -3z-1 \\ y = -2z+3 \end{array}$ $\begin{array}{l} x = -3r-1 \\ y = -2r+3 \\ z = r \end{array}$

d) $\begin{bmatrix} 1 & -2 & -4 & 7 \\ 1 & -1 & -1 & 5 \\ 3 & -4 & -6 & 18 \end{bmatrix}$ $\overset{\sim}{\underset{\text{Add }(-3)R1 \text{ to R3}}{\text{Add }(-1)R1 \text{ to R2}}}$ $\begin{bmatrix} 1 & -2 & -4 & 7 \\ 0 & 1 & 3 & -2 \\ 0 & 2 & 6 & -3 \end{bmatrix}$

$\underset{\text{Add }(-2)R2 \text{ to R3}}{\text{Add }(2)R2 \text{ to R1}}\overset{\sim}{}$ $\begin{bmatrix} 1 & 0 & 2 & 3 \\ 0 & 1 & 3 & 2 \\ 0 & 0 & 0 & -1 \end{bmatrix}$ $(-1)R3\overset{\sim}{}$ $\begin{bmatrix} 1 & 0 & 2 & 3 \\ 0 & 1 & 3 & 2 \\ 0 & 0 & 0 & 1 \end{bmatrix}$

$\underset{\text{Add }(-2)R2 \text{ to R3}}{\text{Add }(-3)R3 \text{ to R1}}\overset{\sim}{}$ $\begin{bmatrix} 1 & 0 & 2 & 0 \\ 0 & 1 & 3 & 0 \\ 0 & 0 & 0 & 1 \end{bmatrix}$ No solution

e) $\begin{bmatrix} 1 & -1 & -3 & 2 \\ 2 & -1 & -2 & 5 \\ 1 & 1 & 5 & 4 \end{bmatrix}$ $\overset{\sim}{\underset{\text{Add }(-1)R1 \text{ to R3}}{\text{Add }(-2)R1 \text{ to R2}}}$ $\begin{bmatrix} 1 & -1 & -3 & 2 \\ 0 & 1 & 4 & 1 \\ 0 & 2 & 8 & 2 \end{bmatrix}$ $\underset{\text{Add }(-2)R2 \text{ to R3}}{\text{Add R2 to R1}}\overset{\sim}{}$

$\begin{bmatrix} 1 & 0 & 1 & 3 \\ 0 & 1 & 4 & 1 \\ 0 & 0 & 0 & 0 \end{bmatrix}$ $\begin{array}{l} x = -z+3 \\ y = -4z+1 \end{array}$ $\begin{array}{l} x = -r+3 \\ y = -4r+1 \\ z = r \end{array}$

(f) $\begin{bmatrix} 1 & 1 & 6 & 4 \\ -1 & 1 & 2 & -2 \\ 3 & 4 & 22 & 14 \end{bmatrix}$ $\overset{\sim}{\underset{\text{Add }(-3)R1 \text{ to R3}}{\text{Add R1 to R2}}}$ $\begin{bmatrix} 1 & 1 & 6 & 4 \\ 0 & 2 & 8 & 2 \\ 0 & 1 & 4 & 2 \end{bmatrix}$ $(1/2)R3\overset{\sim}{}$ $\begin{bmatrix} 1 & 1 & 6 & 4 \\ 0 & 1 & 4 & 1 \\ 0 & 1 & 4 & 2 \end{bmatrix}$

$\underset{\text{Add }(-1)R2 \text{ to R3}}{\text{Add }(-1)R2 \text{ to R1}}\overset{\sim}{}$ $\begin{bmatrix} 1 & 0 & 2 & 3 \\ 0 & 1 & 4 & 1 \\ 0 & 0 & 0 & 1 \end{bmatrix}$ $\underset{\text{Add }(-1)R2 \text{ to R3}}{\text{Add }(-3)R3 \text{ to R1}}\overset{\sim}{}$ $\begin{bmatrix} 1 & 0 & 2 & 0 \\ 0 & 1 & 4 & 0 \\ 0 & 0 & 0 & 1 \end{bmatrix}$

No solution.

11. (a) $x^2-y^2=15$, $x+y=5$. $y=5-x$. $x^2-(5-x)^2=15$, $10x-40=0$. $x=4$. $y=5-x=1$. Solution $(4,1)$.

(b) $2x^2+y^2=9$, $2x-y=3$. $y=2x-3$. $2x^2+(2x-3)^2=9$, $6x^2-12x=0$. $x(x-2)$. $x=0,2$. When $x=0,y=-3$. When $x=2,y=1$. Solns $(0,-3),(2,1)$.

(c) $2x^2+3y^2=35$ (1), $\qquad$ $2x^2+3y^2=35$
$\qquad$ $x^2+y^2=13$ (2). Mult by -2, $-2x^2-2y^2=-26$
$\qquad\qquad\qquad$ Add $\qquad$ $y^2=9$, $y=\pm3$.
$\qquad$ Subst in (2), $x^2+9=13$, $x^2=4$, $x=\pm2$. Solutions $(-2,-3),(-2,3)$,
$\qquad$ $(2,-3),(2,3)$.

(d) $2x^2+y^2=8$, $x+y=4$. $y=4-x$. $2x^2+(4-x)^2=8$, $3x^2-8x+16=8$.
$\qquad$ $3x^2-8x+8=0$. $b^2-4ac=(-8)^2-4(3)(8)=-32<0$, No solution.

(e) $\qquad$ $y^2+2xy=33$, $y^2=9$. $y=\pm3$.
$\qquad$ When $y=3$, $(3)^2+2x(3)=7$, $9+6x=33$, $x=4$.
$\qquad$ When $y=-3$, $(-3)^2+2x(-3)=7$, $9-6x=33$, $x=-4$. solns $(4,3),(-4,-3)$.

(f) $1/x+5/y=3/2$, $2/x-1/y=4/5$. $2y+10x=3xy$ (1), $10y-5x=4xy$ (2).
$\qquad$ Mult (2) by 2, $20y-10x=8xy$ (3). Add (1) and (3). $22y=11xy$.
$\qquad$ $11y(2-x)=0$. Since $y\neq0$, $x=2$. Subst into (1). $2y+20=6y$, $y=5$.
$\qquad$ Solution $(2,5)$.

12. (a) $y=x^2+1$
$\qquad$ $y=x+2$
$\qquad$ $(-.62,1.38)$,
$\qquad$ $(1.62,3.62)$.

(b) $y=x^2+2x+3$
$\qquad$ $y=-x^3-3$
$\qquad$ $(-1.78,2.60)$.

(c) $y=e^{x-1}+1$
$\qquad$ $y=\ln(x-1)+4$
$\qquad$ $(1.16,2.17)$,
$\qquad$ $(2.14,4.13)$.

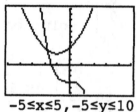

$-5\le x\le5, -1\le y\le10$ $\qquad$ $-5\le x\le5, -5\le y\le10$ $\qquad$ 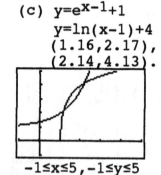 $-1\le x\le5, -1\le y\le5$

13. (a) $(1,5)$, $3x+y\le7$. $3(1)+5=8\not\le7$. Not on graph.

(b) $(-2,3)$, $5x+2y\le9$. $5(-2)+2(3)=-4\le9$. On graph.

(c) $(4,7)$, $-2x+7y\ge6$. $-2(4)+7(7)=41\ge6$. On graph.

(d) $(0,0)$, $x+y\le4$. $0+0=0\le4$. On graph.

(e) $(2,-1)$, $3x-2y\ge10$. $3(2)-2(-1)=8\not\ge10$. Not on graph.

(f) $(4,-3)$, $3x+5y\le-4$. $3(4)+5(-3)=-3\le-4$. Not on graph.

14. (a) $x+2y\le4$ $\qquad$ (b) $3x-y\le4$ $\qquad$ (c) $-4x+5y\le3$ $\qquad$ (d) $7x+12y\ge28$

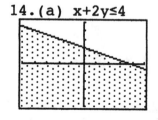 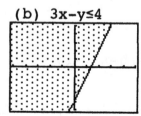 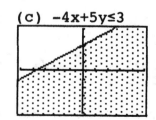 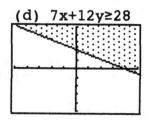

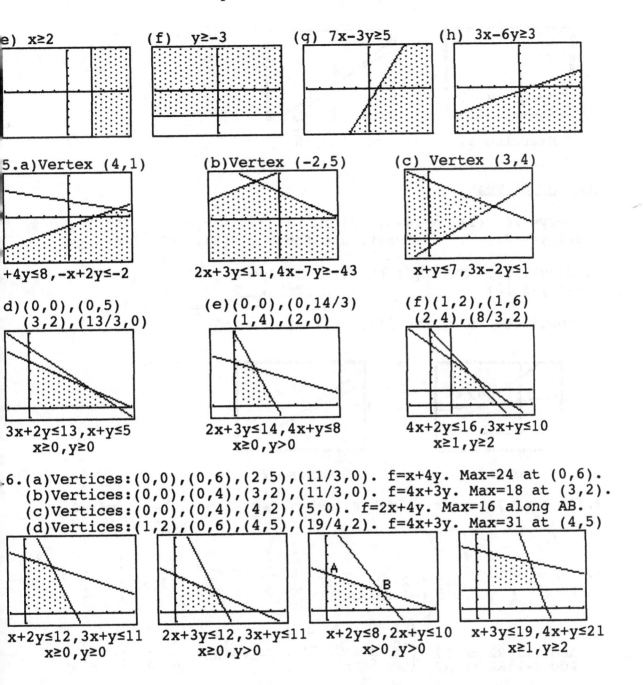

(e) x≥2

(f) y≥-3

(g) 7x-3y≥5

(h) 3x-6y≥3

5.a)Vertex (4,1)

(b)Vertex (-2,5)

(c) Vertex (3,4)

+4y≤8,-x+2y≤-2

2x+3y≤11,4x-7y≥-43

x+y≤7,3x-2y≤1

d)(0,0),(0,5)
  (3,2),(13/3,0)

(e)(0,0),(0,14/3)
   (1,4),(2,0)

(f)(1,2),(1,6)
   (2,4),(8/3,2)

3x+2y≤13,x+y≤5
   x≥0,y≥0

2x+3y≤14,4x+y≤8
   x≥0,y>0

4x+2y≤16,3x+y≤10
   x≥1,y≥2

6.(a)Vertices:(0,0),(0,6),(2,5),(11/3,0). f=x+4y. Max=24 at (0,6).
  (b)Vertices:(0,0),(0,4),(3,2),(11/3,0). f=4x+3y. Max=18 at (3,2).
  (c)Vertices:(0,0),(0,4),(4,2),(5,0). f=2x+4y. Max=16 along AB.
  (d)Vertices:(1,2),(0,6),(4,5),(19/4,2). f=4x+3y. Max=31 at (4,5)

x+2y≤12,3x+y≤11
   x≥0,y≥0

2x+3y≤12,3x+y≤11
   x≥0,y>0

x+2y≤8,2x+y≤10
   x>0,y>0

x+3y≤19,4x+y≤21
   x≥1,y≥2

17. Let the company manufacture x Panorama TVs, y Vision TVs.x≥0,y≥0.
    Time constraint: 4x+y≤2200. Cost constraint: 200x+300y≤150000.
    Profit f=40x+35y. Vertices:A(0,500),B(510,160),C(550,0).
    See fig below. f_A=17500,f_B=26000,f_C=22000. Max f=26000 at B.
    Maximum profit is $26,000 by making 510 Panoramas and 160 Visions.

18. Let the company manufacture x cabinets, y tables.x≥0,y≥0.
    Cost constraint:20x+30y≤1900=>2x+3y≤190. Time constraint:x+2y≤110.
    Profit f=7x+12y. Vertices:A(0,55),B(50,30),C(95,0),see fig.
    See fig below. f_A=660,f_B=710,f_C=665. Max f=710 at B.
    Maximum daily profit is $710 by making 50 cabinets and 30 tables.

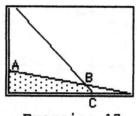

Exercise 17   Exercise 18

Chapter 6   Test

1. x+3y= 14. (1)   x=14-3y.(3) In (2) 2(14-3y)+y=8. -5y=-28. <u>y=4</u>.
   2x+ y= 8. (2)   Substitute for y in (3), x=14-3(4). <u>x=2</u>.

2. x+2y=1 (1)   Mult (1) by -2.   -2x-4y=-2
   2x+3y=3 (2)   2x+3y= 3
                       Add   -y= 1. <u>y=-1</u>.
   Substitute for y into (1). x+2(-1)=1. <u>x=3</u>.

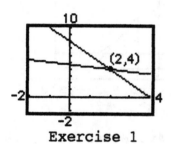

Exercise 1   Exercise 2

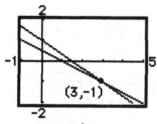

3.(1,4),(2,7). y=mx+b. m+b=4,2m+b=7. -m=-3,m=3. b=1. y=3x+1.

4. $\begin{bmatrix} 1 & 1 & 1 & 2 \\ 2 & 1 & -1 & 5 \\ 1 & 3 & 4 & 3 \end{bmatrix}$ Add (-2)R̃1 to R2 $\begin{bmatrix} 1 & 1 & 1 & 2 \\ 0 & -1 & -3 & 1 \\ 0 & 2 & 3 & 1 \end{bmatrix}$ (-1̃)R2 $\begin{bmatrix} 1 & 1 & 1 & 2 \\ 0 & 1 & 3 & -1 \\ 0 & 2 & 3 & 1 \end{bmatrix}$
      Add (-1)R1 to R3

Add (-1)R̃2 to R1 $\begin{bmatrix} 1 & 0 & -2 & 3 \\ 0 & 1 & 3 & -1 \\ 0 & 0 & -3 & 3 \end{bmatrix}$ (-1/3)R̃3 $\begin{bmatrix} 1 & 0 & -2 & 3 \\ 0 & 1 & 3 & -1 \\ 0 & 0 & 1 & -1 \end{bmatrix}$
Add (-2)R2 to R3

Add (2)R̃3 to R1 $\begin{bmatrix} 1 & 0 & 0 & 1 \\ 0 & 1 & 0 & 2 \\ 0 & 0 & 1 & -1 \end{bmatrix}$ x = 1
Add (-3)R3 to R2   y = 2
                                                   z =-1

5. Points (1,4), (3,14), (-1,10). y = ax² + bx + c.

   a + b + c = 4   a = 2
   9a + 3b + c = 14   b = -3   y = 2x² - 3x + 5.
   a - b + c = 10   c = 5
   A(1,4), B(3,14), C(-1,10) in figure below.

202

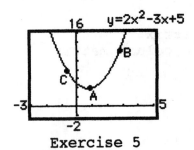

$y=2x^2-3x+5$

Exercise 5

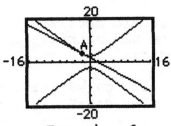

Exercise 6

5. $2x^2-y^2=-8$, $x+y=2$. $y=2-x$. $2x^2-(2-x)^2=-8$, $2x^2-(4-4x+x^2)=-8$.
   $2x^2-4+4x-x^2=-8$. $x^2+4x+4=0$. $(x+2)^2=0$. $x=-2$.
   Substitute back into $y=2-x$ to get $y=4$. One solution $(-2,4)$.
   This is the point A in the figure above.

7.a) $(2,4)$. $3x+2y\le16$. $3(2)+2(4)=14\le16$. On graph.
   b) $(3,-4)$. $-2x+y\le-11$. $-2(3)+(-4)=-10\not\le-11$. Not on graph.

8. $x+3y\le9$, $2x-y\le4$, $x\ge0$, $y\ge0$. Vertices $A(0,3),B(3,2),C(2,0),O(0,0)$.
   $f=2x+3y$. See figure below. $f_A=9$, $f_B=12$, $f_C=4$, $f_O=0$. $f_{max}=12$
   when $x=3$, $y=2$.

9. Let the company manufacture x Cruisers, y Jets. $x\ge0,y\ge0$.
   Time constraint: $5x+4y\le3620$. Cost constraint: $600x+800y\le556000$.
   Profit $f=70x+65y$. Vertices:$A(0,695),B(420,380),C(724,0)$.
   See figure below. $f_A=45175,f_B=54100,f_C=50680$. Max $f=54100$ at B.
   Maximum profit is $54,100 by making 420 Cruisers and 380 Jets.

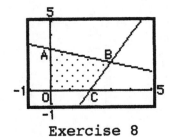

Exercise 8

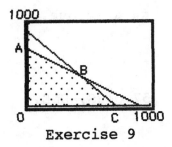

Exercise 9

10. 600 English,320 math for $36240. 50 English,20 math for $2740.
    Let English books cost $x each and math books $y each.
    $600x+320y=36240$. $600x+320y=36240$
    $50x+20y=2740$. Mult by -12 $\underline{-600x-240y=-32880}$
                                Add $80y=3360$. $y=42$.
    Substitute, $50x+20(42)=2740$. $50x=1900$. $x=38$.
    English books cost $38, math books cost $42.

203

Chapter 7

Section 7.1

1. 2x2, a square matrix.   3. 3x2 matrix.
5. 3x1, a column matrix.   7. 2x1, a column matrix.
9. 2x4 a matrix.

11. $a_{12}=4$ 13. $c_{23}=3$ 15. $c_{11}=1$
17. $d_{32}=1$ 19. $d_{33}=5$ 21. $d_{34}=6$

23. $p_{14}=3$ 25. $q_{24}=3$ 27. $q_{46}=-12$
29. $q_{35}=-6$ 31. $p_{13}=-1$ 33. $q_{26}=2$

35. Comparing corresponding elements: $x=2$, $x+y=5$. Soln $x=2,y=3$.

37. $x=2,x+y=3,x-y=1,x+z=4$. Soln $x=2,y=1,z=2$.

39. $x=3,x-y=4,x+y=6,x+z=8$. 1st,2nd,4th give $x=3,y=-1,z=5$.
    2nd not satisfied, no solution.

41. $x+y=3,x+z=4,x+y+z=6,x-1=0$. 4th,1st,2nd give $x=1,y=2,z=3$.
    3rd satisfied, this is the solution.

43. $x-y+z=0,y=3,x+z=3,7=7$. Many solutions, $x=3-z$, $y=3$ or
    $x=3-r,y=3,z=r$.

45. $A+B=\begin{bmatrix} 5 & 4 \\ 4 & 0 \end{bmatrix}$ 47. A+D DNE.   49. $A+2B=\begin{bmatrix} 9 & 6 \\ 9 & -3 \end{bmatrix}$

51. $A+B+C=\begin{bmatrix} 10 & 5 \\ 6 & 0 \end{bmatrix}$ 53. A+5E DNE.

55. $P+Q=\begin{bmatrix} 5 & 7 \\ 4 & 6 \\ -1 & 8 \end{bmatrix}$ 57. Q+3S, DNE.

59. $P-Q=\begin{bmatrix} -3 & -3 \\ 2 & 2 \\ -1 & -4 \end{bmatrix}$ 61. $R+2S-3T=\begin{bmatrix} 8 & -6 & 11 \\ 15 & -3 & 17 \end{bmatrix}$

63. P+2Q+3R-4T, DNE.

65. $x+y+z=8,y-z=1,z=1,x^2=25$. 2nd,3rd give $y=2$. 1st gives $x=5$.
    4th satisfied. Soln $x=5,y=2,z=1$.

67. $x^2=9,x+y=3,4=4,y=-3$. 1st,4th give $x=6$. 1st not satisfied.
    No solution.

69. $x^2-3x=-2,2x=6,y=4,z=3$. 2nd gives $x=3$. 1st not satisfied. No soln

1. $D = 2\begin{bmatrix} 1 & 2 \\ 3 & -1 \end{bmatrix} - 3\begin{bmatrix} 5 & 1 \\ 3 & 2 \end{bmatrix} + 4\begin{bmatrix} -2 & 7 \\ 1 & 4 \end{bmatrix} =$

$$\begin{bmatrix} 2 & 4 \\ 6 & -2 \end{bmatrix} - \begin{bmatrix} 15 & 3 \\ 9 & 6 \end{bmatrix} + \begin{bmatrix} -8 & 28 \\ 4 & 16 \end{bmatrix} = \begin{bmatrix} -21 & 29 \\ 1 & 8 \end{bmatrix}.$$

(a) $d_{12}=29$ (b) $d_{22}=8$ (c) $3d_{21}=3$ (d) $d_{12}+2d_{22}=29+16=45$

3. (a) $i=1, j=2$. $i=4, j=4$.   (b) $i=1,4$.   (c) $j=2,5$.   (d) $i=j=3$.
   (e) $i=1, j=4$. $i=2, j=4$.   (f) $i=3, j=2$.

5. (a) Daily traffic from zone I to zone IV = 200.
   (b) The daily traffic from zone III to zone II = 900.
   (c) Internal traffic in zone II = 6000.
   (d) Internal traffic in zone IV = 1800.
   Such an O-D matrix will always be square:
   # rows = # zones = # columns.

7. (a)   (b)

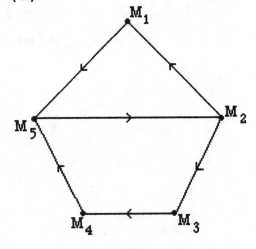

$$D = \begin{bmatrix} 0 & 2 & 3 & 4 & 1 \\ 1 & 0 & 1 & 2 & 3 \\ 4 & 3 & 0 & 1 & 2 \\ 3 & 2 & 3 & 0 & 1 \\ 2 & 1 & 2 & 3 & 0 \end{bmatrix} \begin{matrix} 10 \\ 7 \\ 10 \\ 9 \\ 8 \end{matrix}$$

$$D = \begin{bmatrix} 0 & 2 & 1 & 2 & 1 \\ 1 & 0 & 2 & 3 & 2 \\ 2 & 1 & 0 & 4 & 3 \\ 3 & 2 & 1 & 0 & 4 \\ 4 & 3 & 2 & 1 & 0 \end{bmatrix} \begin{matrix} 6 \\ 8 \\ 10 \\ 10 \\ 10 \end{matrix}$$

M_2, M_5, M_4, M_1 and M_3. M_1, M_2, M_3 and M_4 and M_5.

(c)

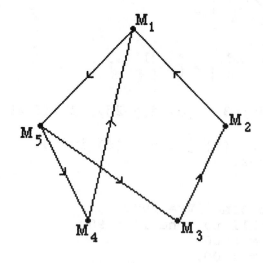

(d)

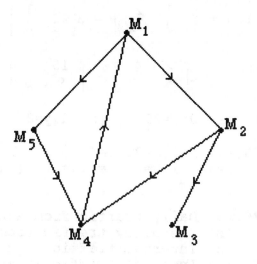

$$D = \begin{bmatrix} 0 & 3 & 2 & 2 & 1 \\ 1 & 0 & 3 & 3 & 2 \\ 2 & 1 & 0 & 4 & 3 \\ 1 & 4 & 3 & 0 & 2 \\ 2 & 2 & 1 & 1 & 0 \end{bmatrix} \begin{matrix} 8 \\ 9 \\ 10 \\ 10 \\ 6 \end{matrix}$$

M_5, M_1, M_2, M_3 and M_4.

$$D = \begin{bmatrix} 0 & 1 & 2 & 2 & 1 \\ 2 & 0 & 1 & 1 & 3 \\ \infty & \infty & 0 & \infty & \infty \\ 1 & 2 & 3 & 0 & 2 \\ 2 & 3 & 4 & 1 & 0 \end{bmatrix} \begin{matrix} 6 \\ 7 \\ \infty \\ 8 \\ 10 \end{matrix}$$

M_1, M_2, M_4, M_5, M_3.

79.

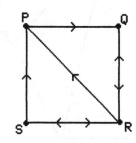

81.

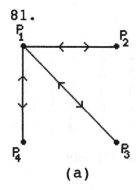

(a)

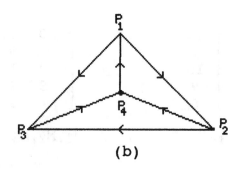

(b)

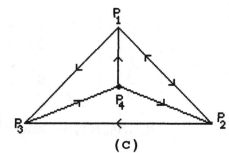

(c)

Section 7.2

$1.\ AB = \begin{bmatrix} 2 & 4 \\ 8 & 10 \end{bmatrix}$ $3.\ AC = \begin{bmatrix} 2 & 0 & 4 \\ -4 & 9 & 10 \end{bmatrix}$ $5.\ AD = \begin{bmatrix} 6 & 0 \\ -9 & 12 \end{bmatrix}$

$.BC = \begin{bmatrix} -1 & 6 & 10 \\ -1 & 12 & 22 \end{bmatrix}$ 　　　$9.ABC = \begin{bmatrix} 2 & 4 \\ 8 & 10 \end{bmatrix}\begin{bmatrix} 1 & 0 & 2 \\ -1 & 3 & 4 \end{bmatrix} = \begin{bmatrix} -2 & 12 & 20 \\ -2 & 30 & 56 \end{bmatrix}$

$1.B^2 = \begin{bmatrix} 7 & 10 \\ 15 & 22 \end{bmatrix}$ 　　　$12.C^2$, DNE. 　　$13.AB+A = \begin{bmatrix} 2 & 4 \\ 8 & 10 \end{bmatrix} + \begin{bmatrix} 2 & 0 \\ -1 & 3 \end{bmatrix} = \begin{bmatrix} 4 & 4 \\ 7 & 13 \end{bmatrix}$

$5.DB = \begin{bmatrix} 3 & 6 \\ 10 & 12 \end{bmatrix}$ 　　　$17.AB = \begin{bmatrix} 12 & 6 \\ 18 & -1 \end{bmatrix}$ 　　　$19.AC = \begin{bmatrix} 5 \\ 5 \end{bmatrix}$

$1.DE = \begin{bmatrix} 10 & 2 \end{bmatrix}$ 　　　$23.2AB-3AE = 2\begin{bmatrix} 12 & 6 \\ 18 & -1 \end{bmatrix} - 3\begin{bmatrix} 8 & 1 \\ 2 & 4 \end{bmatrix} = \begin{bmatrix} 0 & 9 \\ 30 & -14 \end{bmatrix}$

$5.A^2+B = \begin{bmatrix} -1 & 8 \\ -4 & 7 \end{bmatrix} + \begin{bmatrix} 0 & 4 \\ 6 & 1 \end{bmatrix} = \begin{bmatrix} -1 & 12 \\ 2 & 8 \end{bmatrix}$

$7.AB$ 2x2 　　　$29.CA$, DNE 　　$31.AF$, DNE 　　$33.EA$, DNE 　　$35.2DA$, DNE

$7.EFD$ 2x3 　　　$39.A^2+3B$ 2x2 　$41.PQ$ 2x4 　　　$43.QRT$ 2x2

$5.QP$, DNE 　　　$47.RS+T$ 4x2 　　$49.6TQ+8RS$, DNE

$1.ABD = (AB)D = \begin{bmatrix} -1 & -3 \\ 6 & -4 \end{bmatrix}\begin{bmatrix} 2 & 0 & 1 \\ 1 & 4 & -3 \end{bmatrix} = \begin{bmatrix} -5 & -12 & 8 \\ 8 & -16 & 18 \end{bmatrix}$

$3.AC+BC$, DNE since AC DNE.

$5.A^2BC = (AA)(BC)$, DNE since BC DNE.

$7.CBAB = (CB)(AB) = \begin{bmatrix} 3 & -2 \\ 10 & -3 \\ 9 & 5 \end{bmatrix}\begin{bmatrix} -1 & -3 \\ 6 & -4 \end{bmatrix} = \begin{bmatrix} -15 & -1 \\ -28 & -18 \\ 21 & -47 \end{bmatrix}$

$9.DA$, DNE.

$1.(DC)^2+A^2 = (\begin{bmatrix} 1 & 3 \\ 12 & -5 \end{bmatrix})^2 + \begin{bmatrix} -1 & -1 \\ 2 & -2 \end{bmatrix} = \begin{bmatrix} 36 & -13 \\ -46 & 59 \end{bmatrix}$

$3.A^2$, 2x2 　　　$65.ABCD$, 2x2

$7.EBCD$, 4x2 　　$69.(AB)^4$, DNE since AB not square.

$1.BCD+2A^2$, 2x2

$3.AB = \begin{bmatrix} -17.56 & 74.6 \\ 11.02 & -26.1 \end{bmatrix}$ 　　　$75.3.5AB = \begin{bmatrix} -61.46 & 261.1 \\ 38.57 & -91.35 \end{bmatrix}$

$7.AB-BA = \begin{bmatrix} -22.04 & 86.78 \\ 23.18 & 22.04 \end{bmatrix}$

79. $0.0007AB^5 = \begin{bmatrix} -33.5148 & -22.3330 \\ 3.9774 & 15.2089 \end{bmatrix}$, to 4 dec places.

81. $PQ = \begin{bmatrix} 10.71 & -14.28 \\ -31.4 & 59.64 \end{bmatrix}$ 83. $3.7RP$, DNE. 85. R^2, DNE.

87. $A = \begin{bmatrix} 0.99 & 0.02 \\ 0.01 & 0.98 \end{bmatrix}$, $X_0 = \begin{bmatrix} 200 \\ 65 \end{bmatrix}$. $X_1 = AX_0$ etc. Year 2010, get $A^{15}X_0$.

.	1995	'96	'97	'98	'99	2000	...	2010
Metro	200	199.3	198.62	197.96	197.32	196.70		191.44
Nonmetro	65	65.7	66.38	67.04	67.68	68.30		73.56

Long term prediction, 176.22 for metro, 188.33 for non-metro.

Section 7.3

1. $\begin{bmatrix} 3 & 6 \\ 1 & 0 \end{bmatrix}\begin{bmatrix} 1 & 0 \\ 0 & 1 \end{bmatrix} = \begin{bmatrix} 3 & 6 \\ 1 & 0 \end{bmatrix}\begin{bmatrix} 1 & 0 \\ 0 & 1 \end{bmatrix} = \begin{bmatrix} 3 & 6 \\ 1 & 0 \end{bmatrix}$

3. $\begin{bmatrix} 4 & -5 \\ 2 & 3 \end{bmatrix}\begin{bmatrix} 1 & 0 \\ 0 & 1 \end{bmatrix} = \begin{bmatrix} 1 & 0 \\ 0 & 1 \end{bmatrix}\begin{bmatrix} 4 & -5 \\ 2 & 3 \end{bmatrix} = \begin{bmatrix} 4 & -5 \\ 2 & 3 \end{bmatrix}$

5. $\begin{bmatrix} 0 & 7 & 6 \\ -2 & 4 & 3 \\ 1 & 9 & 8 \end{bmatrix}\begin{bmatrix} 1 & 0 & 0 \\ 0 & 1 & 0 \\ 0 & 0 & 1 \end{bmatrix} = \begin{bmatrix} 1 & 0 & 0 \\ 0 & 1 & 0 \\ 0 & 0 & 1 \end{bmatrix}\begin{bmatrix} 0 & 7 & 6 \\ -2 & 4 & 3 \\ 1 & 9 & 8 \end{bmatrix} = \begin{bmatrix} 0 & 7 & 6 \\ -2 & 4 & 3 \\ 1 & 9 & 8 \end{bmatrix}$

7. $\begin{bmatrix} 1 & 2 & -1 & 3 \\ 4 & 6 & 0 & 5 \\ 7 & 1 & 2 & 9 \\ 12 & 3 & 4 & 1 \end{bmatrix}\begin{bmatrix} 1 & 0 & 0 & 0 \\ 0 & 1 & 0 & 0 \\ 0 & 0 & 1 & 0 \\ 0 & 0 & 0 & 1 \end{bmatrix} = \begin{bmatrix} 1 & 0 & 0 & 0 \\ 0 & 1 & 0 & 0 \\ 0 & 0 & 1 & 0 \\ 0 & 0 & 0 & 1 \end{bmatrix}\begin{bmatrix} 1 & 2 & -1 & 3 \\ 4 & 6 & 0 & 5 \\ 7 & 1 & 2 & 9 \\ 12 & 3 & 4 & 1 \end{bmatrix}$

$= \begin{bmatrix} 1 & 2 & -1 & 3 \\ 4 & 6 & 0 & 5 \\ 7 & 1 & 2 & 9 \\ 12 & 3 & 4 & 1 \end{bmatrix}$

9. $\begin{bmatrix} 3 & 7 \\ 2 & 5 \end{bmatrix}\begin{bmatrix} 5 & -7 \\ -2 & 3 \end{bmatrix} = \begin{bmatrix} 5 & -7 \\ -2 & 3 \end{bmatrix}\begin{bmatrix} 3 & 7 \\ 2 & 5 \end{bmatrix} = \begin{bmatrix} 1 & 0 \\ 0 & 1 \end{bmatrix}$. Yes.

11. $\begin{bmatrix} 5 & -2 \\ -7 & 3 \end{bmatrix}\begin{bmatrix} 3 & 2 \\ 7 & 5 \end{bmatrix} = \begin{bmatrix} 3 & 2 \\ 7 & 5 \end{bmatrix}\begin{bmatrix} 5 & -2 \\ -7 & 3 \end{bmatrix} = \begin{bmatrix} 1 & 0 \\ 0 & 1 \end{bmatrix}$. Yes.

13. $\begin{bmatrix} 3 & 2 \\ 4 & 3 \end{bmatrix}\begin{bmatrix} 3 & -2 \\ -4 & 3 \end{bmatrix} = \begin{bmatrix} 3 & -2 \\ -4 & 3 \end{bmatrix}\begin{bmatrix} 3 & 2 \\ 4 & 3 \end{bmatrix} = \begin{bmatrix} 1 & 0 \\ 0 & 1 \end{bmatrix}$. Yes.

5. $\begin{bmatrix} 1 & 0 & 2 \\ -1 & 2 & 3 \\ 1 & -1 & 0 \end{bmatrix}\begin{bmatrix} 3 & -2 & -4 \\ 3 & -2 & -5 \\ -1 & 1 & 2 \end{bmatrix} = \begin{bmatrix} 3 & -2 & -4 \\ 3 & -2 & -5 \\ -1 & 1 & 2 \end{bmatrix}\begin{bmatrix} 1 & 0 & 2 \\ -1 & 2 & 3 \\ 1 & -1 & 0 \end{bmatrix}$

$= \begin{bmatrix} 1 & 0 & 0 \\ 0 & 1 & 0 \\ 0 & 0 & 1 \end{bmatrix}$. **Yes.**

7. $\begin{bmatrix} 4 & -1 & -4 \\ 3 & 0 & -4 \\ 3 & -1 & -3 \end{bmatrix}\begin{bmatrix} 4 & -1 & -4 \\ 3 & 0 & -4 \\ 3 & -1 & -3 \end{bmatrix} = \begin{bmatrix} 4 & -1 & -4 \\ 3 & 0 & -4 \\ 3 & -1 & -3 \end{bmatrix}\begin{bmatrix} 4 & -1 & -4 \\ 3 & 0 & -4 \\ 3 & -1 & -3 \end{bmatrix}$

$= \begin{bmatrix} 1 & 0 & 0 \\ 0 & 1 & 0 \\ 0 & 0 & 1 \end{bmatrix}$. **Yes.**

9. $\begin{bmatrix} 6 & 2 & 3 \\ 5 & 1 & -1 \\ -7 & 8 & 2 \end{bmatrix}\begin{bmatrix} 2 & -1 & 0 \\ -3 & 1 & -6 \\ -2 & 1 & 4 \end{bmatrix} = \begin{bmatrix} 0 & -1 & 0 \\ 9 & -5 & -10 \\ -42 & 17 & -40 \end{bmatrix} \neq \begin{bmatrix} 1 & 0 & 0 \\ 0 & 1 & 0 \\ 0 & 0 & 1 \end{bmatrix}$. **No.**

21. $\begin{bmatrix} 1 & -3 \\ -2 & 5 \end{bmatrix}$. $[A:I_2] = \begin{bmatrix} 1 & -3 & 1 & 0 \\ -2 & 5 & 0 & 1 \end{bmatrix}$ Add $(2)\tilde{R}1$ to R2 $\begin{bmatrix} 1 & -3 & 1 & 0 \\ 0 & 1 & -2 & -1 \end{bmatrix}$

Add $(3)\tilde{R}2$ to R1 $\begin{bmatrix} 1 & 0 & -5 & -3 \\ 0 & 1 & -2 & -1 \end{bmatrix}$. $A^{-1} = \begin{bmatrix} -5 & -3 \\ -2 & -1 \end{bmatrix}$.

23. $\begin{bmatrix} 1/2 & 0 \\ 0 & 1/6 \end{bmatrix}$. $[A:I_2] = \begin{bmatrix} 1/2 & 0 & 1 & 0 \\ 0 & 1/6 & 0 & 1 \end{bmatrix}$ $(2)\tilde{R}1 \begin{bmatrix} 1 & 0 & 2 & 0 \\ 0 & 1/6 & 0 & 1 \end{bmatrix}$

$(6)\tilde{R}2 \begin{bmatrix} 1 & 0 & 2 & 0 \\ 0 & 1 & 0 & 6 \end{bmatrix}$. $A^{-1} = \begin{bmatrix} 2 & 0 \\ 0 & 6 \end{bmatrix}$.

25. $\begin{bmatrix} 2 & 4 \\ 1 & 2 \end{bmatrix}$. $[A:I_2] = \begin{bmatrix} 2 & 4 & 1 & 0 \\ 1 & 2 & 0 & 1 \end{bmatrix}$ $(1/2)\tilde{R}1 \begin{bmatrix} 1 & 2 & 1/2 & 0 \\ 1 & 2 & 0 & 1 \end{bmatrix}$

Add $(-1)\tilde{R}1$ to R2 $\begin{bmatrix} 1 & 2 & 1/2 & 0 \\ 0 & 0 & -1/2 & 1 \end{bmatrix}$. **Inverse does not exist.**

27. $\begin{bmatrix} 3 & 4 \\ 2 & 3 \end{bmatrix}$. $[A:I_2] = \begin{bmatrix} 3 & 4 & 1 & 0 \\ 2 & 3 & 0 & 1 \end{bmatrix}$ $(1/3)\tilde{R}1 \begin{bmatrix} 1 & 4/3 & 1/3 & 0 \\ 2 & 3 & 0 & 1 \end{bmatrix}$

Add $(-2)\tilde{R}1$ to R2 $\begin{bmatrix} 1 & 4/3 & 1/3 & 0 \\ 0 & 1/3 & -2/3 & 1 \end{bmatrix}$ $(3)\tilde{R}2 \begin{bmatrix} 1 & 4/3 & 1/3 & 0 \\ 0 & 1 & -2 & 3 \end{bmatrix}$

Add $(4/3)\tilde{R}2$ to R1 $\begin{bmatrix} 1 & 0 & 3 & -4 \\ 0 & 1 & -2 & 3 \end{bmatrix}$. $A^{-1} = \begin{bmatrix} 3 & -4 \\ -2 & 3 \end{bmatrix}$.

29. $[A:I_3] = \begin{bmatrix} 1 & 2 & 0 & 1 & 0 & 0 \\ 0 & 1 & 1 & 0 & 1 & 0 \\ -1 & 0 & 1 & 0 & 0 & 1 \end{bmatrix}$ Add $(1)\tilde{R}1$ to R3 $\begin{bmatrix} 1 & 2 & 0 & 1 & 0 & 0 \\ 0 & 1 & 1 & 0 & 1 & 0 \\ 0 & 2 & 1 & 1 & 0 & 1 \end{bmatrix}$

Add $(-2)\tilde{R}2$ to R1
Add $(-2)\tilde{R}2$ to R3 $\begin{bmatrix} 1 & 0 & -2 & 1 & -2 & 0 \\ 0 & 1 & 1 & 0 & 1 & 0 \\ 0 & 0 & -1 & 1 & -2 & 1 \end{bmatrix}$ $(-\tilde{1})R3$ $\begin{bmatrix} 1 & 0 & -2 & 1 & -2 & 0 \\ 0 & 1 & 1 & 0 & 1 & 0 \\ 0 & 0 & 1 & -1 & 2 & -1 \end{bmatrix}$

Add $(2)\tilde{R}3$ to R1
Add $(-1)R3$ to R2 $\begin{bmatrix} 1 & 0 & 0 & -1 & 2 & -2 \\ 0 & 1 & 0 & 1 & -1 & 1 \\ 0 & 0 & 1 & -1 & 2 & -1 \end{bmatrix}$. $A^{-1} = \begin{bmatrix} -1 & 2 & -2 \\ 1 & -1 & 1 \\ -1 & 2 & -1 \end{bmatrix}$.

31. $[A:I_3] = \begin{bmatrix} 2 & -1 & 0 & 1 & 0 & 0 \\ 3 & 4 & 2 & 0 & 1 & 0 \\ -1 & 6 & 2 & 0 & 0 & 1 \end{bmatrix}$ $(1/\tilde{2})R2$ $\begin{bmatrix} 1 & -1/2 & 0 & 1/2 & 0 & 0 \\ 3 & 4 & 2 & 0 & 1 & 0 \\ -1 & 6 & 2 & 0 & 0 & 1 \end{bmatrix}$

$(-3)R\tilde{1} + R2$
$(1)R1 + R3$ $\begin{bmatrix} 1 & -1/2 & 0 & 1/2 & 0 & 0 \\ 0 & 11/2 & 2 & -3/2 & 1 & 0 \\ 0 & 11/2 & 2 & 1/2 & 0 & 1 \end{bmatrix}$ $(2/1\tilde{1})R2$ $\begin{bmatrix} 1 & -1/2 & 0 & 1/2 & 0 & 0 \\ 0 & 1 & 4/11 & -3/11 & 2/11 & 0 \\ 0 & 11/2 & 2 & 1/2 & 0 & 1 \end{bmatrix}$

$(1/2)\tilde{R}2 + R1$
$(-11/2)R2 + R3$ $\begin{bmatrix} 1 & 0 & 2/11 & 4/11 & 1/11 & 0 \\ 0 & 1 & 4/11 & -3/11 & 2/11 & 0 \\ 0 & 0 & 0 & 2 & -1 & 1 \end{bmatrix}$. A^{-1} does not exist.

33. $[A:I_3] = \begin{bmatrix} -1 & 4 & 0 & 1 & 0 & 0 \\ -1 & 5 & 1 & 0 & 1 & 0 \\ 1 & 0 & -3 & 0 & 0 & 1 \end{bmatrix}$ $(-\tilde{1})R2$ $\begin{bmatrix} 1 & -4 & 0 & -1 & 0 & 0 \\ -1 & 5 & 1 & 0 & 1 & 0 \\ -1 & 0 & -3 & 0 & 0 & 1 \end{bmatrix}$

$(1)R\tilde{1} + R2$
$(1)R1 + R3$ $\begin{bmatrix} 1 & -4 & 0 & -1 & 0 & 0 \\ 0 & 1 & 1 & -1 & 1 & 0 \\ 0 & -4 & -3 & -1 & 0 & 1 \end{bmatrix}$ $(4)R\tilde{2} + R1$
$(4)R2 + R3$ $\begin{bmatrix} 1 & 0 & 4 & -5 & 4 & 0 \\ 0 & 1 & 1 & -1 & 1 & 0 \\ 0 & 0 & 1 & -5 & 4 & 1 \end{bmatrix}$

$(-4)R\tilde{3} + R1$
$(-1)R3 + R2$ $\begin{bmatrix} 1 & 0 & 0 & 15 & -12 & -4 \\ 0 & 1 & 0 & 4 & -3 & -1 \\ 0 & 0 & 1 & -5 & 4 & 1 \end{bmatrix}$. $A^{-1} = \begin{bmatrix} 15 & -12 & -4 \\ 4 & -3 & -1 \\ -5 & 4 & 1 \end{bmatrix}$.

35. $\begin{bmatrix} 1 & 1 \\ 3 & -1 \end{bmatrix}\begin{bmatrix} x \\ y \end{bmatrix} = \begin{bmatrix} 2 \\ 4 \end{bmatrix}$. 37. $\begin{bmatrix} 2 & 3 \\ 4 & 0 \end{bmatrix}\begin{bmatrix} x \\ y \end{bmatrix} = \begin{bmatrix} 1 \\ 2 \end{bmatrix}$. 39. $\begin{bmatrix} 3 & 2 & -7 \\ 4 & 3 & -2 \\ 1 & 1 & -1 \end{bmatrix}\begin{bmatrix} x \\ y \\ z \end{bmatrix} = \begin{bmatrix} 1 \\ 4 \\ 5 \end{bmatrix}$.

41. $\begin{bmatrix} 1 & 1 & 1 \\ 5 & -1 & 0 \\ 1 & 1 & 2 \end{bmatrix}\begin{bmatrix} x \\ y \\ z \end{bmatrix} = \begin{bmatrix} 4 \\ 6 \\ 9 \end{bmatrix}$.

43. $\begin{bmatrix} 1 & -3 \\ -2 & 5 \end{bmatrix}\begin{bmatrix} x \\ y \end{bmatrix} = \begin{bmatrix} 2 \\ 1 \end{bmatrix}$. $\begin{bmatrix} x \\ y \end{bmatrix} = \begin{bmatrix} 1 & -3 \\ -2 & 5 \end{bmatrix}^{-1}\begin{bmatrix} 2 \\ 1 \end{bmatrix} = \begin{bmatrix} -5 & -3 \\ -2 & -1 \end{bmatrix}\begin{bmatrix} 2 \\ 1 \end{bmatrix} = \begin{bmatrix} -13 \\ -5 \end{bmatrix}$

$x=-13$, $y=-5$.

45. $\begin{bmatrix} 5 & 2 \\ 3 & 1 \end{bmatrix}\begin{bmatrix} x \\ y \end{bmatrix} = \begin{bmatrix} -2 \\ 6 \end{bmatrix}$. $\begin{bmatrix} x \\ y \end{bmatrix} = \begin{bmatrix} 5 & 2 \\ 3 & 1 \end{bmatrix}^{-1}\begin{bmatrix} -2 \\ 6 \end{bmatrix} = \begin{bmatrix} -1 & 2 \\ 3 & -5 \end{bmatrix}\begin{bmatrix} -2 \\ 6 \end{bmatrix} = \begin{bmatrix} 14 \\ -36 \end{bmatrix}$

$x=14$, $y=-36$.

47. $\begin{bmatrix} 1 & 0 & 2 \\ -3 & 3 & -1 \\ 0 & 1 & 2 \end{bmatrix}\begin{bmatrix} x \\ y \\ z \end{bmatrix} = \begin{bmatrix} 1 \\ 2 \\ 3 \end{bmatrix}$. $\begin{bmatrix} x \\ y \\ z \end{bmatrix} = \begin{bmatrix} 1 & 0 & 2 \\ -3 & 3 & -1 \\ 0 & 1 & 2 \end{bmatrix}^{-1}\begin{bmatrix} 1 \\ 2 \\ 3 \end{bmatrix} = \begin{bmatrix} 7 & 2 & -6 \\ 6 & 2 & -5 \\ -3 & -1 & 3 \end{bmatrix}\begin{bmatrix} 1 \\ 2 \\ 3 \end{bmatrix}$

$= \begin{bmatrix} -7 \\ -5 \\ 4 \end{bmatrix}$. $x=-7$, $y=-5$, $z=4$.

49. $\begin{bmatrix} 1 & 0 \\ 2 & 1 \end{bmatrix}^{-1} = \begin{bmatrix} 1 & 0 \\ -2 & 1 \end{bmatrix}$. 51. $\begin{bmatrix} 2 & 1 \\ 4 & 3 \end{bmatrix}^{-1} = \begin{bmatrix} 1.5 & -.5 \\ -2 & 1 \end{bmatrix}$.

53. $\begin{bmatrix} 1 & 2 \\ 3 & 6 \end{bmatrix}^{-1}$. Inverse does not exist.

55. $\begin{bmatrix} 1 & 2 & 3 \\ 0 & 1 & 2 \\ 4 & 5 & 3 \end{bmatrix}^{-1} = \begin{bmatrix} 2.3333 & -3.0000 & -.3333 \\ -2.6667 & 3.0000 & .6667 \\ 1.3333 & -1.0000 & -.3333 \end{bmatrix}$.

57. $\begin{bmatrix} 1 & 2 & -3 \\ 1 & -2 & 1 \\ 5 & -2 & -3 \end{bmatrix}^{-1}$. Inverse does not exist.

59. $\begin{bmatrix} 1 & 2 & 3 \\ 2 & -1 & 4 \\ 0 & -1 & 1 \end{bmatrix}^{-1} = \begin{bmatrix} -.4286 & .7143 & -1.5714 \\ .2857 & -.1429 & -.2857 \\ .2857 & -.1429 & .7143 \end{bmatrix}$.

61. $x=-2$, $y=2$.   63. $x=5$, $y=0$. 65. $x=11$, $y=-4$.

67. $x=0.7778$, $y=0.3333$, $z=-0.5556$.   69. $x=143$, $y=-47$, $z=-16$.

71. $\begin{bmatrix} 60 \\ 40 \end{bmatrix}$ 73. $\begin{bmatrix} 210 \\ 175 \end{bmatrix}$ 75. $\begin{bmatrix} 165 \\ 480 \\ 250 \end{bmatrix}$

77. a_{ij} is the amount from industry i required to produce one unit from industry j. Thus no a_{ij} can be negative. If the amount from industry i required for one unit of industry j were greater than 1, the value going into industry j would be greater than the final value, an economically infeasible situation.

78. The sum of the elements of column j of the input-output matrix

211

is the sum of the amounts from all the industries required to produce one unit from industry j. This sum will certainly be greater than or equal to zero since all elements are, and in an economically feasible situation, the sum of all that goes into producing one unit from industry j should be worth less than one unit of the product from industry j.

Section 7.4

1. $\begin{vmatrix} 1 & 2 \\ 3 & 4 \end{vmatrix} = 4-6 = -2.$

3. $\begin{vmatrix} 5 & 2 \\ 1 & 3 \end{vmatrix} = 15-2 = 13.$

5. $\begin{vmatrix} 2 & 5 \\ 0 & 1 \end{vmatrix} = 2-0 = 2.$

7. $\begin{vmatrix} -2 & -6 \\ -3 & -4 \end{vmatrix} = 8-18 = -10.$

9. $A = \begin{vmatrix} 1 & 2 & -1 \\ 3 & 4 & 2 \\ 5 & 1 & 3 \end{vmatrix}$. Minor of a_{11} is $A_{11} = \begin{vmatrix} 4 & 2 \\ 1 & 3 \end{vmatrix} = 12-2 = 10.$

$A_{12} = \begin{vmatrix} 3 & 2 \\ 5 & 3 \end{vmatrix} = 9-10 = -1.$ $A_{21} = \begin{vmatrix} 2 & -1 \\ 1 & 3 \end{vmatrix} = 6+1 = 7.$

11. $A = \begin{vmatrix} 5 & 0 & 4 \\ 1 & -2 & 6 \\ 3 & 1 & -5 \end{vmatrix}$. Minor of $a_{11} = A_{11} = \begin{vmatrix} -2 & 6 \\ 1 & -5 \end{vmatrix} = 10-6 = 4.$

$A_{31} = \begin{vmatrix} 0 & 4 \\ -2 & 6 \end{vmatrix} = 0+8 = 8.$ $A_{23} = \begin{vmatrix} 5 & 0 \\ 3 & 1 \end{vmatrix} = 5-0 = 5.$

13. $\begin{vmatrix} -1 & 2 & 0 \\ 4 & 1 & 3 \\ 5 & 2 & -1 \end{vmatrix} = -1 \begin{vmatrix} 1 & 3 \\ 2 & -1 \end{vmatrix} - 2 \begin{vmatrix} 4 & 3 \\ 5 & -1 \end{vmatrix} + 0 \begin{vmatrix} 4 & 1 \\ 5 & 2 \end{vmatrix}$

$= -1(-1-6) - 2(-4-15) + 0(8-5) = 7+38 = 45.$

15. $\begin{vmatrix} 0 & 0 & 4 \\ 1 & 2 & 5 \\ 4 & 9 & 1 \end{vmatrix} = 0 \begin{vmatrix} 2 & 5 \\ 9 & 1 \end{vmatrix} - 0 \begin{vmatrix} 1 & 5 \\ 4 & 1 \end{vmatrix} + 4 \begin{vmatrix} 1 & 2 \\ 4 & 9 \end{vmatrix}$

$= 0(2-45) - 0(1-20) + 4(9-8) = 4.$

17. $\begin{vmatrix} 1 & -1 & 1 \\ 2 & 4 & 7 \\ 3 & 5 & 1 \end{vmatrix} = 1 \begin{vmatrix} 4 & 7 \\ 5 & 1 \end{vmatrix} - (-1) \begin{vmatrix} 2 & 7 \\ 3 & 1 \end{vmatrix} + 1 \begin{vmatrix} 2 & 4 \\ 3 & 5 \end{vmatrix}$

$= 1(4-35) + 1(2-21) + 1(10-12) = -31-19-2 = -52.$

19. $\begin{vmatrix} 1 & 0 & 0 \\ 2 & 3 & 4 \\ 1 & 4 & 5 \end{vmatrix}$. Row1: $= 1 \begin{vmatrix} 3 & 4 \\ 4 & 5 \end{vmatrix} - 0 \begin{vmatrix} 2 & 4 \\ 1 & 5 \end{vmatrix} + 0 \begin{vmatrix} 2 & 3 \\ 1 & 4 \end{vmatrix}$

$$= 1(15-16) - 0(10-4) + 0(8-3) = -1.$$

Col3: $= 0\begin{vmatrix} 2 & 3 \\ 1 & 4 \end{vmatrix} - 4\begin{vmatrix} 1 & 0 \\ 1 & 4 \end{vmatrix} + 5\begin{vmatrix} 1 & 0 \\ 2 & 3 \end{vmatrix}$

$$= 0(8-3) - 4(4-0) + 5(3-0) = -16+15 = -1.$$

21. $\begin{vmatrix} 5 & 2 & 1 \\ 0 & 4 & 3 \\ 5 & 1 & 2 \end{vmatrix}$. Col1: $= 5\begin{vmatrix} 4 & 3 \\ 1 & 2 \end{vmatrix} - 0\begin{vmatrix} 2 & 1 \\ 1 & 2 \end{vmatrix} + 5\begin{vmatrix} 2 & 1 \\ 4 & 3 \end{vmatrix}$

$$= 5(8-3) - 0(4-1) + 5(6-4) = 25+10=35.$$

Col2: $= -2\begin{vmatrix} 0 & 3 \\ 5 & 2 \end{vmatrix} + 4\begin{vmatrix} 5 & 1 \\ 5 & 2 \end{vmatrix} - 1\begin{vmatrix} 5 & 1 \\ 0 & 3 \end{vmatrix}$

$$= -2(0-15) + 4(10-5) - 1(15-0) = 30+20-15=35.$$

23. $\begin{vmatrix} 0 & 2 & 4 \\ 0 & 4 & 3 \\ 5 & 1 & 2 \end{vmatrix}$. Row1: $= 5\begin{vmatrix} 2 & 4 \\ 4 & 3 \end{vmatrix} - 1\begin{vmatrix} 0 & 4 \\ 0 & 3 \end{vmatrix} + 2\begin{vmatrix} 0 & 2 \\ 0 & 4 \end{vmatrix}$

$$= 5(6-16) - 1(0-0) + 2(0-0) = -50.$$

Col1: $= 0\begin{vmatrix} 4 & 3 \\ 1 & 2 \end{vmatrix} - 0\begin{vmatrix} 2 & 4 \\ 1 & 2 \end{vmatrix} + 5\begin{vmatrix} 2 & 4 \\ 4 & 3 \end{vmatrix}$

$$= 0(8-3) - 0(4-4) + 5(6-16) = -50.$$

25. $\begin{array}{l} x + 2y = 5 \\ -x + y = 1 \end{array}$. $x = \dfrac{\begin{vmatrix} 5 & 2 \\ 1 & 1 \end{vmatrix}}{\begin{vmatrix} 1 & 2 \\ -1 & 1 \end{vmatrix}} = \dfrac{3}{3} = 1.$ $y = \dfrac{\begin{vmatrix} 1 & 5 \\ -1 & 1 \end{vmatrix}}{\begin{vmatrix} 1 & 2 \\ -1 & 1 \end{vmatrix}} = \dfrac{6}{3} = 2.$

27. $\begin{array}{l} x + y = 0 \\ 2x + 3y = 1 \end{array}$. $x = \dfrac{\begin{vmatrix} 0 & 1 \\ 1 & 3 \end{vmatrix}}{\begin{vmatrix} 1 & 1 \\ 2 & 3 \end{vmatrix}} = \dfrac{-1}{1} = -1.$ $y = \dfrac{\begin{vmatrix} 1 & 0 \\ 2 & 1 \end{vmatrix}}{\begin{vmatrix} 1 & 1 \\ 2 & 3 \end{vmatrix}} = \dfrac{1}{1} = 1.$

29. $\begin{array}{l} 6x - 2y = 5 \\ -3x + y = -4 \end{array}$. Determinant of matrix of coeff, $|A| = \begin{vmatrix} 6 & -2 \\ -3 & 1 \end{vmatrix} = 0.$ Cramer's rule cannot be used for this system.

31. $\begin{array}{l} 3x - y = 5 \\ 2x + 2y = 6 \end{array}$. $x = \dfrac{\begin{vmatrix} 5 & -1 \\ 6 & 2 \end{vmatrix}}{\begin{vmatrix} 3 & -1 \\ 2 & 2 \end{vmatrix}} = \dfrac{16}{8} = 2.$ $y = \dfrac{\begin{vmatrix} 3 & 5 \\ 2 & 6 \end{vmatrix}}{\begin{vmatrix} 3 & -1 \\ 2 & 2 \end{vmatrix}} = \dfrac{8}{8} = 1.$

33. $\begin{array}{l} x + 4y = 9 \\ 3x - y = 14 \end{array}$. $x = \dfrac{\begin{vmatrix} 9 & 4 \\ 14 & -1 \end{vmatrix}}{\begin{vmatrix} 1 & 4 \\ 3 & -1 \end{vmatrix}} = \dfrac{-65}{-13} = 5$. $y = \dfrac{\begin{vmatrix} 1 & 9 \\ 3 & 14 \end{vmatrix}}{\begin{vmatrix} 1 & 4 \\ 3 & -1 \end{vmatrix}} = \dfrac{-13}{-13} = 1$.

35. $\begin{array}{l} x + y = 7 \\ x - y = 1 \end{array}$. $x = \dfrac{\begin{vmatrix} 7 & 1 \\ 1 & -1 \end{vmatrix}}{\begin{vmatrix} 1 & 1 \\ 1 & -1 \end{vmatrix}} = \dfrac{-8}{-2} = 4$. $y = \dfrac{\begin{vmatrix} 1 & 7 \\ 1 & 1 \end{vmatrix}}{\begin{vmatrix} 1 & 1 \\ 1 & -1 \end{vmatrix}} = \dfrac{-6}{-2} = 3$.

37. $\begin{array}{l} x + 2y + 3z = 6 \\ x - 2y + z = 0 \\ 3x + y - z = 3 \end{array}$. $|A| = \begin{vmatrix} 1 & 2 & 3 \\ 1 & -2 & 1 \\ 3 & 1 & -1 \end{vmatrix} = 30$. $x = \dfrac{\begin{vmatrix} 6 & 2 & 3 \\ 0 & -2 & 1 \\ 3 & 1 & -1 \end{vmatrix}}{30} = \dfrac{30}{30} = 1$.

$y = \dfrac{\begin{vmatrix} 1 & 6 & 3 \\ 1 & 0 & 1 \\ 3 & 3 & -1 \end{vmatrix}}{30} = \dfrac{30}{30} = 1$. $z = \dfrac{\begin{vmatrix} 1 & 2 & 6 \\ 1 & -2 & 0 \\ 3 & 1 & 3 \end{vmatrix}}{30} = \dfrac{30}{30} = 1$.

39. $\begin{array}{l} 2x - y + z = -1 \\ 3x + y - z = -4 \\ x + 2y + 3z = 7 \end{array}$. $|A| = \begin{vmatrix} 2 & -1 & 1 \\ 3 & 1 & -1 \\ 1 & 2 & 3 \end{vmatrix} = 25$. $x = \dfrac{\begin{vmatrix} -1 & -1 & 1 \\ -4 & 1 & -1 \\ 7 & 2 & 3 \end{vmatrix}}{25} = \dfrac{-25}{25} = -1$.

$y = \dfrac{\begin{vmatrix} 2 & -1 & 1 \\ 3 & -4 & -1 \\ 1 & 7 & 3 \end{vmatrix}}{25} = \dfrac{25}{25} = 1$. $z = \dfrac{\begin{vmatrix} 2 & -1 & -1 \\ 3 & 1 & -4 \\ 1 & 2 & 7 \end{vmatrix}}{25} = \dfrac{50}{25} = 2$.

41. $\begin{array}{l} x - 3y + z = 6 \\ 2x - 5y + 3z = 15 \\ -x + 3y + z = -2 \end{array}$. $|A| = \begin{vmatrix} 1 & -3 & 1 \\ 2 & -5 & 3 \\ -1 & 3 & 1 \end{vmatrix} = 2$. $x = \dfrac{\begin{vmatrix} 6 & -3 & 1 \\ 15 & -5 & 3 \\ -2 & 3 & 1 \end{vmatrix}}{2} = \dfrac{14}{2} = 7$.

$y = \dfrac{\begin{vmatrix} 1 & 6 & 1 \\ 2 & 15 & 3 \\ -1 & 2 & 1 \end{vmatrix}}{2} = \dfrac{2}{2} = 1$. $z = \dfrac{\begin{vmatrix} 1 & -3 & 6 \\ 2 & -5 & 15 \\ -1 & 3 & -2 \end{vmatrix}}{2} = \dfrac{4}{2} = 2$.

43. $\begin{array}{l} 2x - y + z = 1 \\ 3x + 2y + 3z = 15 \\ -x - 2y + 4z = 1 \end{array}$. $|A| = \begin{vmatrix} 2 & -1 & 1 \\ 3 & 2 & 3 \\ -1 & -2 & 4 \end{vmatrix} = 39$. $x = \dfrac{\begin{vmatrix} 1 & -1 & 1 \\ 15 & 2 & 3 \\ 1 & -2 & 4 \end{vmatrix}}{39} = \dfrac{39}{39} = 1$.

$$y = \frac{\begin{vmatrix} 2 & 1 & 1 \\ 3 & 15 & 3 \\ -1 & 1 & 4 \end{vmatrix}}{39} = \frac{117}{39} = 3. \quad z = \frac{\begin{vmatrix} 2 & -1 & 1 \\ 3 & 2 & 15 \\ -1 & -2 & 1 \end{vmatrix}}{39} = \frac{78}{39} = 2.$$

45. $\begin{aligned} 6x + 2y + z &= 3 \\ 3x - y - z &= 5 \\ 2x + 2y - 3z &= 3 \end{aligned}$. $|A| = \begin{vmatrix} 6 & 2 & 1 \\ 3 & -1 & -1 \\ 2 & 2 & -3 \end{vmatrix} = 52.$ $x = \frac{\begin{vmatrix} 3 & 2 & 1 \\ 5 & -1 & -1 \\ 3 & 2 & -3 \end{vmatrix}}{52} = \frac{52}{52} = 1.$

$$y = \frac{\begin{vmatrix} 6 & 3 & 1 \\ 3 & 5 & -1 \\ 2 & 3 & -3 \end{vmatrix}}{52} = \frac{-52}{52} = -1. \quad z = \frac{\begin{vmatrix} 6 & 2 & 3 \\ 3 & -1 & 5 \\ 2 & 2 & 3 \end{vmatrix}}{52} = \frac{-52}{52} = -1.$$

47. $\begin{aligned} 3x + y - z &= 3 \\ -x - y + 2z &= 5 \\ 5x + 2y - z &= 8 \end{aligned}$. $|A| = \begin{vmatrix} 3 & 1 & -1 \\ -1 & -1 & 2 \\ 5 & 2 & -1 \end{vmatrix} = -3.$ $x = \frac{\begin{vmatrix} 3 & 1 & -1 \\ 5 & -1 & 2 \\ 8 & 2 & -1 \end{vmatrix}}{-3} = \frac{-6}{-3} = 2.$

$$y = \frac{\begin{vmatrix} 3 & 3 & -1 \\ -1 & 5 & 2 \\ 5 & 8 & -1 \end{vmatrix}}{-3} = \frac{-3}{-3} = 1. \quad z = \frac{\begin{vmatrix} 3 & 1 & 3 \\ -1 & -1 & 5 \\ 5 & 2 & 8 \end{vmatrix}}{-3} = \frac{-12}{-3} = 4.$$

49. -2. 51. 2.13 53. -460. 55. 438. 57. 0

Section 7.5

Group Discussion: (i) The current model does not distinguish between "ties" in the locations of graves. The graves may not have well defined ordering according to pottery content. e.g.

Consider the graves/pottery situation defined by $A = \begin{bmatrix} 0 & 1 & 0 \\ 1 & 1 & 1 \\ 1 & 1 & 1 \\ 0 & 0 & 1 \end{bmatrix}$.

$$G = AA^t = \begin{bmatrix} 0 & 1 & 0 \\ 1 & 1 & 1 \\ 1 & 1 & 1 \\ 0 & 0 & 1 \end{bmatrix} \begin{bmatrix} 0 & 1 & 1 & 0 \\ 1 & 1 & 1 & 0 \\ 0 & 1 & 1 & 1 \end{bmatrix} = \begin{bmatrix} 1 & 1 & 1 & 0 \\ 1 & 3 & 3 & 1 \\ 1 & 3 & 3 & 1 \\ 0 & 1 & 1 & 1 \end{bmatrix}.$$

$g_{12} = 1$, $g_{13} = 1$, $g_{23} = 3$, $g_{24} = 1$, $g_{34} = 1$. Ordering 1 - (2&3) - 4. We cannot decide between graves 2 and 3 on the basis of this information. There could be large such groups in a real situation. The mathematics does give information, but not complete information. Archaeologists may have other information that enables them to resolve such cases. (ii) There may be inconsistencies in information given by pottery content. e.g.

$$A = \begin{bmatrix} 1 & 0 & 0 & 1 \\ 1 & 1 & 0 & 0 \\ 0 & 1 & 1 & 0 \\ 0 & 0 & 1 & 1 \end{bmatrix}. \; G = AA^t = \begin{bmatrix} 1 & 0 & 0 & 1 \\ 1 & 1 & 0 & 0 \\ 0 & 1 & 1 & 0 \\ 0 & 0 & 1 & 1 \end{bmatrix}\begin{bmatrix} 1 & 1 & 0 & 0 \\ 0 & 1 & 1 & 0 \\ 0 & 0 & 1 & 1 \\ 1 & 0 & 0 & 1 \end{bmatrix} = \begin{bmatrix} 2 & 1 & 0 & 1 \\ 1 & 2 & 1 & 0 \\ 0 & 1 & 2 & 1 \\ 1 & 0 & 1 & 2 \end{bmatrix}.$$

$g_{12}=1, g_{14}=1, g_{23}=1, g_{34}=1$. Ordering 1-2-3-4, but $g_{14}=1$ spoils this. The model does not give consistent information. The order is probably 1-2-3-4 with the pottery common to graves 1 and 4 being misleading? The model gives partial information. Archaeologists would investigate this possibility further.

1. $A = \begin{bmatrix} -1 & 2 \\ 2 & -3 \end{bmatrix}$. $A^t = \begin{bmatrix} -1 & 2 \\ 2 & -3 \end{bmatrix} = A$. Symmetric.

3. $C = \begin{bmatrix} 3 & -1 \\ 2 & 4 \end{bmatrix}$. $C^t = \begin{bmatrix} 3 & 2 \\ -1 & 4 \end{bmatrix} \neq C$. Not symmetric.

5. $E = \begin{bmatrix} 4 & 5 & 6 \\ -1 & 2 & 3 \\ 0 & 1 & 2 \end{bmatrix}$. $E^t = \begin{bmatrix} 4 & -1 & 0 \\ 5 & 2 & 1 \\ 6 & 3 & 2 \end{bmatrix} \neq E$. Not symmetric.

7. $G = \begin{bmatrix} -2 & 4 & 5 & 7 \\ 1 & 0 & 3 & -7 \end{bmatrix}$. $G^t = \begin{bmatrix} -2 & 1 \\ 4 & 0 \\ 5 & 3 \\ 7 & -7 \end{bmatrix} \neq G$. Not symmetric.

9. $\begin{bmatrix} -1 & * \\ 2 & -3 \end{bmatrix} = \begin{bmatrix} -1 & 2 \\ * & -3 \end{bmatrix}$. Thus $A = \begin{bmatrix} -1 & 2 \\ 2 & -3 \end{bmatrix}$.

11. $\begin{bmatrix} 3 & 5 & * \\ * & 8 & 4 \\ -3 & * & 3 \end{bmatrix} = \begin{bmatrix} 3 & * & -3 \\ 5 & 8 & * \\ * & 4 & 3 \end{bmatrix}$. Thus $C = \begin{bmatrix} 3 & 5 & -3 \\ 5 & 8 & 4 \\ -3 & 4 & 3 \end{bmatrix}$.

13. $G = AA^t = \begin{bmatrix} 1 & 0 & 0 & 1 \\ 0 & 0 & 1 & 1 \\ 1 & 1 & 0 & 0 \\ 0 & 1 & 0 & 0 \end{bmatrix}\begin{bmatrix} 1 & 0 & 1 & 0 \\ 0 & 0 & 1 & 1 \\ 0 & 1 & 0 & 0 \\ 1 & 1 & 0 & 0 \end{bmatrix} = \begin{bmatrix} 2 & 1 & 1 & 0 \\ 1 & 2 & 0 & 0 \\ 1 & 0 & 2 & 1 \\ 0 & 0 & 1 & 1 \end{bmatrix}$.

$g_{12}=1, g_{13}=1, g_{34}=1$. Ordering 2 - 1 - 3 - 4 (either way)

15. $G = AA^t = \begin{bmatrix} 0 & 1 & 0 & 0 \\ 1 & 1 & 0 & 0 \\ 1 & 0 & 0 & 1 \\ 1 & 0 & 1 & 1 \end{bmatrix}\begin{bmatrix} 0 & 1 & 1 & 1 \\ 1 & 1 & 0 & 0 \\ 0 & 0 & 0 & 1 \\ 0 & 0 & 1 & 1 \end{bmatrix} = \begin{bmatrix} 1 & 1 & 0 & 0 \\ 1 & 2 & 1 & 1 \\ 0 & 1 & 2 & 2 \\ 0 & 1 & 2 & 3 \end{bmatrix}$.

$g_{12}=1, g_{23}=1, g_{24}=1, g_{34}=2$. Ordering 1 - 2 - 3 - 4.

17. $G = AA^t = \begin{bmatrix} 0 & 1 & 1 & 0 \\ 0 & 1 & 0 & 1 \\ 1 & 0 & 0 & 0 \\ 0 & 0 & 1 & 0 \\ 1 & 0 & 0 & 1 \end{bmatrix}\begin{bmatrix} 0 & 0 & 1 & 0 & 1 \\ 1 & 1 & 0 & 0 & 0 \\ 1 & 0 & 0 & 1 & 0 \\ 0 & 1 & 0 & 0 & 1 \end{bmatrix} = \begin{bmatrix} 2 & 1 & 0 & 1 & 0 \\ 1 & 2 & 0 & 0 & 1 \\ 0 & 0 & 1 & 0 & 1 \\ 1 & 0 & 0 & 1 & 0 \\ 0 & 1 & 1 & 0 & 2 \end{bmatrix}$.

$g_{12}=1, g_{14}=1, g_{25}=1, g_{35}=1$. Ordering 4 - 1 - 2 - 5 - 3.

19. The element g_{ij} of the matrix G (=AA^t) is equal to the number of types of pottery common to graves i and j. g_{ji} is equal to the number of pottery types common to graves j and i. Thus $g_{ij} = g_{ji}$ implying that the matrix G is symmetric.

21. (a) $F = AA^t$. f_{ij} = element in row i, column j of F
$$= (\text{row } i \text{ of } A) \times (\text{column } j \text{ of } A^t)$$

$$= [\; a_{i1} \quad a_{i2} \quad \cdots \quad a_{in} \;] \begin{bmatrix} a_{j1} \\ a_{j2} \\ \vdots \\ a_{jn} \end{bmatrix}$$

$$= a_{i1}a_{j1} + a_{i2}a_{j2} + \cdots + a_{in}a_{jn}$$

Each term in this sum will be either 1 or 0. For example the term $a_{i2}a_{j2}$ will be 1 if a_{i2} and a_{j2} are both 1; that is if person 2 is a friend to both i and j. It will be 0 if person 2 is not a friend of both i and j. Thus the number of 1's in this expression for f_{ij} , (the actual value of f_{ij}), is the number of friends common to people i and j.
(b) If friendship is not mutual then person i could consider person j to be a friend, but not vice-versa. The matrix F would not then be symmetric.

Chapter 7 Project
(b) Groups are group$_A$, group$_B$, and group$_C$ with encoding matrices A, B, and C. Let message be in the form of the matrix X. Then group$_B$ receives the message in coded form AX. Group$_C$ receives the messsage in coded form BAX. Thus to decode the matrix group$_C$ needs the decoding matrix $(BA)^{-1}$. A can send and receive messages from C using the coding matrix AB and the decoding matrix $(BA)^{-1}$.

Chapter 7 Review Exercises

1. (a) $\begin{bmatrix} 1 & 5 \\ 6 & 9 \end{bmatrix}$, 2x2 matrix.   (b) $\begin{bmatrix} 4 & 7 & 3 \\ -2 & 8 & -1 \end{bmatrix}$, 2x3 matrix.

(c) $\begin{bmatrix} 7 & 4 & -8 & 5 \\ 5 & -2 & 6 & 8 \\ 4 & 6 & -3 & 7 \end{bmatrix}$, 3x4 matrix.

(d) $[\; 2 \quad 4 \quad -2 \quad 7 \;]$, 1x4 matrix, a row matrix.

2. $A = \begin{bmatrix} 4 & 7 & 3 \\ -2 & 8 & -1 \end{bmatrix}$, $B = \begin{bmatrix} 7 & 4 & -8 & 5 & 9 \\ 5 & -2 & 6 & 8 & -5 \\ 4 & 6 & -3 & 7 & 12 \end{bmatrix}$

Chapter 7 Review Exercises

(a) $a_{13}=3$ (b) $a_{23}=-1$ (c) $a_{22}=8$ (d) $b_{31}=4$
(e) $b_{22}=-2$ (f) $b_{34}=7$ (g) $b_{25}=-5$

3. (a) $\begin{bmatrix} x & 2y \\ x-y & x+3z \end{bmatrix} = \begin{bmatrix} 5 & 8 \\ 1 & 11 \end{bmatrix}$. Compare corresponding elements: $x=5$, $2y=8$, $x-y=1$, $x+3z=11$. Thus $x=5$, $y=4$, $z=2$.

(b) $\begin{bmatrix} x^2 & x-y \\ z & x+y+z \end{bmatrix} = \begin{bmatrix} 9 & 1 \\ 2 & -5 \end{bmatrix}$. Compare corresponding elements: $x^2=9$, $x-y=1$, $z=2$, $x+y+z=-5$. Thus $x=\pm3$. $y=x-1=2$ or -4. $z=2$. But must have $x+y+z=-5$. Only $x=-3$, $y=-4$, $z=2$ satisfies this.

4. (a) $A+B=\begin{bmatrix} 1 & 8 \\ 1 & 8 \end{bmatrix}$.   (b) $A+3D$, DNE. A is 2x2, 3D is 2x3

(c) $5A-2B=\begin{bmatrix} 33 & 5 \\ 19 & 26 \end{bmatrix}$,   (d) $C+3D=\begin{bmatrix} 25 & 19.4 & -21.1 \\ 18 & 2 & 18.5 \end{bmatrix}$.

5. (a) $AB=\begin{bmatrix} -8 & 29 \\ -12 & 21 \end{bmatrix}$.   (b) $AC=\begin{bmatrix} 14 & 33 & -9 \\ 12 & 27 & -9 \end{bmatrix}$.

(c) AD, DNE, A is 2x2 & D is 3x3. #cols in A ≠ #rows in D.

(d) $A^2=\begin{bmatrix} -11 & 28 \\ -21 & 24 \end{bmatrix}$.   (e) $2AB+A^3=\begin{bmatrix} -16 & 58 \\ -24 & 42 \end{bmatrix}+\begin{bmatrix} -95 & 124 \\ -93 & 60 \end{bmatrix}=\begin{bmatrix} -111 & 182 \\ -117 & 102 \end{bmatrix}$

(f) $AC-3BC=\begin{bmatrix} 14 & 33 & -9 \\ 12 & 27 & -9 \end{bmatrix}-\begin{bmatrix} 45 & 105 & -30 \\ 42 & 96 & -30 \end{bmatrix}=\begin{bmatrix} -31 & -72 & 21 \\ -30 & -69 & 21 \end{bmatrix}$.

6. (a) AB is 2x2.   (b) AC is 2x3.   (c) ABC is 2x3.
(d) CD is 2x3.   (e) AE, DNE.   (f) CDE is 2x1.
(g) A^2C+2BD is 2x3.   (h) $D^3E + 4E$ is 3x1.

7. (a) $\begin{bmatrix} 2 & 0 \\ 0 & 6 \end{bmatrix}$. $[A:I_2] = \begin{bmatrix} 2 & 0 & 1 & 0 \\ 0 & 6 & 0 & 1 \end{bmatrix} \underset{(1/6)R2}{\overset{(1/2)R1}{\sim}} \begin{bmatrix} 1 & 0 & 1/2 & 0 \\ 0 & 1 & 0 & 1/6 \end{bmatrix}$

$A^{-1} = \begin{bmatrix} 1/2 & 0 \\ 0 & 1/6 \end{bmatrix}$.

(b) $\begin{bmatrix} 1 & 2 \\ 1 & 3 \end{bmatrix}$. $[A:I_2] = \begin{bmatrix} 1 & 2 & 1 & 0 \\ 1 & 3 & 0 & 1 \end{bmatrix}$ Add $(-1)\tilde{R}1$ to R2 $\begin{bmatrix} 1 & 2 & 1 & 0 \\ 0 & 1 & -1 & 1 \end{bmatrix}$

Add $(-2)\tilde{R}2$ to R1 $\begin{bmatrix} 1 & 0 & 3 & -2 \\ 0 & 1 & -1 & 1 \end{bmatrix}$. $A^{-1} = \begin{bmatrix} 3 & -2 \\ -1 & 1 \end{bmatrix}$.

(c) $\begin{bmatrix} 2 & 4 \\ 1 & 2 \end{bmatrix}$. $[A:I_2] = \begin{bmatrix} 2 & 4 & 1 & 0 \\ 1 & 2 & 0 & 1 \end{bmatrix} \underset{(1/2)R1}{\sim} \begin{bmatrix} 1 & 2 & 1/2 & 0 \\ 1 & 2 & 0 & 1 \end{bmatrix}$

Add $(-1)\tilde{R}1$ to R2 $\begin{bmatrix} 1 & 2 & 1/2 & 0 \\ 0 & 0 & -1/2 & 1 \end{bmatrix}$. Inverse does not exist.

(d) $[A:I_3] = \begin{bmatrix} 1 & 1 & 5 & 1 & 0 & 0 \\ 1 & 2 & 8 & 0 & 1 & 0 \\ -1 & 1 & 2 & 0 & 0 & 1 \end{bmatrix}$ Add $(-1)\tilde{R}1$ to R2 $\begin{bmatrix} 1 & 1 & 5 & 1 & 0 & 0 \\ 0 & 1 & 3 & -1 & 1 & 0 \\ 0 & 2 & 7 & 1 & 0 & 1 \end{bmatrix}$
Add R1 to R3

Add $(-1)\tilde{R}2$ to R1 $\begin{bmatrix} 1 & 0 & 2 & 2 & -1 & 0 \\ 0 & 1 & 3 & -1 & 1 & 0 \\ 0 & 0 & 1 & 3 & -2 & 1 \end{bmatrix}$
Add $(-2)R2$ to R3

Add $(-2)\tilde{R}3$ to R1 $\begin{bmatrix} 1 & 0 & 0 & -4 & 3 & -2 \\ 0 & 1 & 0 & -10 & 7 & -3 \\ 0 & 0 & 1 & 3 & -2 & 1 \end{bmatrix}$. $A^{-1} = \begin{bmatrix} -4 & 3 & -2 \\ -10 & 7 & -3 \\ 3 & -2 & 1 \end{bmatrix}$.
Add $(-3)R3$ to R2

(e) $[A:I_3] = \begin{bmatrix} 1 & 2 & -1 & 1 & 0 & 0 \\ 3 & 1 & 4 & 0 & 1 & 0 \\ 5 & 5 & 2 & 0 & 0 & 1 \end{bmatrix}$ Add $(-3)\tilde{R}1$ to R2
Add $(-5)R1$ to R3

$\begin{bmatrix} 1 & 2 & -1 & 1 & 0 & 0 \\ 0 & -5 & 7 & -3 & 1 & 0 \\ 0 & -5 & 7 & -5 & 0 & 1 \end{bmatrix}$ $(-1/5)R2$ $\begin{bmatrix} 1 & 2 & -1 & 1 & 0 & 0 \\ 0 & 1 & -7/5 & 3/5 & -1/5 & 0 \\ 0 & -5 & 7 & -5 & 0 & 1 \end{bmatrix}$

Add $(-2)\tilde{R}2$ to R1 $\begin{bmatrix} 1 & 0 & 9/5 & -1/5 & 2/5 & 0 \\ 0 & 1 & -7/5 & 3/5 & -1/5 & 0 \\ 0 & 0 & 0 & -2 & -1 & 1 \end{bmatrix}$. Inverse DNE.
Add $(5)R2$ to R3

8. (a) $\begin{bmatrix} 1 & 2 \\ 1 & 3 \end{bmatrix}\begin{bmatrix} x \\ y \end{bmatrix} = \begin{bmatrix} 8 \\ 11 \end{bmatrix}$. $\begin{bmatrix} x \\ y \end{bmatrix} = \begin{bmatrix} 1 & 2 \\ 1 & 3 \end{bmatrix}^{-1}\begin{bmatrix} 8 \\ 11 \end{bmatrix} = \begin{bmatrix} 3 & -2 \\ -1 & 1 \end{bmatrix}\begin{bmatrix} 8 \\ 11 \end{bmatrix} = $

$\begin{bmatrix} 2 \\ 3 \end{bmatrix}$. $x=2$, $y=3$.

(b) $\begin{bmatrix} 1 & 1 & 1 \\ 0 & 1 & 2 \\ 2 & 1 & -4 \end{bmatrix}\begin{bmatrix} x \\ y \\ z \end{bmatrix} = \begin{bmatrix} 3 \\ 5 \\ 3 \end{bmatrix}$. $\begin{bmatrix} x \\ y \\ z \end{bmatrix} = \begin{bmatrix} 1 & 1 & 1 \\ 0 & 1 & 2 \\ 2 & 1 & -4 \end{bmatrix}^{-1}\begin{bmatrix} 3 \\ 5 \\ 3 \end{bmatrix} = $

$\begin{bmatrix} 3/2 & -5/4 & -1/4 \\ -1 & 3/2 & 1/2 \\ 1/2 & -1/4 & -1/4 \end{bmatrix}\begin{bmatrix} 3 \\ 5 \\ 3 \end{bmatrix} = \begin{bmatrix} -5/2 \\ 6 \\ -1/2 \end{bmatrix}$. $x=-5/2$, $y=6$, $z=-1/2$.

9. (a) $\begin{vmatrix} 2 & 3 \\ 1 & 6 \end{vmatrix} = 12-3=9$. (b) $\begin{vmatrix} -3 & 5 \\ -1 & 2 \end{vmatrix} = -6+5=-1$.

(c) $\begin{vmatrix} 1 & 5 & 7 \\ 2 & 3 & -2 \\ 4 & 2 & 1 \end{vmatrix} = 1\begin{vmatrix} 3 & -2 \\ 2 & 1 \end{vmatrix} - 5\begin{vmatrix} 2 & -2 \\ 4 & 1 \end{vmatrix} + 7\begin{vmatrix} 2 & 3 \\ 4 & 2 \end{vmatrix} =$

$1(3+4) - 5(2+8) + 7(4-12) = 7-50-56 = -99.$

(d) Expand using 2nd row: $\begin{vmatrix} -3 & 5 & 7 \\ 1 & 0 & 0 \\ 4 & 2 & 9 \end{vmatrix} = -1\begin{vmatrix} 5 & 7 \\ 2 & 9 \end{vmatrix} = -1(45-14) = -31.$

(e) Expand using 3rd col: $\begin{vmatrix} 2 & 7 & 2 \\ 3 & 1 & 0 \\ 0 & 2 & 0 \end{vmatrix} = 2\begin{vmatrix} 3 & 1 \\ 0 & 2 \end{vmatrix} = 2(6-0) = 12.$

10. (a) $\begin{array}{l} x + 2y = 8 \\ 2x + y = 7 \end{array}$. $x = \dfrac{\begin{vmatrix} 8 & 2 \\ 7 & 1 \end{vmatrix}}{\begin{vmatrix} 1 & 2 \\ 2 & 1 \end{vmatrix}} = \dfrac{-6}{-3} = 2.$ $y = \dfrac{\begin{vmatrix} 1 & 8 \\ 2 & 7 \end{vmatrix}}{\begin{vmatrix} 1 & 2 \\ 2 & 1 \end{vmatrix}} = \dfrac{-9}{-3} = 3.$

(b) $\begin{array}{l} x + 2y - z = 3 \\ 2x - y + 4z = 7 \\ -x + 3y + 5z = 6 \end{array}$. $|A| = \begin{vmatrix} 1 & 2 & -1 \\ 2 & -1 & 4 \\ -1 & 3 & 5 \end{vmatrix} = -50.$

$x = \dfrac{\begin{vmatrix} 3 & 2 & -1 \\ 7 & -1 & 4 \\ 6 & 3 & 5 \end{vmatrix}}{-50} = \dfrac{-100}{-50} = 2.$ $y = \dfrac{\begin{vmatrix} 1 & 3 & -1 \\ 2 & 7 & 4 \\ -1 & 6 & 5 \end{vmatrix}}{-50} = \dfrac{-50}{50} = 1.$

$z = \dfrac{\begin{vmatrix} 1 & 2 & 3 \\ 2 & -1 & 7 \\ -1 & 3 & 6 \end{vmatrix}}{-50} = \dfrac{-50}{-50} = 1.$

11. (a) $A = \begin{bmatrix} -1 & 2 \\ 2 & -3 \end{bmatrix}$. $A^t = \begin{bmatrix} -1 & 2 \\ 2 & -3 \end{bmatrix}$. Symmetric.

(b) $B = \begin{bmatrix} -2 & -3 \\ 3 & 4 \end{bmatrix}$. $B^t = \begin{bmatrix} -2 & 3 \\ -3 & 4 \end{bmatrix}$. Not symmetric.

(c) $C = \begin{bmatrix} 4 & 5 \\ -2 & 3 \\ 7 & 0 \end{bmatrix}$. $C^t = \begin{bmatrix} 4 & -2 & 7 \\ 5 & 3 & 0 \end{bmatrix}$. Not symmetric.

(d) $D = \begin{bmatrix} 2 & 3 & -5 \\ 3 & 0 & 4 \\ -5 & 4 & 6 \end{bmatrix}$. $D^t = \begin{bmatrix} 2 & 3 & -5 \\ 3 & 0 & 4 \\ -5 & 4 & 6 \end{bmatrix}$. Symmetric.

12. $P = \begin{bmatrix} .96 & .01 & .015 \\ .03 & .98 & .005 \\ .01 & .01 & .98 \end{bmatrix}$. $X_0 = \begin{bmatrix} 60 \\ 140 \\ 65 \end{bmatrix}$. $X_i = PX_{i-1}$, and $X_n = P^n X_0$.

1995 is X_0. 1996 is X_1, $= PX_0$. etc. In 2010, $X_{15} = P^{15} X_0$.

	1995	1996	1997	1998	1999	2000
City	60	59.98	59.95	59.94	59.93	59.92
Suburb	140	139.33	138.67	138.02	137.40	136.78
Nonmetro	65	65.7	66.38	67.04	67.68	68.30

In 2010, $X_{15} = P^{15} X_0$ = City:60, Suburb:131.45, NonMetro:73.56. It is interesting that the city population dropped then started rising again and has reached its original level. Long term: City:61.83, Suburb:114.83 NonMetro:88.33.

13. $P = AA^t = \begin{bmatrix} 0 & 1 & 0 & 1 & 0 & 1 \\ 0 & 0 & 0 & 0 & 1 & 0 \\ 1 & 0 & 0 & 1 & 1 & 1 \\ 1 & 1 & 0 & 0 & 0 & 0 \end{bmatrix} \begin{bmatrix} 0 & 0 & 1 & 1 \\ 1 & 0 & 0 & 1 \\ 0 & 0 & 0 & 0 \\ 1 & 0 & 1 & 0 \\ 0 & 1 & 1 & 0 \\ 1 & 0 & 1 & 0 \end{bmatrix} = \begin{bmatrix} 3 & 0 & 2 & 1 \\ 0 & 1 & 1 & 0 \\ 2 & 1 & 4 & 1 \\ 1 & 0 & 1 & 2 \end{bmatrix}$.

$p_{13} = 2$, producers 1 and 3 have 2 customers in common.
$p_{14} = 1$, producers 1 and 4 have 1 customer in common.
$p_{23} = 1$, producers 2 and 3 have 1 customer in common.
$p_{34} = 1$, producers 3 and 4 have 1 customer in common.
Thus producers 1 and 3 are in most direct competition.

14.

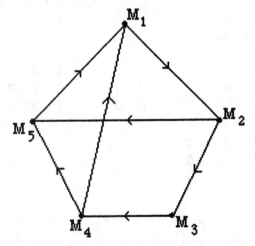

$D = \begin{bmatrix} 0 & 1 & 2 & 3 & 2 \\ 2 & 0 & 1 & 2 & 1 \\ 2 & 3 & 0 & 1 & 2 \\ 1 & 2 & 3 & 0 & 1 \\ 1 & 2 & 3 & 4 & 0 \end{bmatrix} \begin{matrix} 8 \\ 6 \\ 8 \\ 7 \\ 10 \end{matrix}$

$M_2, M_4, (M_1 \text{ and } M_3), M_5$.

Chapter 7 Test

1. (a) $\begin{bmatrix} 3 & 2 \\ 4 & 5 \\ 6 & 8 \end{bmatrix}$ 3x2 matrix. (b) $\begin{bmatrix} 1 & 0 & -4 \\ 6 & 8 & 7 \end{bmatrix}$, 2x3.

(c) $\begin{bmatrix} 2 & 1 & 3 \\ 4 & 5 & 7 \\ 6 & 0 & 8 \end{bmatrix}$, 3x3. (d) [3 2 4 5 6], 1x5.

2. $A = \begin{bmatrix} 2 & 0 & 1 & -7 & 4 \\ 9 & 7 & -2 & 0 & 6 \\ 8 & 3 & 5 & 1 & 2 \end{bmatrix}$ (a) $a_{14}=-7$. (b) $a_{22}=7$. (c) $a_{31}=8$.

3. $\begin{bmatrix} x & 3y \\ x+y & x+y+z \end{bmatrix} = \begin{bmatrix} 1 & 6 \\ 3 & 8 \end{bmatrix}$. Compare corresponding elements: $x=1$, $3y=6$, $x+y=3$, $x+y+z=8$. Thus $x=1$, $y=2$, $z=5$.

4. $A = \begin{bmatrix} 0 & -1 & 3 \\ 4 & 2 & 8 \end{bmatrix}$, $B = \begin{bmatrix} 2 & -5 & 2 \\ 5 & 8 & 9 \end{bmatrix}$ (a) $A+B = \begin{bmatrix} 2 & -6 & 5 \\ 9 & 10 & 17 \end{bmatrix}$

   (b) $3A-2B = \begin{bmatrix} 0 & -3 & 9 \\ 12 & 6 & 24 \end{bmatrix} - \begin{bmatrix} 4 & -10 & 4 \\ 10 & 16 & 18 \end{bmatrix} = \begin{bmatrix} -4 & 7 & 5 \\ 2 & -10 & 6 \end{bmatrix}$

5. $A = \begin{bmatrix} 1 & 3 \\ -4 & 2 \end{bmatrix}$, $B = \begin{bmatrix} 0 & 5 & -1 \\ 3 & 2 & 6 \end{bmatrix}$ (a) $AB = \begin{bmatrix} 9 & 11 & 17 \\ 6 & -16 & 16 \end{bmatrix}$.

   (b) BA, B is 2x3 & A is 2x2, #cols in $B \neq$ #rows in A. DNE.

   (c) $A^2 = \begin{bmatrix} 1 & 3 \\ -4 & 2 \end{bmatrix}\begin{bmatrix} 1 & 3 \\ -4 & 2 \end{bmatrix} = \begin{bmatrix} -11 & 9 \\ -12 & -8 \end{bmatrix}$.

6. A 2x2, B 2x3, C 3x1. (a) AB is 2x3 (b) ABC is 2x1

   (c) BA DNE. (d) A^2B+2AB is 2x3 (e) C^tB^tA is 1x2.

7. $AB = \begin{bmatrix} 1 & 2 \\ 2 & 5 \end{bmatrix}\begin{bmatrix} 5 & -2 \\ -2 & 1 \end{bmatrix} = \begin{bmatrix} 1 & 0 \\ 0 & 1 \end{bmatrix}$, $BA = \begin{bmatrix} 5 & -2 \\ -2 & 1 \end{bmatrix}\begin{bmatrix} 1 & 2 \\ 2 & 5 \end{bmatrix} = \begin{bmatrix} 1 & 0 \\ 0 & 1 \end{bmatrix}$.

   $AB=BA=I_2$. Thus B is the inverse of A.

8. $[A:I_3] = \begin{bmatrix} 1 & 2 & 0 & 1 & 0 & 0 \\ 0 & 1 & -1 & 0 & 1 & 0 \\ -1 & 1 & -2 & 0 & 0 & 1 \end{bmatrix}$ Add R1 to R2 $\overset{\approx}{}$ $\begin{bmatrix} 1 & 2 & 0 & 1 & 0 & 0 \\ 0 & 1 & -1 & 0 & 1 & 0 \\ 0 & 3 & -2 & 1 & 0 & 1 \end{bmatrix}$

   Add (-2)R2 to R1 $\overset{\approx}{}$ $\begin{bmatrix} 1 & 0 & 2 & 1 & -2 & 0 \\ 0 & 1 & -1 & 0 & 1 & 0 \\ 0 & 0 & 1 & 1 & -3 & 1 \end{bmatrix}$
   Add (-3)R2 to R2

   Add (-2)R3 to R1 $\overset{\approx}{}$ $\begin{bmatrix} 1 & 0 & 0 & -1 & 4 & -2 \\ 0 & 1 & 0 & 1 & -2 & 1 \\ 0 & 0 & 1 & 1 & -3 & 1 \end{bmatrix}$. $A^{-1} = \begin{bmatrix} -1 & 4 & -2 \\ 1 & -2 & 1 \\ 1 & -3 & 1 \end{bmatrix}$.
   Add R3 to R2

9. $\begin{array}{l} x - 4y = -1 \\ 2x - 7y = -1 \end{array}$. $\begin{bmatrix} 1 & -4 \\ 2 & -7 \end{bmatrix}\begin{bmatrix} x \\ y \end{bmatrix} = \begin{bmatrix} -1 \\ -1 \end{bmatrix}$. $\begin{bmatrix} x \\ y \end{bmatrix} = \begin{bmatrix} 1 & -4 \\ 2 & -7 \end{bmatrix}^{-1}\begin{bmatrix} -1 \\ -1 \end{bmatrix}$.

   $\begin{bmatrix} 1 & -4 & 1 & 0 \\ 2 & -7 & 0 & 1 \end{bmatrix} \overset{\approx}{} \begin{bmatrix} 1 & -4 & 1 & 0 \\ 0 & 1 & -2 & 1 \end{bmatrix} \overset{\approx}{} \begin{bmatrix} 1 & 0 & -7 & 4 \\ 0 & 1 & -2 & 1 \end{bmatrix}$. $\begin{bmatrix} 1 & -4 \\ 2 & -7 \end{bmatrix}^{-1} = \begin{bmatrix} -7 & 4 \\ -2 & 1 \end{bmatrix}$.

$$\begin{bmatrix} x \\ y \end{bmatrix} = \begin{bmatrix} -7 & 4 \\ -2 & 1 \end{bmatrix}\begin{bmatrix} -1 \\ -1 \end{bmatrix} = \begin{bmatrix} 3 \\ 1 \end{bmatrix}. \quad x=3, \ y=1.$$

10. (a) $\begin{vmatrix} 2 & -1 \\ 4 & 3 \end{vmatrix} = 6-(-4) = 6+4 = 10.$

(b) $\begin{vmatrix} 3 & 0 & 5 \\ 3 & 2 & -1 \\ 7 & 2 & 8 \end{vmatrix} = 3\begin{vmatrix} 2 & -1 \\ 2 & 8 \end{vmatrix} - 0\begin{vmatrix} 3 & -1 \\ 7 & 8 \end{vmatrix} + 5\begin{vmatrix} 3 & 2 \\ 7 & 2 \end{vmatrix} =$

$3(16+2) - 0(24+7) + 5(6-14) = 54-0-40 = 14.$

11. $\begin{array}{l} x + 3y = -1 \\ 4x + 2y = 6 \end{array} \cdot \ |A| = \begin{vmatrix} 1 & 3 \\ 4 & 2 \end{vmatrix} = -10.$

$x = \dfrac{\begin{vmatrix} -1 & 3 \\ 6 & 2 \end{vmatrix}}{-10} = \dfrac{-20}{-10} = 2. \quad y = \dfrac{\begin{vmatrix} 1 & -1 \\ 4 & 6 \end{vmatrix}}{-10} = \dfrac{10}{-10} = -1.$

12. (a) $A = \begin{bmatrix} 1 & 2 \\ 3 & 4 \end{bmatrix}. \ A^t = \begin{bmatrix} 1 & 3 \\ 2 & 4 \end{bmatrix} \neq A.$ Not symmetric.

(b) $B = \begin{bmatrix} 2 & 3 & 1 \\ 3 & 4 & -5 \\ 1 & -5 & 0 \end{bmatrix}. \ B^t = \begin{bmatrix} 2 & 3 & 1 \\ 3 & 4 & -5 \\ 1 & -5 & 0 \end{bmatrix} = B.$ Symmetric.

13.

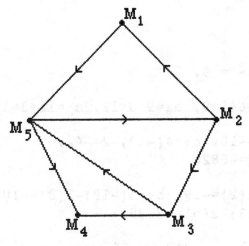

$$D = \begin{bmatrix} 0 & 2 & 3 & 2 & 1 \\ 1 & 0 & 1 & 2 & 2 \\ 3 & 2 & 0 & 1 & 1 \\ \infty & \infty & \infty & 0 & \infty \\ 2 & 1 & 2 & 1 & 0 \end{bmatrix} \begin{matrix} 8 \\ 6 \\ 7 \\ \infty \\ 6 \end{matrix}$$

$(M_2$ and $M_5), M_3, M_1, M_4.$

14. (a) Daily traffic from zone II to zone IV = 700.
(b) The daily traffic from zone III to zone I = 500.
(c) Internal traffic in zone II = 1000.
(d) Largest traffic, 3400 vehicles, from zone IV to zone III.

Chapter 8

1. $a_n=2n+3$. $a_1=2(1)+3=5$, $a_2=2(2)+3=7$, $a_3=2(3)+3=9$, $a_4=2(4)+3=11$.

3. $a_n=n^3+3n+2$. $a_1=(1)^3+3(1)+2=6$, $a_2=(2)^3+3(2)+2=16$, $a_3=(3)^3+3(3)+2=38$, $a_4=(4)^3+3(4)+2=78$.

5. $a_n=2^n$. $a_1=2^1=2$, $a_2=2^2=4$, $a_3=2^3=8$, $a_4=2^4=16$.

7. $a_n=2n(n+1)$. $a_1=2(1)(1+1)=4$. $a_2=2(2)(2+1)=12$. $a_3=2(3)(3+1)=24$. $a_4=2(4)(4+1)=40$.

9. $a_n=2n-5$. $a_1=2(1)-5=-3$, $a_2=2(2)-5=-1$, $a_3=2(3)-5=1$, $a_4=2(4)-5=3$, $a_5=2(5)-5=5$.

11. $a_n=2n(4n-5)$. $a_1=2(1)(4(1)-5)=-2$, $a_2=2(2)(4(2)-5)=12$, $a_3=2(3)(4(4)-5)=42$, $a_4=2(4)(4(4)-5)=88$, $a_5=2(5)(4(5)-5)=150$.

13. $a_n = \dfrac{n-3}{2+3n}$. $a_1 = \dfrac{1-3}{2+3(1)} = -\dfrac{2}{5}$, $a_2 = \dfrac{2-3}{2+3(2)} = -\dfrac{1}{8}$, $a_3 = \dfrac{3-3}{2+3(3)} = 0$, $a_4 = \dfrac{4-3}{2+3(4)} = \dfrac{1}{14}$, $a_5 = \dfrac{5-3}{2+3(5)} = \dfrac{2}{17}$.

15. $a_n=1+(\frac{1}{2})^n$. $a_1=1+(\frac{1}{2})^1 = \dfrac{3}{2}$, $a_2=1+(\frac{1}{2})^2 = \dfrac{5}{4}$, $a_3=1+(\frac{1}{2})^3 = \dfrac{9}{8}$, $a_4=1+(\frac{1}{2})^4 = \dfrac{17}{16}$, $a_5=1+(\frac{1}{2})^5 = \dfrac{33}{32}$.

17. $a_n=5n-1$. 4th term $= 5(4)-1=19$.

19. $a_n=\dfrac{n-7}{2n+3}$. 7th term $= \dfrac{7-7}{2(7)+3} = 0$.

21. $a_n=3^{n-10}$. 12th term $= 3^{12-10} = 3^2 = 9$.

23. $a_1=3$, $a_n=a_{n-1} + 3$. $a_2=3+3=6$, $a_3=6+3=9$, $a_4=9+3=12$, $a_2=12+3=15$.

25. $a_1=-2$, $a_n=4a_{n-1} - 2$. $a_2=4(-2)-2=-10$, $a_3=4(-10)-2=-42$, $a_4=4(-42)-2=-170$, $a_5=4(-170)-2=-682$.

27. $a_1=-3$, $a_n=5a_{n-1} - 2n$. $a_2=5(-3)-2(2)=-19$, $a_3=5(-19)-2(3)=-101$, $a_4=5(-101)-2(4)=-513$, $a_5=5(-513)-2(5)=-2575$.

29. $a_1=2$, $a_n= \dfrac{n+2}{a_{n-1} + 3}$. $a_2= \dfrac{2+2}{2+3} = \dfrac{4}{5}$, $a_3= \dfrac{3+2}{4/5+3} = \dfrac{25}{19}$, $a_4= \dfrac{4+2}{25/19+3} = \dfrac{57}{41}$, $a_5= \dfrac{5+2}{57/41+3} = \dfrac{287}{180}$.

Section 8.1

31. $a_1=1$, $a_n=(a_{n-1})^2$. $a_2=(1)^2 = 1$, $a_3=(1)^2 = 1$, $a_4=(1)^2 = 1$, $a_5=(1)^2 = 1$.

33. $\sum\limits_{k=1}^{4}(2k+3)= 2(1)+3 + 2(2)+3 + 2(3)+3 + 2(4)+3 = 5+7+9+11 = 32.$

35. $\sum\limits_{k=1}^{6} k(k+3)= 1(1+3) + 2(2+3) + 3(3+3) + 4(4+3) + 5(5+3) + 6(6+3) =$
$4+10+18+28+40+54 = 154.$

37. $\sum\limits_{k=1}^{4}(2k^2+3k-4)= (2(1)^2+3(1)-4) + (2(2)^2+3(2)-4) + (2(3)^2+3(3)-4)+$
$(2(4)^2+3(4)-4) = 1+10+23+40 = 74.$

39. $\sum\limits_{k=1}^{4}\frac{4}{2k+1} = \frac{4}{2(1)+1} + \frac{4}{2(2)+1} + \frac{4}{2(3)+1} + \frac{4}{2(4)+1} = \frac{4}{3}+\frac{4}{5}+\frac{4}{7}+\frac{4}{9} = \frac{992}{315}.$

41. $\sum\limits_{k=1}^{10}(-1)^k= (-1)^1 + (-1)^2 + (-1)^3 + (-1)^4 + (-1)^5 + (-1)^6 + (-1)^7+$
$(-1)^8 + (-1)^9 + (-1)^{10} = 0.$

43. $\sum\limits_{k=1}^{8}27 = 27+27+27+27+27+27+27+27=216.$

45. $1 + 2 + 3 + 4 + 5 + 6 = \sum\limits_{k=1}^{6} k.$

47. $1 - 2 + 3 - 4 + 5 - 6 = \sum\limits_{k=1}^{6} (-1)^{k+1}k.$

49. $1 + 3 + 5 + 7 + 9 + 11 = \sum\limits_{k=1}^{6} (2k+1).$

51. $\frac{1}{4} + \frac{2}{5} + \frac{3}{6} + \frac{4}{7} = \sum\limits_{k=1}^{4} \frac{k}{k+3}.$

225

53. S_3, $a_k=2k$. $S_3 = 2(1)+2(2)+2(3) = 2+4+6 = 12$.

55. $a_k = k^2+3$. $S_5 = 1^2+3 + 2^2+3 + 3^2+3 + 4^2+3 + 5^2+3 =$
$$4+7+12+19+28 = 70.$$

57. $a_k = k$. $S_6 = 1+2+3+4+5+6 = 21$.

59. $a_k = 1/k$. $S_5 = 1/1 + 1/2 + 1/3 + 1/4 + 1/5 = 137/60$.

61. $a_n = 2n+3$, first five terms. $S_5 = 2(1)+3 + 2(2)+3 + 2(3)+3 +$
$$2(4) + 3 + 2(5) + 3 = 5+7+9+11+13 = 45. \quad \overline{a}=45/5=9.$$

63. $a_n = n(n-3)$, first five terms. $S_5 = 1(1-3) + 2(2-3) + 3(3-3)+$
$$4(4-3) + 5(5-3) = -2-2+0+4+10 = 10. \quad \overline{a}=10/5=2.$$

65. $a_n = (-1)^n$, first fifty terms. $S_{50} = (-1)^1 + (-1)^2 + (-1)^3 +$
$$(-1)^4 + (-1)^5 + \dots \quad (-1)^{49} + (-1)^{50} = 0. \quad \overline{a}=0/50=0.$$

67. $a_n = 2n-4$, first six terms. $f(n)=2n-4$. $-2, 0, 2, 4, 6, 8$.

69. $a_n = 7n^{-1} + n$, first seven terms. $f(n)=7n^{-1} + n$.
$8, 5.5, 5.3333, 5.75, 6.4, 7.1616, 8$.

71. $a_n = (n+1)(n-3)$, a_8 through a_{14}. $f(n)=(n+1)(n-3)$.
$45, 60, 77, 96, 117, 140, 165$.

73. $a_n = \dfrac{2n+3}{n+1}$, a_5 through a_{12}. $f(n)=\dfrac{2n+3}{n+1}$.
$2.1667, 2.1429, 2.125, 2.1111, 2.1, 2.0909, 2.0833, 2.0769$.

Section 8.2

1. $1, 3, 5, 7,\dots$ $a_2-a_1=2$, $a_3-a_2=2$, $a_4-a_3=2$. Common diff. 2.

3. $-1, 3, 7, 11,\dots$ $a_2-a_1=4$, $a_3-a_2=4$, $a_4-a_3=4$. Common diff. 4.

5. $\dfrac{1}{2}, \dfrac{3}{2}, \dfrac{5}{2},\dots$ $a_2-a_1=1$, $a_3-a_2=1$, $a_4-a_3=1$. Common difference 1.

7. $0, 5, 10, 15,\dots$ $a_2-a_1=5$, $a_3-a_2=5$, $a_4-a_3=5$. A.S. with common
difference 5.

8. $0, \dfrac{3}{4}, \dfrac{3}{2}, \dfrac{5}{4},\dots$ $a_2-a_1=\dfrac{3}{4}$, $a_3-a_2=\dfrac{3}{4}$, $a_4-a_3=-\dfrac{1}{4}$. Not an A.S.

9. $\frac{1}{2}$, $-\frac{1}{4}$, -1, $-\frac{7}{4}$,... $a_2-a_1=-\frac{3}{4}$, $a_3-a_2=-\frac{3}{4}$, $a_4-a_3=-\frac{3}{4}$. A.S. with common difference $-\frac{3}{4}$.

11. -5, 1, 7, 13,... $a_2-a_1=6$, $a_3-a_2=6$, $a_4-a_3=6$. A.S. with common difference 6.

13. 0, 3, 6, 9,... $d=3$. $a_5=12$, $a_6=15$.

15. $-\frac{1}{2}$, 0, $\frac{1}{2}$, 1,... $d=\frac{1}{2}$. $a_5=\frac{3}{2}$, $a_6=2$.

17. 41, 48, 55, 62,... $d=7$. $a_5=69$, $a_6=76$.

19. $a_1=2$, $d=3$. $2,5,8,11$.

21. $a_1=-3$, $d=-1$. $-3,-4,-5,-6$.

23. $a_1=-6$, $d=\frac{3}{4}$. -6, $-\frac{21}{4}$, $-\frac{9}{2}$, $-\frac{15}{4}$.

25. $a_1=2.36$, $d=3.2$. Sequence: 2.36, 5.56, 8.76, 11.96, 15.16, 18.36, 21.56, 24.76, 27.96, 31.16.

27. $a_1=2.7689$, $d=5.23$. Sequence: 2.7869, 7.9989, 13.2289, 18.4589, 23.6889, 28.9189, 34.1489, 39.3789, 44.6089, 49.8389.

29. $a_1=67.92$, $d=-5.6$. Sequence: 67.92, 62.32, 56.72, 51.12, 45.52, 39.92, 34.32, 28.72, 23.12, 17.52.

31. $a_1=3$, $d=4$. 5th term. $a_n=a_1+(n-1)d$. $a_5=3+(5-1)4=19$.

33. $a_1=-1$, $d=8$. 4th term. $a_n=a_1+(n-1)d$. $a_4=-1+(4-1)(8)=23$.

35. $a_1=-5$, $d=2$. 20th term. $a_n=a_1+(n-1)d$. $a_{20}=-5+(20-1)2=33$.

37. $a_2=3$, $a_3=5$. 5th term. $d=a_3-a_2=2$. $a_n=a_1+(n-1)d$ with $n=2$, $a_2=3$, $d=2$ gives $a_1=1$. Thus $a_n=a_1+(n-1)d$ gives $a_5=1+(5-1)(2)=9$.

39. $a_8=-4$, $a_9=4$. 1st term. $d=a_9-a_8=8$. $a_n=a_1+(n-1)d$ with $n=8$, $a_8=-4$, $d=8$ gives $a_1=-60$.

41. $a_3=0$, $a_4=-5$. 15th term. $d=a_4-a_3=-5$. $a_n=a_1+(n-1)d$ with $n=3$, $a_3=0$, $d=-5$ gives $a_1=10$. Thus $a_n=a_1+(n-1)d$ gives $a_{15}=0+(15-1)(-5)=-60$.

43. $a_2=4$, $a_4=8$. 7th term. $a_n=a_1+(n-1)d$ with $n=2$ and $n=4$ gives $4=a_1+1d$ and $8=a_1+3d$. Solving gives $a_1=2$, $d=2$. Thus $a_n=a_1+(n-1)d$ now gives $a_7=2+(7-1)(2)=14$.

45. $a_4 = \frac{1}{2}$, $a_{11} = 4$. 6th term. $a_n = a_1 + (n-1)d$ with $n=4$ and $n=11$ gives $\frac{1}{2} = a_1 + 3d$ and $4 = a_1 + 10d$. Solving gives $a_1 = -1$, $d = \frac{1}{2}$. Thus $a_n = a_1 + (n-1)d$ now gives $a_6 = -1 + (6-1)(\frac{1}{2}) = \frac{3}{2}$.

47. $a_1 = 27$, $a_9 = -4$. 20th term. $a_n = a_1 + (n-1)d$ with $n=1$ and $n=9$ gives $27 = a_1$ and $-4 = a_1 + 8d$. Solving gives $d = -\frac{31}{8}$. Thus $a_n = a_1 + (n-1)d$ now gives $a_{20} = 27 + (20-1)(-\frac{31}{8}) = -\frac{373}{8}$.

In exercises 49–53 use $S_n = \frac{n}{2}[2a_1 + (n-1)d]$.

49. $a_1 = 3$, $d = 2$. $S_5 = \frac{5}{2}[2(3) + (5-1)2] = 35$.

51. $a_1 = -5$, $d = 2$. $S_4 = 2[2(-5) + (4-1)2] = -8$.

53. $a_1 = 0$, $d = \frac{3}{2}$. $S_5 = \frac{5}{2}[2(0) + (5-1)(\frac{3}{2})] = 15$.

54. $a_1 = -\frac{1}{3}$, $d = -\frac{3}{4}$. $S_{12} = 6[2(-\frac{1}{3}) + (12-1)(-\frac{3}{4})] = -\frac{107}{2} = -53.5$.

In exercises 55–59 use $S_n = \frac{n}{2}(a_1 + a_n)$.

55. $a_n = 2n+1$. $S_8 = \frac{8}{2}(a_1 + a_8) = 4(2(1)+1 + 2(8)+1) = 80$.

57. $a_n = n-6$. $S_{15} = \frac{15}{2}(a_1 + a_{15}) = \frac{15}{2}(1-6 + 15-6) = 30$.

59. $a_n = -4n+5$. $S_4 = \frac{4}{2}(a_1 + a_4) = 2(-4(1)+5 + -4(4)+5) = -20$.

61. $a_1 = 2$, $a_n = 10$, $S_n = 30$. $S_n = \frac{n}{2}(a_1 + a_n)$ gives $30 = \frac{n}{2}(2+10)$, $60 = 12n$, $n = 5$.

63. $a_1 = 4$, $a_n = 99$, $S_n = 1030$. $S_n = \frac{n}{2}(a_1 + a_n)$ gives $1030 = \frac{n}{2}(4+99)$, $2060 = 103n$, $n = 20$.

65. $S_8 = 48$, $d = -4$. $S_n = \frac{n}{2}[2a_1 + (n-1)d]$ gives $48 = \frac{8}{2}[2a_1 + (8-1)(-4)]$. $12 = 2a_1 - 28$, $a_1 = 20$. $a_n = a_1 + (n-1)d = 20 + (n-1)(-4)$, $a_n = 24 - 4n$.

67. $a_1 = -3$, $a_{14} = 75$. $a_n = a_1 + (n-1)d$ gives $75 = -3 + (14-1)d$, $78 = 13d$, $d = 6$. Thus $a_n = a_1 + (n-1)d = -3 + (n-1)6$, $a_n = 6n-9$.

69. \$2500, 6% simple interest. $P = 2500$, $r = .06$. $d = Pr = (2500)(.06) = 150$. Amounts for first 5 yr are \$2500, \$2650, \$2800, \$2950, \$3100.

71. $1000 deposited at 7% simple interest. How long before $1490?
P=1000, r=.07. d=Pr=70. $a_n=a_1+(n-1)d$ gives 1490=1000+(n-1)70.
490=70n-70, 70n=560, n=8. it will take 8 years.

73. Acceleration 12ft/sec^2. Starts from rest, u=0. $v_n=u+an$.
$v_n=0+12n$, $v_n=12n$. letting n=1,2,3,..,8 velocities for the
first eight seconds are 12, 24, 36, 48, 60, 72, 84, 96 ft/sec.

75. Clock chimes on the hour, once at one o'clock, twice at two
o'clock, ..., twelve times at twelve o'clock. Chimes 1+2+...+12
in a twelve hour period. $a_1=1, a_2=2,..., a_{12}=12$.
$S_n = \frac{n}{2}(a_1+a_n)$ gives $S_{12} = 6(1+12)=78$. Chimes 78 times.

Section 8.3

Group Discussion: Ancestor 4 generations back who was Greek.
Parent would be 1/2, grandparent $(1/4) = (1/2)^2$, great-
grandparent $(1/2)^4$, 4 generations $(1/2)^8$. Thus 1/16 Greek.
In general, n generations back, $(1/2)^n$. nth term in a
geometrical sequence with first term (1/2) and ratio 1/2.

1. 3, 9, 27, 81,.. 9/3=3, 27/9=3, 81/27=3. G.S. with r=3.

3. 1, 5, 25, 125,.. 5/1=5, 25/5=5, 125/25=5. G.S. with r=5.

5. 2,-2,2,-2,.. -2/2=-1, 2/(-2)=-1, -2/2=-1. G.S. with r=-1.

7. 2, 10, 20, 40,.. 10/2=5, 20/10=2. Not a G.S.

9. 1, 3, 9, 27,.. 3/1=3, 9/3=3, 27/9=3. G.S. with r=3.

11. 15,-5, $\frac{5}{3}$, $-\frac{5}{9}$,.. -5/15=-1/3, $(\frac{5}{3})/(-5)$=-1/3, $(-\frac{5}{9})/(\frac{5}{3})$=-1/3. G.S.
with r=-1/3.

13. 1, 4, 16, 64,.. 4/1=16/4=64/16=4, r=4. $a_5=64(4)=256$.
$a_6=256(4)=1024$.

15. 27, 9, 3, 1,.. r=1/3. $a_5=1(1/3)=1/3$. $a_6=(1/3)(1/3)=1/9$.

17. -12, -6, -3, $-\frac{3}{2}$,.. $r=\frac{1}{2}$. $a_5=(-\frac{3}{2})(\frac{1}{2})=-\frac{3}{4}$. $a_6=(-\frac{3}{4})(\frac{1}{2})=-\frac{3}{8}$.

19. $a_1=2$, r=4. Sequence: 2, 8, 32, 128, 512.

21. $a_1=5$, $r=\frac{1}{2}$. Sequence: 5, $\frac{5}{2}$, $\frac{5}{4}$, $\frac{5}{8}$, $\frac{5}{16}$.

23. $a_1=1$, r=5. Sequence: 1, 5, 25, 125, 625.

25. a_1=4.76, r=3.8. Sequence: 4.7600, 18.0880, 68.7344, 261.1907, 992.5247, 3771.5940.

27. a_1=2.981, r=3.25. Sequence: 2.9810, 9.6883, 31.4868, 102.3321, 332.5795, 1080.8832.

29. a_1=93.26, r=-0.6. Sequence: 93.2600, -55.9560, 33.5736, -20.1442, 12.0865, -7.2519.

In exercises 31-47 use $a_n = a_1(r)^{n-1}$.

31. a_1=3,r=4. $a_6 = 3(4)^5 = 3072$. 　　　33. a_1=63,r=$-\frac{1}{3}$. $a_7 = 63(-\frac{1}{3})^6 = \frac{7}{81}$.

35. $a_1 = \frac{1}{3}$,r=$\frac{1}{2}$. $a_5 = \frac{1}{3}(\frac{1}{2})^4 = \frac{1}{48}$.

37. a_2=6, a_3=12. r=12/6=2. $a_n = a_1(r)^{n-1}$ with n=2 gives $a_2 = a_1(r)^{2-1}$, $6 = a_1(2)^1$, a_1=3.

39. a_7=2916,a_8=8748. r=8748/2916=3. $a_7 = a_1(r)^{7-1}$, $2916 = a_1(3)^6$, a_1=4.

41. $a_2 = -\frac{11}{3}$,$a_3 = \frac{11}{9}$. r=$(\frac{11}{9}/-\frac{11}{3}) = -\frac{1}{3}$. $a_2 = a_1(r)^{2-1}$, $-\frac{11}{3} = a_1(-\frac{1}{3})^1$, a_1=11.

43. a_3=8, a_6=64. 4th term. $a_n = a_1(r)^{n-1}$ gives $8 = a_1r^2$, $64 = a_1r^5$. Thus $64/8 = a_1r^5/a_1r^2$, $8 = r^3$, r=2. $a_1 = 8/r^2 = 8/2^2 = 2$. $a_4 = a_1r^3 = 2(2)^3 = 16$.

45. a_5=32, a_{10}=1. 12th term. $a_n = a_1(r)^{n-1}$ gives $32 = a_1r^4$, $1 = a_1r^9$. Thus $1/32 = a_1r^9/a_1r^4$, $1/32 = r^5$, r=1/2. $a_1 = 32/(1/2)^4 = 512$. $a_{12} = a_1r^{11} = 512(1/2)^{11} = 1/4$.

47. a_7=-1, a_9=-1. 2nd term. $a_n = a_1(r)^{n-1}$ gives $-1 = a_1r^6$, $-1 = a_1r^8$. Thus $(-1)/(-1) = a_1r^8/a_1r^6$, $1 = r^2$, r=$\pm$1. $a_1 = -1/r^6 = -1$. $a_2 = a_1r = \pm 1$.

In exercise 49-53 use $S_n = \dfrac{a_1(1-r^n)}{(1-r)}$.

49. $a_n = 3(5)^{n-1}$. S_4. Compare with $a_n = a_1(r)^{n-1}$. a_1=3, r=5. Thus $S_4 = \dfrac{3(1-5^4)}{(1-5)} = 468$.

51. $a_n = -2(2)^{n-1}$. a_1=-2, r=2. $S_3 = \dfrac{-2(1-2^3)}{(1-2)} = -14$.

53. $a_n = (\frac{1}{81})(-3)^{n-1}$. $a_1 = \frac{1}{81}$, r=-3. $S_4 = \dfrac{\frac{1}{81}(1-(-3)^4)}{(1-(-3))} = -\dfrac{20}{81}$.

In exercises 55-59 use $S = \dfrac{a_1}{(1-r)}$.

55. $a_1=2, r=\dfrac{1}{3}$. $S = \dfrac{2}{(1-\frac{1}{3})} = 3$.

57. $a_1=13, r=\dfrac{1}{8}$. $S = \dfrac{13}{(1-\frac{1}{8})} = \dfrac{104}{7}$.

59. $a_1=11, r=-\dfrac{1}{5}$. $S = \dfrac{11}{(1+\frac{1}{5})} = \dfrac{55}{6}$.

61. $.32\overline{42} = .32 + .0042 + .000042 + .00000042 + .0000000042 + \ldots$
$= .32 + .0042 + .0042(10^{-2}) + .0042(10^{-4}) + .0042(10^{-6}) + \ldots$
Use $a_1 = .0042$, $r = 10^{-2}$.

$.32\overline{42} = .32 + \dfrac{a_1}{(1-r)} = .32 + \dfrac{.0042}{(1-10^{-2})} = .32 + \dfrac{.0042}{.99} = \dfrac{32}{100} + \dfrac{42}{9900}$

$\qquad = \dfrac{3210}{9900} = \dfrac{107}{330}$.

63. $.58\overline{365} = .58 + .00365 + .0000365 + .000000365 + \ldots$
$= .58 + .00365 + .00365(10^{-3}) + .00365(10^{-6}) + .00365(10^{-9}) \ldots$
Use $a_1 = .00365$, $r = 10^{-3}$.

$.58\overline{365} = .58 + \dfrac{a_1}{(1-r)} = .58 + \dfrac{.00365}{(1-10^{-3})} = \dfrac{58307}{99900}$.

65. $.7\overline{345} = .7 + .0345 + .0000345 + .000000345 + \ldots$
$= .7 + .0345 + .0345(10^{-3}) + .0345(10^{-6}) + .0345(10^{-9}) \ldots$
Use $a_1 = .0345$, $r = 10^{-3}$.

$.7\overline{345} = .7 + \dfrac{a_1}{(1-r)} = .7 + \dfrac{.0345}{(1-10^{-3})} = \dfrac{1223}{1665}$.

67. $a_1=4.53$, $r=0.85$. $f(x) = \dfrac{a_1(1-r^x)}{(1-r)} = \dfrac{4.53(1-.85^x)}{(1-.85)}$.

$S = \dfrac{a_1}{(1-r)} = \dfrac{4.53}{(1-.85)} = 30.2000$.

69. $a_1=-9.867$, $r=0.5$. $f(x) = \dfrac{a_1(1-r^x)}{(1-r)} = \dfrac{-9.867(1-.5^x)}{(1-.5)}$.

$S = \dfrac{a_1}{(1-r)} = \dfrac{-9.867}{(1-.5)} = -19.7340$.

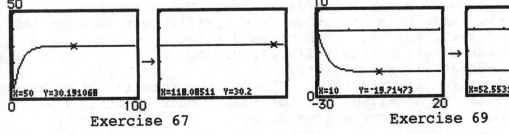

Exercise 67   Exercise 69

71. $a_1=5$, $r=1.25$. Exceed 18. $f(x)=5(1.25)^{x-1}$. 7th term.

73. $a_1=1000$, $r=0.85$. Fall below 50. $f(x)=1000(.85)^{x-1}$. 20th term.

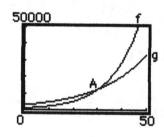

X	Y1
1	5
2	6.25
3	7.8125
4	9.7656
5	12.207
6	15.259
7	19.073

X	Y1
16	87.354
17	74.251
18	63.113
19	53.646
20	45.599
21	38.76
22	32.946

Exercise 71   Exercise 73

75. Initial count of bacterium = 100. Doubling time = 1hr.
    Let $a_1=100$, $r=2$. Thus $a_n=a_1 r^{n-1}$, $a_n=100 \cdot 2^{n-1}$.
    $a_5=100 \cdot 2^{5-1}=1600$.

77. How many ancestors going back 6 generations? G.S. with $a_1=2$,
    $r=2$. $a_n=a_1 r^{n-1}$, $a_n=2(2)^{n-1}$. $a_6=2(2)^{6-1} = 64$ ancestors.

79. Population decreasing at rate of 5% per annum. Current
    population = 75,000. Get G.S. where $a_{n+1}=a_n(1-5/100)$. Let
    $a_1=75000$, $r=.95$. $a_n=a_1 r^{n-1}$, $a_n=75000(.95)^{n-1}$. In 5 yrs time
    n=5. $a_3=75000(.95)^{5-1} = 61087.96875$. Predicted population is
    about 61,088.

81. 1st account, $a_1=1000$, ratio=1.09. $a_n=a_1 r^{n-1}$, $a_n=1000(1.09)^{n-1}$.
    2nd account, $a_1=3000$, ratio=1.05. $a_n=a_1 r^{n-1}$, $a_n=1000(1.05)^{n-1}$.
    Let $f(x)=1000(1.09)^{n-1}$, $g(x)=1000(1.05)^{n-1}$.
    A(30.384456,12582.22), below. It will take 30.38 years for the
    9% account to catch up with the 5% account.

50000   f

g

A

0   50

Section 8.4

1. $2+4+6+\ldots+2n = n(n+1)$. (A)
    1. Let n=1. Formula gives $2=1(1+1)$, true. Valid for n=1.
    2. Let n=k. Formula is $2+4+6+\ldots+2k = k(k+1)$. Next term on left
    would be $2(k+1)$. Add this term to each side.
    $2+4+6+\ldots+2k+2(k+1) = k(k+1)+2(k+1)$, giving
    $2+4+6+\ldots+2k+2(k+1) = (k+1)(k+2)$.
    This is formula (A) with n=k+1. Thus if the formula holds for
    n=k it holds for n=k+1. By mathematical induction it is true
    for all positive integer values of n.

. $2+5+8+\ldots+(3n-1) = \dfrac{n(3n+1)}{2}$. (A)

1. Let n=1. Formula gives $2 = \dfrac{1(3(1)+1)}{2}$, true. Valid for n=1.

2. Let n=k. Formula is $2+5+8+\ldots+(3k-1) = \dfrac{k(3k+1)}{2}$. Next term on left would be $3(k+1)-1$, i.e. $3k+2$. Add this term to each side.
$2+5+8+\ldots+(3k-1)+(3k+2) = \dfrac{k(3k+1)}{2} +(3k+2) = \dfrac{k(3k+1)+2(3k+2)}{2}$

$= \dfrac{3k^2+7k+4}{2} = \dfrac{(k+1)(3k+4)}{2}$.

This is formula (A) with n=k+1. Thus if the formula holds for n=k it holds for n=k+1. By mathematical induction it is true for all positive integer values of n.

. $3+9+15+\ldots+(6n-3) = 3n^2$. (A)

1. Let n=1. Formula gives $3 = 3(1)^2$, true. Valid for n=1.
2. Let n=k. Formula is $3+9+15+\ldots+(6k-3) = 3k^2$. Next term on left would be $6(k+1)-3$, i.e. $6k+3$. Add this term to each side.
$3+9+15+\ldots+(6k-3)+(6k+3) = 3k^2+6k+3=3(k^2+2k+1)=3(k+1)^2$.
This is formula (A) with n=k+1. Thus if the formula holds for n=k it holds for n=k+1. By mathematical induction it is true for all positive integer values of n.

. $1+3^2+5^2+\ldots+(2n-1)^2 = \dfrac{n(2n+1)(2n-1)}{3}$. (A)

1. Let n=1. Formula is $1 = \dfrac{(1)(2(1)+1)(2(1)-1)}{3}$, true. Valid n=1.

2. Let n=k. Formula is $1+3^2+5^2+\ldots+(2k-1)^2 = \dfrac{k(2k+1)(2k-1)}{3}$.
Next term on left would be $(2(k+1)-1)^2$, $(2k+1)^2$. Add to each side.
$1+3^2+5^2+\ldots+(2k-1)^2+(2k+1)^2 = \dfrac{k(2k+1)(2k-1)}{3} + (2k+1)^2$

$= \dfrac{k(2k+1)(2k-1)+3(2k+1)^2}{3} = \dfrac{(2k+1)(2k^2+5k+3)}{3}$

$= \dfrac{(2k+1)(k+1)(2k+3)}{3} = \dfrac{(k+1)(2k+3)(2k+1)}{3}$.

This is formula (A) with n=k+1. Thus if the formula holds for n=k it holds for n=k+1. By mathematical induction it is true for all positive integer values of n.

. $1+4^1+4^2++\ldots+4^{n-1} = \dfrac{4^n-1}{3}$. (A)

1. Let n=1. Formula gives $1 = \dfrac{4^1-1}{3}$, true. Valid for n=1.

2. Let n=k. Formula is $1+4^1+4^2++\ldots+4^{k-1} = \dfrac{4^k-1}{3}$. Next term on left would be $4^{(k+1)-1}$, i.e. 4^k. Add this term to each side.

$$1+4^1+4^2++\ldots+4^{k-1}+4^k = \frac{4^k-1}{3} + 4^k = \frac{4^k-1+3(4^k)}{3} = \frac{4(4^k)-1}{3} = \frac{4^{k+1}-1}{3}$$

This is formula (A) with n=k+1. Thus if the formula holds for n=k it holds for n=k+1. By mathematical induction it is true for all positive integer values of n.

11. $1\cdot3+2\cdot5+3\cdot6+\cdots+n(2n+1) = \dfrac{n(n+1)(4n+5)}{6}$. (A)

1. Let n=1. Formula is $3 = \dfrac{1(1+1)(4(1)+5)}{6}$, true. Valid n=1.

2. Let n=k. Formula is $1\cdot3+2\cdot5+3\cdot6+\cdots+k(2k+1) = \dfrac{k(k+1)(4k+5)}{6}$.
Next term on left would be (k+1)(2k+3). Add to each side.
$1\cdot3+2\cdot5+3\cdot6+\cdots+k(2k+1)+(k+1)(2k+3)= \dfrac{k(k+1)(4k+5)}{6} +(k+1)(2k+3)=$
$\dfrac{k(k+1)(4k+5)+6(k+1)(2k+3)}{6} = \dfrac{(k+1)(4k^2+5k+12k+18)}{6} =$
$\dfrac{(k+1)(4k^2+17k+18)}{6} = \dfrac{(k+1)(k+2)(4k+9)}{6}$.

This is formula (A) with n=k+1. Thus if the formula holds for n=k it holds for n=k+1. By mathematical induction it is true for all positive integer values of n.

13. $(xy)^n = x^ny^n$. (A)
   1. Let n=1. Formula gives $(xy)^1 = x^1y^1$, true. Valid for n=1.
   2. Let n=k. Formula is $(xy)^k = x^ky^k$. Assume this to be true.
   We now show that the formula is valid for n=k+1. We have that
   $(xy)^{k+1} = x^ky^k(xy)= x^kxy^ky = x^{k+1}y^{k+1}$. Thus $(xy)^{k+1} = x^{k+1}y^{k+1}$.
   This is formula (A) with n=k+1. Thus if the formula holds for n=k it holds for n=k+1. By mathematical induction it is true for all positive integer values of n.

15. If $x > 1$ then $x^n > x^{n-1}$. (A)
   1. Let n=1. Statement gives: If x>1, $x^1>x^{1-1}$ ($x>x^0$), true. Valid for n=1.
   2. Let n=k. Statement is: If $x > 1$ then $x^k > x^{k-1}$. Assume this to be true. We now show that the statement is valid for n=k+1. Since x>1, $x^kx > x^{k-1}x$, $x^{k+1} > x^k$.
   This is statement (A) with n=k+1. Thus if the statement holds for n=k it holds for n=k+1. By mathematical induction it is true for all positive integer values of n.

17. $4^n - 1$ is divisible by 3. (A)
   1. Let n=1. Statement gives: 4^1-1 is divisible by 3., true. Valid for n=1.
   2. Let n=k. Statement is: 4^k-1 is divisible by 3. Assume this to be true. We now show that the statement is valid for n=k+1. Have that
   $4^{k+1}-1 = 4^k4^1 - 1 = 4^k4 - 1 = 4^k(3+1) - 1 = 4^k3 + 4^k - 1$.
   4^k3 is divisible by 3, and $4^k - 1$ is divisible by 3

(asumption). Thus $4^{k+1}-1$ is divisible by 3.
This is statement (A) with n=k+1. Thus if the statement holds for n=k it holds for n=k+1. By mathematical induction it is true for all positive integer values of n.

19. n^3-n+3 is divisible by 3. (A)

1. Let n=1. Statement gives: 1^3-1+3 is divisible by 3, true. Valid for n=1.

2. Let n=k. Statement is: k^3-k+3 is divisible by 3. Assume this to be true. We now show that the statement is valid for n=k+1. Have that
$(k+1)^3-(k+1)+3 = k^3+3k^2+3k+1-k-1+3 = k^3+3k^2+2k+3 = (k^3-k+3) + (3k^2+3k)$. Both terms are divisible by 3. Thus $(k+1)^3-(k+1)+3$ is divisible by 3.
This is statement (A) with n=k+1. Thus if the statement holds for n=k it holds for n=k+1. By mathematical induction it is true for all positive integer values of n.

21. G.S. sequence, 1st term a_1 & ratio r. To show $a_n = a_1r^{n-1}$.(A)

1. Let n=1. Statement gives: $a_1 = a_1r^{1-1}$, true since $r^1=1$.
   Valid for n=1.

2. Let n=k. Statement is: $a_k = a_1r^{k-1}$. Assume this to be true. Now show that the statement is valid for n=k+1. Have that
$a_{k+1} = a_kr = a_1r^{k-1}r = a_1r^k$.
This is statement (A) with n=k+1. Thus if the statement holds for n=k it holds for n=k+1. By mathematical induction it is true for all positive integer values of n.

23. $n^3 > n^2+100$. (A)

1. Table below shows that $n^3 > n^2+100$ first for n=6.

2. Let n=k, k≥6. Statement is: $k^3 > k^2+100$. Assume this to be true. Now show that the statement is valid for n=k+1. Have that
$(k+1)^3 = k^3+3k^2+3k+1 > k^2+100+3k^2+3k+1 = k^2+2k+1+100+3k^2+k = (k+1)^2+100+3k^2+k > (k+1)^2+100$. Thus $(k+1)^3 > (k+1)^2+100$.
This is statement (A) with n=k+1. Thus if the statement holds for n=k it holds for n=k+1. By mathematical induction it is true for n=1 and all positive integer values of n≥5.
true for n=1 and all positive integer values of n≥5.

$x^3 \quad x^2+100$

X	Y₁	Y₂
1	1	101
2	8	104
3	27	109
4	64	116
5	125	125
6	216	136
7	343	149

Exercise 23

Section 8.5

1. $(a+b)^5 = a^5 + 5a^4b + 10a^3b^2 + 10a^2b^3 + 5ab^4 + b^5$.

3. $(1+2x)^7 = (1)^7 + 7(1)^6(2x) + 21(1)^5(2x)^2 + 35(1)^4(2x)^3 + 35(1)^3(2x)^4 + 21(1)^2(2x)^5 + 7(1)(2x)^6 + (2x)^7 = 1 + 14x + 84x^2 + 280x^3 + 560x^4 + 672x^5 + 448x^6 + 128x^7$.

5. $(3x-2y)^5 = (3x)^5 + 5(3x)^4(-2y) + 10(3x)^3(-2y)^2 + 10(3x)^2(-2y)^3 + 5(3x)(-2y)^4 + (-2y)^5 = 243x^5 - 810x^4y + 1080x^3y^2 - 720x^2y^3 + 240xy^4 - 32y^5$.

7. $(x+y^2)^6 = (x)^6 + 6(x)^5(y^2) + 15(x)^4(y^2)^2 + 20(x)^3(y^2)^3 + 15(x)^2(y^2)^4 + 6(x)(y^2)^5 + (y^2)^6 = x^6 + 6x^5y^2 + 15x^4y^4 + 20x^3y^6 + 15x^2y^8 + 6xy^{10} + y^{12}$.

9. 10th: 1   9   36   84   126   126   84   36   9   1

11. $6! = 6 \cdot 5 \cdot 4 \cdot 3 \cdot 2 \cdot 1 = 720$.   13. $3! = 3 \cdot 2 \cdot 1 = 6$.

15. $\dfrac{12!}{10!} = \dfrac{12 \cdot 11 \cdot 10!}{10!} = 12 \cdot 11 = 132$.   17. $\dfrac{100!}{99!} = \dfrac{100 \cdot 99!}{99!} = 100$.

19. $\binom{10}{6} = \dfrac{10!}{4!6!} = \dfrac{10 \cdot 9 \cdot 8 \cdot 7 \cdot 6!}{4 \cdot 3 \cdot 2 \cdot 1 \cdot 6!} = 210$.   21. $\binom{5}{5} = \dfrac{5!}{0!5!} = 1$.

23. $(x+y)^4 = x^4 + \binom{4}{1}x^3y + \binom{4}{2}x^2y^2 + \binom{4}{3}xy^3 + y^4 = x^4 + 4x^3y + 6x^2y^2 + 4xy^3 + y^4$.

25. $(2+2y)^3 = 2^3 + \binom{3}{1}2^2(2y) + \binom{3}{2}2(2y)^2 + (2y)^3 = 8 + 24y + 24y^2 + 8y^3$.

27. $(3-4y)^5 = 3^5 + \binom{5}{1}3^4(-4y) + \binom{5}{2}3^3(-4y)^2 + \binom{5}{3}3^2(-4y)^3 + \binom{5}{4}3(-4y)^4 + (-4y)^5 = 243 - 1620y + 4320y^2 - 5760y^3 + 3840y^4 - 1024y^5$.

29. $(x + 1/y)^4 = x^4 + \binom{4}{1}x^3(1/y) + \binom{4}{2}x^2(1/y)^2 + \binom{4}{3}x(1/y)^3 + (1/y)^4 =$

$x^4 + 4x^3/y + 6x^2/y^2 + 4x/y^3 + 1/y^4$.

31. $(x+y^{1/2})^5 = x^5 + \binom{5}{1}x^4(y^{1/2}) + \binom{5}{2}x^3(y^{1/2})^2 + \binom{5}{3}x^2(y^{1/2})^3 + \binom{5}{4}x(y^{1/2})^4 + (y^{1/2})^5 = x^5 + 5x^4y^{1/2} + 10x^3y + 10x^2y^{3/2} + 5xy^2 + y^{5/2}$.

33. $(x^{1/2}+y^{3/2})^4 = (x^{1/2})^4 + \binom{4}{1}(x^{1/2})^3(y^{3/2}) + \binom{4}{2}(x^{1/2})^2(y^{3/2})^2 + \binom{4}{3}(x^{1/2})(y^{3/2})^3 + (y^{3/2})^4 = x^2 + 4x^{3/2}y^{3/2} + 6xy^3 + 4x^{1/2}y^{9/2} + y^6$.

35. $(2x-y)^9$, 6th term. $n=9, r=6$. $\binom{9}{5}(2x)^4(-y)^5 = -2016x^4y^5$.

236

7. $(x + 2y^3)^{10}$, 8th term. $n=10, r=8$. $\binom{10}{7}(x)^3(2y^3)^7 = 15360x^3y^{21}$.

9. $(x^2+5y)^8$, 7th term. $n=8, r=7$. $\binom{8}{6}(x^2)^2(5y)^6 = 437500x^4y^6$.

1. $(5x-y)^5$, 4th term. $n=5, r=4$. $\binom{5}{3}(5x)^2(-y)^3 = -250x^2y^3$.

hapter 8 Project

a) Let a_n be the number of messages in the Inbox at the start of morning n. The realtionship between the number of messages in the inbox for two consecutive days is
$$a_n = (\tfrac{1}{2})a_{n-1} + 8$$
We get $a_n = (\tfrac{1}{2})[(\tfrac{1}{2})a_{n-2} + 8]+8 = (\tfrac{1}{2})^2 a_{n-2} + (\tfrac{1}{2})8+8$
$$= (\tfrac{1}{2})[(\tfrac{1}{2})\{(\tfrac{1}{2})a_{n-3} +8\}+8]+8 = (\tfrac{1}{2})^3 a_{n-3} +(\tfrac{1}{2})^2 8+(\tfrac{1}{2})8+8$$
. . .
Using the properties of geometric series this leads to
$$a_n = (\tfrac{1}{2})^{n-1}a_1 + 8\{\frac{1 - (\tfrac{1}{2})^{(n-1)}}{1 - (\tfrac{1}{2})}\}$$

Enter this expression into the calculator with $a_1 = 208$ and examine the table.

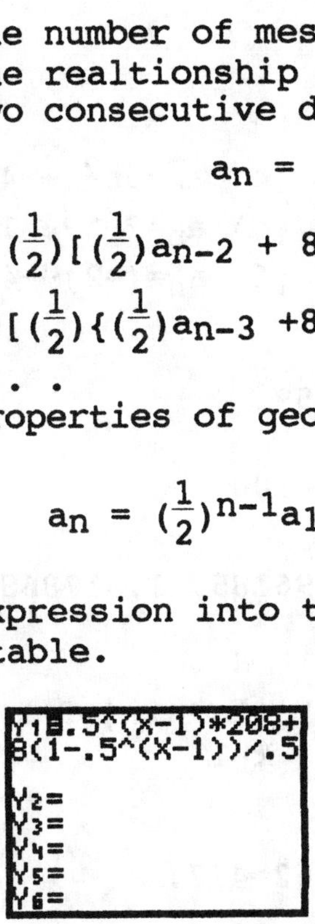

```
Y1≡.5^(X-1)*208+
8(1-.5^(X-1))/.5
Y2=
Y3=
Y4=
Y5=
Y6=
```

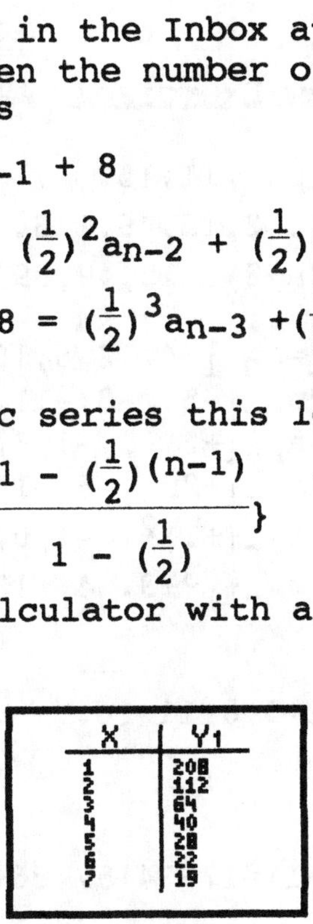

```
X   | Y1
1   | 208
2   | 112
3   | 64
4   | 40
5   | 28
6   | 22
7   | 19
```

We get the following picture of the procedure.

Day	1	2	3	4	5	6	7
start of day	208	112	64	40	28	22	19
	<-----purge, look at new mail, sort, save---->						
end of day		112	64	40	28	22	19

The contents of the Inbox will be reduced to below 20 in six days according to this plan.

(b) Effect of reducing the number of messages added daily to the Inbox to 4 instead of 8 obtained by changing the 8 to a 4 in the equation window would be to give the following:

237

Day	1	2	3	4
start of day	208	108	58	20.5

<purge, look at new mail, sort, save>

| end of day | 108 | 58 | 33 | 14.25 |

Round the 20.5 and 14.25 up. Drops below 20 after 4 days.

(c) Initial number 1552 messages. 8 messages added daily.

Day	1	2	3	4	5	6	7	8	9	10
start of day	1552	784	400	208	112	64	40	28	22	19

<-----purge, look at new mail, sort, save---->

| end of day | | 784 | 400 | 208 | 112 | 64 | 40 | 28 | 22 | 19 |

Drops below 20 after 9 days.

Chapter 8 Review Exercises

1. (a) $a_n=4n-1$. 3,7,11,15.   (b) $a_n=3n^2 + 4$. 7,16,31,52.
   (c) $a_n=2n^3-n^2+1$. 2,13,46,113.   (d) $a_n=2^{n-3}$. 1/4,1/2,1,2.
   (e) $a_n=(4n-1)(2n+3)$. 15,49,99,165.   (f) $a_n=(-2)^n+4n+3$. 5,15,7,35.

2. (a) $a_1=2$, $a_n=a_{n-1}+4$. 2,6,10,14.
   (b) $a_1=-3$, $a_n=2a_{n-1}-5$. -3,-11,-27,-59.
   (c) $a_1=1$, $a_n=-3a_{n-1}+2$. 1,-1,5,-13.
   (d) $a_1=-6$, $a_n=3a_{n-1}+2n$. -6,-14,-36,-100.
   (e) $a_1=-1$, $a_n=(a_{n-1}+1)^2$. -1,0,1,4.
   (f) $a_1=4$, $a_n=2(a_{n-1})^3+3$. 4, 131, 4496185, $1.817868704 \times 10^{20}$.

3. (a) $\sum_{k=1}^{4}(3k-1)=2+5+8+11=26$.   (b) $\sum_{k=1}^{6}(1+2k)=3+5+7+9+11+13=48$.

   (c) $\sum_{k=2}^{7}(3k^2+2k+1)=17+34+57+86+121+162=477$.

   (d) $\sum_{k=3}^{6}(2^k-3k+1)=5+18+47=70$.

4. (a) $a_k=3k$. $S_3 = 3(1)+3(2)+3(3) = 3+6+9 = 18$.
   (b) $a_k=4k+2$. $S_5 = 4(1)+2+4(2)+2+4(3)+2+4(4)+2+4(5)+2=70$.
   (c) $a_k=k^2+3k-1$. $S_2 = 1^2+3(1)-1 + 2^2+3(2)-1 = 12$.

5. (a) 1,4,7,11,... . $a_2-a_1=3$, $a_3-a_2=3$, $a_4-a_3=4$. Not an A.S.

   (b) -10,-2,6,14,... . $a_2-a_1=8$, $a_3-a_2=8$, $a_4-a_3=8$.
   A.S. with common difference 8.

(c) $5, 8, 12, 17, \ldots$. $a_2 - a_1 = 3$, $a_3 - a_2 = 4$, $a_4 - a_3 = 5$. Not an A.S.

(d) $3.5, 4.75, 6, 7.25, \ldots$. $a_2 - a_1 = 1.25$, $a_3 - a_2 = 1.25$, $a_4 - a_3 = 1.25$. A.S. with common difference 1.25.

. (a) $a_1 = 2$, $d = 5$. 4th term. $a_n = a_1 + (n-1)d$. $a_4 = 2 + (4-1)5 = 17$.

(b) $a_1 = -3$, $d = 6$. 5th term. $a_n = a_1 + (n-1)d$. $a_5 = -3 + (5-1)6 = 21$.

(c) $a_1 = 8$, $d = -3$. 6th term. $a_n = a_1 + (n-1)d$. $a_6 = 8 + (6-1)(-3) = -7$.

(d) $a_1 = 0$, $d = 5$. 50th term. $a_n = a_1 + (n-1)d$. $a_{50} = 0 + (50-1)5 = 245$.

. (a) $a_3 = 7$, $a_7 = 19$. 9th term. $a_n = a_1 + (n-1)d$ with $n=3$ and $n=7$ gives $7 = a_1 + 2d$ and $19 = a_1 + 6d$. Solving gives $a_1 = 1$, $d = 3$. Thus $a_n = a_1 + (n-1)d$ now gives $a_9 = 1 + (9-1)(3) = 25$.

(b) $a_7 = 21$, $a_2 = 1$. 5th term. $a_n = a_1 + (n-1)d$ with $n=7$ and $n=2$ gives $21 = a_1 + 6d$ and $1 = a_1 + 1d$. Solving gives $a_1 = -3$, $d = 4$. Thus $a_n = a_1 + (n-1)d$ now gives $a_5 = -3 + (5-1)(4) = 13$.

(c) $a_5 = 11$, $a_8 = 23$. 2nd term. $a_n = a_1 + (n-1)d$ with $n=5$ and $n=8$ gives $11 = a_1 + 4d$ and $23 = a_1 + 7d$. Solving gives $a_1 = -5$, $d = 4$. Thus $a_n = a_1 + (n-1)d$ now gives $a_2 = -5 + (2-1)(4) = -1$.

(d) $a_4 = 18$, $a_9 = 43$. 100th term. $a_n = a_1 + (n-1)d$ with $n=4$ and $n=43$, $18 = a_1 + 3d$ and $43 = a_1 + 8d$. Solving gives $a_1 = 3$, $d = 5$. Thus $a_n = a_1 + (n-1)d$ now gives $a_{100} = 3 + (100-1)(5) = 498$.

8. (a) $a_1 = 5$, $d = 4$. $S_n = \frac{n}{2}[2a_1 + (n-1)d]$, $S_3 = \frac{3}{2}[2(5) + (3-1)4] = 27$.

(b) $a_1 = -3$, $d = 6$. $S_4 = \frac{4}{2}[2(-3) + (4-1)6] = 24$.

(c) $a_n = 2n + 1$. S_6. $a_1 = 3, a_6 = 13$. $S_n = \frac{n}{2}(a_1 + a_n)$. $S_6 = \frac{n}{2}(a_1 + a_6) = 3(3 + 13) = 48$.

(d) $a_n = 3n - 4$. S_5. $a_1 = -1, a_5 = 11$. $S_5 = \frac{n}{2}(a_1 + a_5) = \frac{5}{2}(-1 + 11) = 25$.

9. $a_1 = 2$, $a_n = 17$, $S_n = 57$. $S_n = \frac{n}{2}(a_1 + a_n)$, $57 = \frac{n}{2}(2 + 17)$, $19n = 114$, $n = 6$.

10. $S_6 = -12$, $d = -2$. $S_n = \frac{n}{2}[2a_1 + (n-1)d]$, $-12 = \frac{6}{2}[2a_1 + (6-1)(-2)]$, $-12 = 3[2a_1 - 10]$, $-4 = 2a_1 - 10$, $2a_1 = 6$, $a_1 = 3$.

11. $S_{12} = 210$, $a_1 = 1$. $S_n = \frac{n}{2}[2a_1 + (n-1)d]$, $210 = \frac{12}{2}[2(1) + (12-1)d]$, $210 = 6[2 + 11d]$, $35 = 2 + 11d$, $d = 3$.

12. (a) $a_1=3$, $r=2$. 4th term. $a_n=a_1(r)^{n-1}$, $a_4=3(2)^{4-1}=24$.

(b) $a_1=-1$, $r=3$. 5th term. $a_n=a_1(r)^{n-1}$, $a_5=-1(3)^{5-1}=-81$.

(c) $a_1=1$, $r=\frac{1}{2}$. 3rd term. $a_n=a_1(r)^{n-1}$, $a_3=1(\frac{1}{2})^{3-1}=\frac{1}{4}$.

(d) $a_1=64$, $r=-\frac{1}{2}$. 8th term. $a_n=a_1(r)^{n-1}$, $a_8=64(-\frac{1}{2})^{8-1}=-\frac{1}{2}$.

13. (a) $a_4=324$, $a_7=8748$. 5th term. $a_n=a_1(r)^{n-1}$ gives $324=a_1r^3$, $8748=a_1r^6$. Thus $8748/324=a_1r^6/a_1r^3$, $r^3=27$, $r=3$. $a_1=324/r^3=324/27=12$. $a_5=a_1r^{n-1}=12(3)^4=972$.

(b) $a_5=16$, $a_7=64$. 2nd term. $a_n=a_1(r)^{n-1}$ gives $16=a_1r^4$, $64=a_1r^6$. Thus $64/16=a_1r^6/a_1r^4$, $r^2=4$, $r=\pm2$. $a_1=16/r^4=16/16=1$. $a_2=a_1r=\pm2$.

14. (a) $a_n=2(3)^{n-1}$. Compare with $a_n=a_1(r)^{n-1}$. $a_1=2$, $r=3$. Thus
$$S_n = \frac{a_1(1-r^n)}{(1-r)}, \quad S_5 = \frac{2(1-3^5)}{(1-3)} = 242.$$

(b) $a_n=2(-3)^{n-1}$. Compare with $a_n=a_1(r)^{n-1}$. $a_1=2$, $r=-3$. Thus
$$S_n = \frac{a_1(1-r^n)}{(1-r)}, \quad S_4 = \frac{2(1-(-3)^4)}{(1-(-3))} = -40.$$

15. (a) $a_1=3$, $r=\frac{1}{2}$. $S = \frac{a_1}{(1-r)} = \frac{3}{(1-\frac{1}{2})} = \frac{3}{1/2} = 6$.

(b) $a_1=-1$, $r=-\frac{1}{2}$. $S = \frac{a_1}{(1-r)} = \frac{-1}{(1-(-\frac{1}{2}))} = \frac{-1}{3/2} = -\frac{2}{3}$.

16. (a) $.42\overline{56} = .42+.0056+.000056+.00000056+\ldots$
$$= .42+.0056+.0056(10^{-2})+.0056(10^{-4})+\ldots$$
Use $a_1=.0056$, $r=10^{-2}$.

$.42\overline{56} = .42+\frac{a_1}{(1-r)} = .42 + \frac{.0056}{(1-10^{-2})} = \frac{42}{(100)} + \frac{.0056}{(.99)}$

$= \frac{(42)(99)}{(9900)} + \frac{56}{(9900)} = \frac{4214}{9900} = \frac{2107}{4950}$.

(b) $.93\overline{521} = .93+.00521+.00000521+.00000000521+\ldots$
$$= .93+.00521(\quad)+.00521(^{-3})+.00521(^{-6})+\ldots$$
Use $a_1=.00521$, $r=10^{-3}$.

$$..93\overline{521} = .93 + \frac{a_1}{(1-r)} = .93 + \frac{.00521}{(1-10^{-3})} = \frac{93}{(100)} + \frac{.00521}{(.999)}$$

$$= \frac{(93)(999)}{(99900)} + \frac{521}{(99900)} = \frac{93428}{99900} = \frac{23357}{24975}.$$

17. (a) $3+6+9+...+3n = \frac{3n(n+1)}{2}$. (A)

1. Let n=1. Formula is $3 = \frac{3(1)(1+1)}{2}$, true. Valid for n=1.

2. Let n=k. Formula is $3+6+9+...+3k = \frac{3k(k+1)}{2}$. Next term on left would be $3(k+1)$. Add this term to each side.
$3+6+9+...+3k+3(k+1) = \frac{3k(k+1)}{2} + 3(k+1) = \frac{3k(k+1)+6(k+1)}{2} = \frac{3k^2+9k+6}{2} = \frac{3(k+1)(k+2)}{2}$.
This is formula (A) with n=k+1. Thus if the formula holds for n=k it holds for n=k+1. By mathematical induction it is true for all positive integer values of n.

(b) $1^3+2^3+3^3+...+n^3 = \frac{n^2(n+1)^2}{4}$. (A)

1. Let n=1. Formula is $1 = \frac{1^2(1+1)^2}{4}$, true. Valid for n=1.

2. Let n=k. Formula is $1^3+2^3+3^3+...+k^3 = \frac{k^2(k+1)^2}{4}$. Next term on left would be $(k+1)^3$. Add this term to each side.
$1^3+2^3+3^3+...+k^3+(k+1)^3 = \frac{k^2(k+1)^2}{4} + (k+1)^3 = \frac{k^2(k+1)^2+4(k+1)^3}{4} = \frac{(k+1)^2(k^2+4k+4)}{4} = \frac{(k+1)^2(k+2)^2}{4}$. This is formula (A) with n=k+1. Thus if the formula holds for n=k it holds for n=k+1. By mathematical induction it is true for all positive integer values of n.

(c) $1\cdot2+2\cdot4+3\cdot8+...+n\cdot2^n = (n-1)2^{n+1} + 2$. (A)
1. Let n=1. Formula is $2 = (1-1)2^{2+1} + 2$, true. Valid for n=1.
2. Let n=k. Formula is $1\cdot2+2\cdot4+3\cdot8+...+k\cdot2^k = (k-1)2^{k+1} + 2$. Next term on left would be $(k+1)\cdot2^{k+1}$. Add to each side.
$1\cdot2+2\cdot4+3\cdot8+...+k\cdot2^k+(k+1)\cdot2^{k+1} = (k-1)2^{k+1} +2+(k+1)\cdot2^{k+1} = 2^{k+1}(k-1+k+1)+2 = 2^{k+1}(2k)+2 = (2^{k+1}\cdot2^1)k+2 = k2^{k+2} + 2$.
This is formula (A) with n=k+1. Thus if the formula holds for n=k it holds for n=k+1. By mathematical induction it is true for all positive integer values of n.

18. (a) $(a+b)^4 = a^4 + 4a^3b + 6a^2b^2 + 4ab^3 + b^4$.
(b) $(2-3x)^5 = 2^5 + 5\cdot2^4(-3x)+ 10\cdot2^3(-3x)^2 + 10\cdot2^2(-3x)^3 + 5\cdot2(-3x)^4 + (-3x)^5 = 32 - 240x + 720x^2 -1040x^3 + 810x^4 -243x^5$.

(c) $(2x-y)^6 = (2x)^6 + 6(2x)^5(-y) + 5(2x)^4(-y)^2 + 20(2x)^3(-y)^3 +$
$15(2x)^2(-y)^4 + 6(2x)(-y)^5 + (-y)^6 = 64x^6 - 192x^5y + 240x^4y^2 -$
$160x^3y^3 + 60x^2y^4 - 12xy^5 + y^6.$

19. (a) $5! = 5 \cdot 4 \cdot 3 \cdot 2 \cdot 1 = 120.$ (b) $\dfrac{12!}{10!} = \dfrac{12 \cdot 11 \cdot 10!}{10!} = 12 \cdot 11 = 132.$

(c) $\binom{7}{4} = \dfrac{7!}{3!4!} = \dfrac{7 \cdot 6 \cdot 5 \cdot 4!}{3!4!} = 35.$ (d) $\binom{127}{125} = \dfrac{127 \cdot 126 \cdot 125!}{2!125!} = \dfrac{127 \cdot 126}{2} = 8001.$

20. (a) $(x+2y)^3 = (x)^3 + \binom{3}{1}(x)^2(2y) + \binom{3}{2}(x)(2y)^2 + (2y)^3 =$
$$x^3 + 6x^2y + 12xy^2 + 8y^3.$$

(b) $(x-3y)^5 = x^5 + \binom{5}{1}x^4(-3y) + \binom{5}{2}x^3(-3y)^2 + \binom{5}{3}x^2(-3y)^3 +$
$\binom{5}{4}x(-3y)^4 + (-3y)^5 = x^5 - 15x^4y + 90x^3y^2 - 270x^2y^3 + 405xy^4 - 243y^5.$

(c) $(2x+4y)^3 = (2x)^3 + \binom{3}{1}(2x)^2(4y) + \binom{3}{2}(2x)(4y)^2 + (4y)^3 =$
$$8x^3 + 48x^2y + 96xy^2 + 64y^3.$$

21. (a) $(2x+5y)^7$, 4th term. $n=7, r=4.$ $\binom{7}{3}(2x)^4(5y)^3 = 70000x^4y^3.$

(b) $(x-2y)^9$, 5th term. $n=9, r=5.$ $\binom{9}{5}(x)^5(-2y)^4 = 2016x^5y^4.$

(c) $(5x-3y)^{17}$, 6th term. $n=17, r=6.$ $\binom{17}{5}(5x)^{12}(-3y)^5 =$
$(-3.671103516 \times 10^{14})x^{12}y^5.$

22. $a_1 = 4.27, d = 4.63.$ Sequence: 4.27, 8.9, 13.53, 18.16, 22.79, 27.42, 32.05, 36.68.

23. $a_1 = 5.82, r = 1.93.$ Sequence: 5.82, 11.2326, 21.678918, 41.84031174, 80.75180166, 155.8509772, 300.792386, 580.529305, 1120.421559, 2162.413608.

24. $a_1 = 2.97, r = 0.45.$ $f(x) = \dfrac{a_1(1-r^x)}{(1-r)} = \dfrac{2.97(1-.45^x)}{(1-.45)} \rightarrow 5.4$
Sum = 5.4. $S = \dfrac{a_1}{(1-r)} = \dfrac{2.97}{(1-.45)} = 5.4.$

Chapter 8 Test

Chapter 8   Test

1. $a_n=2n^2 - 3n$. 1st four terms: -1, 2, 9, 20.

2. $a_1=3$, $a_n=2(-1)^n a_{n-1}+ 5$. 1st four terms: 3, 11, -17, -29, 63.

3. $\sum\limits_{k=2}^{5}(k^2-3k+2)= 0+2+6+12 = 20$.

4. $a_n=3n+4$. $S_n= \dfrac{n}{2} [a_1+a_n]$. $S_6= \dfrac{6}{2} [a_1+a_6] = 3[7+22]=87$.

5. $a_1=3$, $d=4$. 5th term. $a_n=a_1+(n-1)d$. $a_5=3+(5-1)4 = 19$.

6. $a_2=4$, $a_8=40$. 10th term. $a_n=a_1+(n-1)d$ with $n=2$ and $n=8$ gives $4=a_1+d$ and $40=a_1+7d$. Solving gives $a_1=-2$, $d=6$. Thus $a_n=a_1+(n-1)d$ now gives $a_{10}=-2+(10-1)(6)=52$.

7. $a_1=3$, $a_n=42$, $S_n=315$. $S_n= \dfrac{n}{2}(a_1+a_n)$, $315= \dfrac{n}{2}(3+42)$, $45n=630$, $n=14$.

8. $S_8=36$, $d=3$. $S_n= \dfrac{n}{2} [2a_1+(n-1)d]$, $36= \dfrac{8}{2} [2a_1+(8-1)(3)]$, $36=4[2a_1+21]$, $9=2a_1+21$, $2a_1=-12$, $a_1=-6$.

9. $S_{10}=345$, $a_1=3$. $S_n= \dfrac{n}{2} [2a_1+(n-1)d]$, $345= \dfrac{10}{2} [2(3)+(10-1)d]$, $345=5[6+9d]$, $69=6+9d$, $63=9d$, $d=7$.

10. $a_1=4$, $r=5$. 6th term. $a_n=a_1(r)^{n-1}$, $a_6=4(5)^{6-1}=12500$.

11. $a_5=112$, $a_9=1792$. 12th term. $a_n=a_1(r)^{n-1}$ gives $112=a_1 r^4$, $1792=a_1 r^8$. Thus $1792/112=a_1 r^8/a_1 r^4$, $r^4=16$, $r=2$. $a_1=112/r^4=112/16=7$. $a_{12}=a_1 r^{n-1}=7(2)^{11}=14336$.

12. $a_n=5(2)^{n-1}$. Compare with $a_n=a_1(r)^{n-1}$. $a_1=5$, $r=2$. Thus $S_n = \dfrac{a_1(1-r^n)}{(1-r)}$, $S_7 = \dfrac{5(1-2^7)}{(1-2)} = 635$.

13. $a_1=7$, $r=\dfrac{1}{3}$. $S = \dfrac{a_1}{(1-r)} = \dfrac{7}{(1-\frac{1}{3})} = \dfrac{7}{2/3} = \dfrac{21}{2} = 10.5$.

14. $.7\overline{48} = .73+.0048+.000048+.00000048+\ldots$
$$= .73+.0048+.0048(10^{-2})+.0048(10^{-4})+\ldots$$
Use $a_1=.0048$, $r=10^{-2}$.

$$.73\overline{48} = .73 + \frac{a_1}{(1-r)} = .73 + \frac{.0048}{(1-10^{-2})} = \frac{73}{(100)} + \frac{.0048}{(.99)}$$

$$= \frac{(73)(99)}{(9900)} + \frac{48}{(9900)} = \frac{97}{132}.$$

15. $(1+3x)^5 = 1^5 + 5 \cdot 1^4(3x) + 10 \cdot 1^3(3x)^2 + 10 \cdot 1^2(3x)^3 + 5 \cdot 1(3x)^4 + (3x)^5 = 1 + 15x + 90x^2 + 270x^3 + 405x^4 + 243x^5$.

16. (a) $7! = 7 \cdot 6 \cdot 5 \cdot 4 \cdot 3 \cdot 2 \cdot 1 = 5040$.

(b) $\binom{8}{3} = \frac{8!}{5!3!} = \frac{8 \cdot 7 \cdot 6 \cdot 5!}{5!3!} = \frac{8 \cdot 7 \cdot 6}{3!} = 8 \cdot 7 = 56$.

(c) $\binom{243}{241} = \frac{243 \cdot 242 \cdot 241!}{2!241!} = \frac{243 \cdot 242}{2} = 29403$.

17. $(2x+y)^5 = (2x)^5 + \binom{5}{1}(2x)^4(y) + \binom{5}{2}(2x)^3(y)^2 + \binom{5}{3}(2x)^2(y)^3 + \binom{5}{4}(2x)(y)^4 + (y)^5 = 32x^5 + 80x^4y + 80x^3y^2 + 40x^2y^3 + 10xy^4 + y^5$.

18. $(3x-2y)^8$, 5th term. $n=8, r=5$. $\binom{8}{4}(3x)^4(-2y)^4 = 90720x^4y^4$.

19. $a_1 = 5.36$, $d = 3.27$. Sequence: 5.36, 8.63, 11.9, 15.17, 18.44.

20. $a_1 = 4.91$, $r = 1.25$. Sequence: 4.91, 6.1375, 7.671875, 9.58984375, 11.98730469, 14.98413086.

Chapter 9

1. $4! = 4 \cdot 3 \cdot 2 \cdot 1 = 24$.

3. $2! = 2 \cdot 1 = 2$.

5. $_4P_2 = 4 = 3 = 12$.

7. $_8P_1 = 8$.

9. $_6P_5 = 6 \cdot 5 \cdot 4 \cdot 3 \cdot 2 = 720$.

11. $_7C_3 = \dfrac{7 \cdot 6 \cdot 5}{3 \cdot 2 \cdot 1} = 35$.

13. $_{12}C_2 = \dfrac{12 \cdot 11}{2 \cdot 1} = 66$

15. $_2C_1 = \dfrac{2}{1} = 2$

17. Smallest positive integer n such that $n! > 10,000$.
    $7! = 5,040$ and $8! = 40,320$. Thus $n = 8$.

19. The values of n for which $_{10}C_n > 200$.
    $_{10}C_4 = 210$, $_{10}C_5 = 252$, $_{10}C_6 = 210$. Thus $n = 4,5,6$.

21. Student takes one biology course, one chemistry course and one physics course. There are three biology classes, four chemistry classes and two physics classes to choose from.
    \# Possibilities $= 3 \cdot 4 \cdot 2 = 24$.

23. Examination with three parts A, B and C. One question has to be answered from each part. There are five questions to choose from in A, four in B and two in C.
    \# ways questions can be answered $= 5 \cdot 4 \cdot 2 = 40$.

25. Eight children line up to have their picture taken.
    \# ways $= 8! = 8 \cdot 7 \cdot 6 \cdot 5 \cdot 4 \cdot 3 \cdot 2 \cdot 1 = 40,320$.

27. Manager arranges the batting order in a nine-man baseball team. \# arrangements $= 9! = 362,880$.

29. Number ways can a winner and runner-up be selected from seven entries $= _7P_2 = 7 \cdot 6 = 42$.

31. Eight runners in the Olympic 100 meters final. Gold, silver and bronze medals be given in $_8P_3 = 336$ ways.

33. Three-digit numbers to be formed using the digits 1,2,3,4.
    (a) No repetition allowed. \# ways $= _4P_3 = 24$.
    (b) Repetition allowed. \# ways $= 4 \cdot 4 \cdot 4 = 64$.

35. Room for 25 students in a class. 30 students wish to enroll.
    Number possibilities $= _{30}C_{25} = 142,506$.

37. Ten faculty members and a secretary in a math department.
    Dept. to be given four new computers (all alike). Secretary to be given one. Thus 3 computers between 10 faculty. Computers be distributed in $_{10}C_3 = 120$ ways.

Section 9.2

39. Meal consists of 1 meat, 2 vegetable, 1 salad and 1 desert. Three meats, five vegetables, two salads, four deserts availble. Number different meals = $3 \cdot {}_5C_2 \cdot 2 \cdot 4 = 480$.

41. Three different contracts to be awarded to six different firms. No firm to get more than two contracts.
No restriction: 6 ways for 1st contract, 6 for 2nd, 6 for 3rd. Gives 6x6x6 ways. Number of ways all go to one firm: 6 ways. Thus # ways no firm gets all three is 6x6x6 - 6 = 210 ways.

43. 6 people selected from 20 people in ${}_{20}C_6 = 38,760$ ways.

45. 7-digit telephone numbers, 0 cannot be used for the first digit. 9 choices for 1st digit, 10 for each of remaining 6 digits. # telephone numbers = $9 \cdot 10^6 = 9$ million.

47. Committee consists of three seniors, two juniors, one sophomore. Six senior, five junior, seven sophomore candidates.
committees = ${}_6C_3 \cdot {}_5C_2 \cdot {}_7C_1 = 1,400$.

49. Foreign Language Department has 200 majors. 100 Spanish, 75 French, 25 German. Each language elects a president and vice president. # ways = ${}_{100}P_2 \cdot {}_{75}P_2 \cdot {}_{25}P_2 = 32,967,000,000$ ways.

51. Three departments, Biology 80 majors, Chemistry 20 majors, Physics 10 majors. Each department to select two representatives for committee. Committee then to elect a chairman and secretary. Total number of students = 110. Biology can select its two students in ${}_{80}C_2$ ways, Chemistry in ${}_{20}C_2$ ways, and Physics in ${}_{10}C_2$ ways. Thus # ways the six people can be selected is ${}_{80}C_2 \cdot {}_{20}C_2 \cdot {}_{10}C_2$ ways. The chair and secretary can then be chosen in ${}_6P_2$ ways. Thus total number of ways = ${}_{80}C_2 \cdot {}_{20}C_2 \cdot {}_{10}C_2 \cdot {}_6P_2 = 810,540,000$.

Section 9.2

Group Discussion: Probability at least two people have birthday on same day of month is $1 - \dfrac{{}_{31}P_n}{31^n}$. With 6 people, $p = 1 - \dfrac{{}_{31}P_6}{31^6} = $ .40268 to 5 decimal. With 7 people, $p = 1 - \dfrac{{}_{31}P_7}{31^7} = $ .51829. Probability becomes more than .5 with 7 people.

1. Coin is tossed. Two possible outcomes, H,T. (a) P(H)=1/2. (b) P(T)=1/2.

3. Word Mississippi. Has 11 letters.
    (a) Four s. P(s)=4/11.   (b) Two p. P(p)=2/11.
    (c) Four i. P(i)=4/11.   (d) No a. P(a)=0.
    (e) One m. P(m)=1/11.   (f) Ten s,p,i. P(s/p/i)=10/11

5. Person with birthday in April. 30 days in April.
   (a) P(April 28)=1/30.   (b) P(2nd week of April)=7/30.
   (c) P(first half of April)=15/30=1/2.

7. Die tossed twice.See sample space in example 2, this
   section;{(1,1),(1,2),...,(6,5),(6,6)}. 36 elements in sample
   space.   (a) P(5,5)=1/36.
   (b) P((3,4)or(4,3))=2/36=1/18.
   (c) P((a,b) with a+b>9)=P((4,6)(5,5),(5,6),(6,4),(6,5),(6,6))
       =6/36=1/6.

9. Coin tossed three times. Sample space{(H,H,H,),(H,H,T),(H,T,H),
   (H,T,T),(T,H,H),(T,H,T),(T,T,H),(T,T,T)}. 8 elements.
   P(H,H,H)=1/8.

11. A box with two black balls B_1,B_2 and two white balls W_1,W_2.
    Three balls drawn in succession without replacement. Sample
    space {$B_1B_2W_1,B_1B_2W_2,B_1W_1B_2,B_1W_1W_2,B_1W_2B_2,B_1W_2W_1,B_2B_1W_1,B_2B_1W_2,$
    $B_2W_1B_1,B_2W_1W_2,B_2W_2B_1,B_2W_2W_1,W_1W_2B_1,W_1W_2B_2,W_1B_1W_2,W_1B_1B_2,W_1B_2W_2,$
    $W_1B_2B_1,W_2W_1B_1,W_2W_1B_2,W_2B_1W_1,W_2B_1B_2,W_2B_2W_1,W_2B_2B_1$}. 24 elements
    in sample space. $P(B_iB_jW_k)$=4/24=1/6.

13. Die tossed. (a) P(odd number less than 5)=P(1or3)=2/6=1/3.
    (b) P(even number or number less than five)=P(2or4,6or1,3)=5/6.

15. Family has four children. Sample space:
    {bbbb,bbbg,bbgb,bbgg,bgbb,bgbg,bggb,bggg,
     gbbb,gbbg,gbgb,gbgg,ggbb,ggbg,gggb,gggg}.16 elements.
    (a) P(either four boys or four girls)=2/16=1/8.
    (b) P(at least one boy & one girl)=14/16=7/8.

17. Die tossed. F is the event of throwing an even number, G is
    the event of throwing a number less than 3. F={2,4,6}, G={1,2}.
    F∩G={2}≠∅. F and G not mutually exclusive.

19. Die is tossed twice. F is the event that the sum of the
    numbers is 9 and G is the event that one of the numbers is 6. F
    and G are not mutually exclusive. e.g. (6, 3) is in F and in G.

21. Die is tossed twice. Sample space has 36 elements.
    P(sum = 8 or 9)=P((2,6),(6,2),(3,5),(5,3),(3,6),(6,3),(4,4),
    (4,5),(5,4)}=9/36=1/4.

23. Two cards from 52 playing cards.
    (a) P(same number or a face card)? # ways selecting 1st
    card=52. # ways selecting 2nd card=3. # outcomes of interest
    =52·3. Total # outcomes=52·51. Thus P=(52·3)/(52·51)=3/51.

    (b) P(of the same suit)? # ways selecting 1st card=52. # ways
    selecting 2nd card=12. Total # outcomes=52·12. Thus
    P=(52·12)/(52·51)=12/51.

25. Twenty lots. 2 swampy and no trees, 3 swampy with trees, 3 dry with no trees, 12 dry with trees. Thus 20 lots.
    (a) P(dry)=((3+12)/20)=3/4.
    (b) P(swampy or has no trees)=P((2+3+3)/20)=8/20=2/5
    (c) P(has trees)=((3+12)/20)=15/20=3/4.

27. Garage has 20 tires, one is defective. Four tires placed on a car, what is probability one is defective? # ways selecting 4 from 20 is $_{20}C_4$. One is defective. $_{19}C_3$ ways of selecting the other three tires. Thus $P=_{19}C_3/_{20}C_4=0.2$.

29. 15 people apply for jobs; 10 men,5 women. 6 people hired.
    P(4 men, 2 women) = $(_{10}C_4 \cdot _5C_2)/(_{15}C_6)=0.4196$ to 4 dec places.

31. Building with 30 floors. 4 people use elevator. Each person can get of at one of 30 floors. Thus # ways they can get off is $30 \cdot 30 \cdot 30 \cdot 30$. (a) 30 ways they can all get off at same floor. $P=30/(30 \cdot 30 \cdot 30 \cdot 30)=1/27000$.
    (b) 1 way all get off at 20th. $P=1/(30 \cdot 30 \cdot 30 \cdot 30)=1/810000$.
    (c) Get off different floors. 1st person choice of 30, 2nd 29, 3rd 28 ... $P=(30 \cdot 29 \cdot 28 \cdot 27)/(30 \cdot 30 \cdot 30 \cdot 30)=0.812$.

33. Channel 2: 10 min(com), 50 min(prog). Channel 6: 12 min,48 min Channel 9: 11.5 min, 48.5 min.
    (a) Channel 6, P(12 min comm in 60min)=12/60=1/5.
    (b) P(commercial)=(10+12+11.5)/(60+60+60)=0.1861.

35. A^+ from A^+, O^+, O^-, A^-. A^- from A^-, O^-.
    AB^- from AB^-, O^-, A^-, B^-. O^- from O^- only.

39. Probability that a B^+ person will be able to give blood to a person (of unknown group) injured in a street accident. B^+ can give blood to B^+ and AB^+. $P(B^+)+P(AB^+)=.092+.033=0.125$.

41. Probability at least two people in room of 45 people have the same birthday = $1-(_{365}P_{45}/365^{45})=1-(365 \cdot 364 \cdot 363 \cdot ... \cdot 321)/365^{45}=$ $1-((365 \cdot 364 \cdot ... \cdot 331)/365^{35})((330 \cdot 329 \cdot ... \cdot 321)/365^{10})=$ 0.94097758995. [Get an overflow as it stands. Split up the arithmetic.]

Chapter 9   Review Exercises

1. (a) 6!=720.   (b) 3!=6.   (c) $_5P_2=20$.   (d) $_7P_1=7$.
   (e) $_{12}P_4=11,880$.   (f) $_7C_3=35$.   (g) $_9C_4=126$.   (h) $_8C_4=70$.

2. 3 from London to Birmingham, 4 Birmingham to Chester, 5 Chester to Colwyn Bay. Thus $3 \cdot 4 \cdot 5 = 60$ from London to Colwyn Bay.

Chapter 9 Review Exercises

3. A senior,junior,sophomore & freshman from 5 seniors,3 juniors, 2 sophomores and 6 freshmen. # ways =$5 \cdot 3 \cdot 2 \cdot 6 = 180$.

4. 4 shirts, 3 shorts, 5 socks. # different uniforms = $4 \cdot 3 \cdot 5 = 60$.

5. 5 carpets,4 papers,3 paints,6 vinyls.# selections=$5 \cdot 4 \cdot 3 \cdot 6 = 360$.

6. 8 books on a shelf. # arrangements=$8!=40,320$.

7. Family of five line up for a picture. # ways = $5! = 120$.

8. Five pairs at a round table. Men and women are to alternate. Treat the pair as a unit, MW. Let pairs be A,B,C,D,E. Place A, anywhere. It is unimportant where. Starting with the left of A we get 4 possibilities for this seat, then 3 to that person's left etc. # seating arrangements = $4!=24$. But can also have pairs as WM. Gives another 24, Thus total =48.

9. 4 history, 3 math, 5 biology books on a shelf. All history books together, all math together, all biology together. History books can be arranged in 4! ways, math in 3! and biology in 5! The 3 groups can be arranged in 3! ways. # arrangements = $(3!)(4!)(3!)(5!)=103,680$.

10. 3 different contracts be awarded to seven different firms. No firm to get more than two contracts. $7 \cdot 7 \cdot 7$ ways if no restriction. 7 ways all contracts can go to 1 firm. Thus $7 \cdot 7 \cdot 7 - 7 = 336$ ways if no firm to get more than two contracts.

11. Recital of 6 classics and 4 moderns. 4 classics and 2 moderns before the intermission. 2 classics, 2 moderns after. Before: Can choose the classics in $_6C_4$ ways, moderns in $_4C_2$. But can then play those in 6! ways. # ways before=$6!(_6C_4)(_4C_2)$. After: The four pieces can be selected in 4! ways. Thus total number arrangements = $4!6!(_6C_4)(_4C_2)=1,555,200$.

12. License plates using two letters, three numbers between 0 & 9, then single letter. # ways=$26 \cdot 26 \cdot 10 \cdot 10 \cdot 10 \cdot 26 = 17,576,000$.

13. Four-digit numbers using 1,2,3,4,5 (a) no repetition. # ways=$5 \cdot 4 \cdot 3 \cdot 2 = 120$. (b) repetition allowed. # ways=$5 \cdot 5 \cdot 5 \cdot 5 = 625$.

14. A={1,2,3,4,5}. # subsets=(# with 1 element)+(# with 2)+...+ (# with 5)+ empty set = $_5C_1+_5C_2+_5C_3+_5C_4+_5C_5+1=32$

15. Test with 16 questions. To answer 10 questions.
    (a) # ways = $_{16}C_{10} = 8,008$.
(b) 6 of first 9 questions,and 4 of the last 7. # ways = $_9C_6 \cdot _7C_4 = 2,940$.

16. To choose 4 cards from 52 cards. # ways = $_{52}C_4$ = 270,725.

17. President, v. president, sec, treasurer and financial sec to be selected from 16 people. # ways=$_{16}P_5$ = 524,160.

18. Single card drawn 52 playing cards. (a) P(red card)=26/52=1/2. (b) P(number less than 5)=P(2,3,4 of C,D,H,S)=12/52=3/13. (c) P(king)=4/52=1/13. (d)P(ace,king,queen or jack)=16/52=4/13.

19. Die is tossed. (a) P(6)=1/6. (b) P(1,2 or 6)=3/6=1/2. (c) P(even number)=3/6=1/2. (d) P(# less than 4)=3/6=1/2.

20. Toss two dice. Sample space given in exampe 2, section 10.2. (a)P(two sixes)=P(6,6)=1/36. (b)P(1&2)=P((1,2),(2,1))=2/36=1/18. (c) P(diff numbers is 3)=P((1,4),(4,1),(2,5),(5,2),(3,6),(6,3)) =6/36=1/6. (d) P(sum of the two numbers is greater than 7)= P((2,6),(6,2),(3,5),(5,3),(3,6),(6,3),(4,4),(4,5),(5,4),(4,6), (6,4),(5,5),(5,6),(6,5),(6,6))=15/36=5/12.

21. Box with 1 red ball and 3 blue balls. Two balls are drawn in succession without replacement. Sample space is S={(R,B1),(R,B2),(R,B3),(B1,R),(B2,R),(B3,R),(B1,B2),(B1,B3), (B2,B3),(B2,B1),(B3,B1),(B3,B2)). (a) P(red then blue ball)=3/12=1/4. (b) P(red and blue ball in any order)=6/12=1/2.

22. Single card drawn from a pack of 52 cards. 4 aces, 13 clubs, 1 ace of clubs. 16 cards of interest. P(ace or club)=16/52=4/13.

23. 16 people apply for jobs. 9 men and 7 women. 5 people hired. P(2 men and 3 women hired)=$(_9C_2)(_7C_3)/(_{16}C_5)$=.2885.

24. 50 CB sets. 5 defective. 10 already sold. # ways selecting 10 from 50 is $_{50}C_{10}$. Assume 2 defective. $_5C_2$ ways of selecting these 2 defective ones. $_{45}C_8$ ways of selecting the remaining 8. P=$(_5C_2 \cdot _{45}C_8)/_{50}C_{10}$=.2098.

25. A die is tossed. F is the event of tossing an even number. G is the event of tossing a number less than 2. F={2,4,6}, G={1}. F∩G=∅. F and G are mutually exclusive.

26. A coin is tossed twice. S={(H,H),(H,T),(T,H),(T,T)}. F={(H,T),(T,T)}. G={(H,H),(T,T)}. F∩G={(T,T)}≠∅. F and G are not mutually exclusive.

27. Telephone busy 105 minutes in 360 minutes. P(busy)=105/360. P(not busy)=1 - 105/360= 51/72.

Chapter 9 Test

28. $11!$ ways the fox and 10 mice can arrive at te log.
(a) Four mice eaten. Thus fox arrives and then 4 mice. 4 mice in $_{10}P_4$ ways. Prob$=(_{10}P_4)/11!=1.262626263 \times 10^{-4}$.
(b) Five mice eaten. Fox arrives 1st,2nd,3rd,4th,5th,6th.
\# ways $= _{10}P_{10}+_{10}P_9+_{10}P_8+_{10}P_7+_{10}P_6+_{10}P_5$.
Prob$=(_{10}P_{10}+_{10}P_9+_{10}P_8+_{10}P_7+_{10}P_6+_{10}P_5)/11!=0.246969697$.
(c) No mice eaten. Fox arrives last. Prob$=1/11!=2.505210 \times 10^{-8}$.

29. 58 letters. (a) 7 gs. P(g)=7/58. (b) 11 l's. P(l)=11/58.
(c) 3 a's, 5 y's. P(a or y)=8/58.
(d) 11 ls, 7 g's, 3 a's. P(l, g or a)=21/58.
(e) English vowels, 3 a's, 1 e, 3 i's, 6 o's, 0 u's.
P(English vowel)=13/58.
(f) Welsh vowel, 13 English + 4 w's + 5 y's.
P(Welsh vowel)=22/58.
(g) Welsh alphabet: ll and ch count as single letters. The "llll" is two ll's. \# letters is 51.
Five ll's, P(ll)=5/51.
(h) Two ch's, P(ch)=2/51.

Chapter 9 Test

1. (a) $_8P_2=8 \times 7=56$.   (b) $_9C_3=9!/(6!3!)=(9 \times 8 \times 7)/(1 \times 2 \times 3)=84$.

2. 3 from Gainesville to Ocala, 4 Ocala to DeLand, 2 DeLand to Daytona. Thus $3 \cdot 4 \cdot 2 = 24$ from Gainesville to Daytona.

3. A senior,junior,sophomore & freshman from 4 seniors,5 juniors, 3 sophomores and 2 freshmen. \# ways $=4 \cdot 5 \cdot 3 \cdot 2=180$.

4. 12 books on a shelf. \# arrangements$=12!=479,001,600$.

5. 2 different contracts be awarded to eight different firms. $8 \cdot 8$ ways if no restriction. 8 ways both contracts can go to 1 firm. Thus $8 \cdot 8-8=56$ ways if no firm to get both contracts.

6. License plates using three letters, three numbers between 0 & 9, then one letter. \# ways$=26 \cdot 26 \cdot 26 \cdot 10 \cdot 10 \cdot 10 \cdot 26=456,976,000$.

7. Test with 10 questions. To answer 10 questions.
(a) \# ways $= _{10}C_7 = 120$. (b) 4 of first 6 questions,and 3 of the last 4. \# ways $= _6C_4 \cdot _4C_3=60$.

8. President, v. president, sec, treasurer and financial sec to be selected from 20 people. \# ways$=_{20}P_5 = 1,860,480$.

9. Die is tossed. (a) P(1)=1/6. (b) P(1,3 or 5)=3/6=1/2.
(c) P(\# less than 5)=4/6=2/3.

10. Box with 2 red ball and 3 blue balls. Two balls are drawn in succession without replacement. Sample space is
S={(R1,B1),(R1,B2),(R1,B3),(B1,R1),(B2,R1),(B3,R1),
   (R2,B1),(R2,B2),(R2,B3),(B1,R2),(B2,R2),(B3,R2)
   (R1,R2),(R2,R1)(B1,B2),(B1,B3),(B2,B3),(B2,B1),(B3,B1),
   (B3,B2)).
   (a) P(red then blue ball)=6/20=3/10.
   (b) P(red and blue ball in any order)=12/20=3/5.

11. 20 people apply for jobs. 12 men and 8 women. 6 people hired. P(5 men and 1 women hired)=$({}_{12}C_5)({}_8C_1)/({}_{20}C_6)$=0.1635.

12. 120 sets. 6 defective. 10 already sold. # ways selecting 10 from 120 is ${}_{120}C_{10}$. Assume 3 defective. ${}_6C_3$ ways of selecting these 3 defective ones. ${}_{117}C_7$ ways of selecting the remaining 7. P=$({}_6C_3 \cdot {}_{120}C_7)/{}_{120}C_{10}$=0.0103.

13. A die is tossed. F is the event of tossing an odd number. G is the event of tossing a number greater than 4. F={1,3,5}, G={5,6}.F∩G={5}≠∅. F and G are not mutually exclusive.

14. Building with 16 floors. 5 people use elevator. Each person can get off at one of 16 floors. Thus # ways they can get off is $16 \cdot 16 \cdot 16 \cdot 16 \cdot 16$. (a) 16 ways they can all get off at same floor. P=$16/(16 \cdot 16 \cdot 16 \cdot 16 \cdot 16)$=1/65536.
   (b) 1 way all get off at 10th. P=$1/(16 \cdot 16 \cdot 16 \cdot 16 \cdot 16)$=1/1048576.
   (c) Get off different floors. 1st person choice of 16, 2nd 15, 3rd 14 ... P=$(16 \cdot 15 \cdot 14 \cdot 13 \cdot 12)/(16 \cdot 16 \cdot 16 \cdot 16 \cdot 16)$=0.4999.

Cumulative Test Chapters 6,7,8,9

1. x−2y= 10. (1) x=10+2y.(3) In (2) 3(10+2y)+y=2. 6y=−28. y=−4.
   3x+ y= 2. (2)   Substitute for y in (3), x=10+2(−4). x=2.

2. $\begin{bmatrix} 1 & 2 & 3 & 3 \\ 1 & 3 & 5 & 4 \\ 2 & 7 & 11 & 8 \end{bmatrix}$ Add $(-1)\tilde{R}1$ to R2 Add $(-2)R1$ to R3 $\begin{bmatrix} 1 & 2 & 3 & 3 \\ 0 & 1 & 2 & 1 \\ 0 & 3 & 5 & 2 \end{bmatrix}$

Add $(-2)\tilde{R}2$ to R1 Add $(-3)R2$ to R3 $\begin{bmatrix} 1 & 0 & -1 & 1 \\ 0 & 1 & 2 & 1 \\ 0 & 0 & -1 & -1 \end{bmatrix}$ $(-\overset{\sim}{1})R3$ $\begin{bmatrix} 1 & 0 & -1 & 1 \\ 0 & 1 & 2 & 1 \\ 0 & 0 & 1 & 1 \end{bmatrix}$

Add $R\overset{\sim}{3}$ to R1 Add $(-2)R3$ to R2 $\begin{bmatrix} 1 & 0 & 0 & 2 \\ 0 & 1 & 0 & -1 \\ 0 & 0 & 1 & 1 \end{bmatrix}$ x = 2 y = −1 z = 1

3. $5x^2+y^2=21$, $2x+y=3$. $y=3-2x$. $5x^2+(3-2x)^2=21$, $5x^2+(9-12x+4x^2)=21$.
   $9x^2-12x-12=0$. $(9x+6)(x-2)=0$. $x=-2/3, 2$. Substitute back into
   $y=3-2x$ to get $y=13/3, -1$. Two solutions $(-2/3,13/3)$ and $(2,-1)$.
   Points A, B in the figure below.

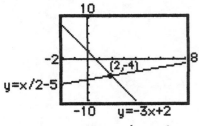

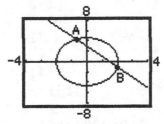

Exercise 1   Exercise 3

4. $A = \begin{bmatrix} 5 & 2 & 0 & -4 & 7 \\ 8 & -6 & 3 & 0 & 1 \\ 3 & 9 & 7 & -5 & 4 \\ 0 & 6 & -9 & -2 & 3 \end{bmatrix}$ (a) $a_{24}=0$. (b) $a_{12}=2$. (c) $a_{45}=3$.

5. $A = \begin{bmatrix} 1 & 0 \\ -2 & 4 \end{bmatrix}$, $B = \begin{bmatrix} 2 & -1 \\ 3 & 0 \end{bmatrix}$, $C = \begin{bmatrix} 1 & 0 & -1 \\ 2 & 1 & 1 \end{bmatrix}$

   (a) $A+2B = \begin{bmatrix} 5 & -2 \\ 4 & 4 \end{bmatrix}$ (b) $AC = \begin{bmatrix} 1 & 0 & -1 \\ 6 & 4 & 6 \end{bmatrix}$

6. $\begin{array}{l} x + 3y = 13 \\ 2x + 5y = 21 \end{array}$. $\begin{bmatrix} 1 & 3 \\ 2 & 5 \end{bmatrix}\begin{bmatrix} x \\ y \end{bmatrix} = \begin{bmatrix} 13 \\ 21 \end{bmatrix}$. $\begin{bmatrix} x \\ y \end{bmatrix} = \begin{bmatrix} 1 & 3 \\ 2 & 5 \end{bmatrix}^{-1}\begin{bmatrix} 13 \\ 21 \end{bmatrix}$.

   $\begin{bmatrix} 1 & 3 & 1 & 0 \\ 2 & 5 & 0 & 1 \end{bmatrix} \approx \begin{bmatrix} 1 & 3 & 1 & 0 \\ 0 & -1 & -2 & 1 \end{bmatrix} \approx \begin{bmatrix} 1 & 3 & 1 & 0 \\ 0 & 1 & 2 & -1 \end{bmatrix} \approx \begin{bmatrix} 1 & 0 & -5 & 3 \\ 0 & 1 & 2 & -1 \end{bmatrix}$.

   $\begin{bmatrix} 1 & 3 \\ 2 & 5 \end{bmatrix}^{-1} = \begin{bmatrix} -5 & 3 \\ 2 & -1 \end{bmatrix}$. $\begin{bmatrix} x \\ y \end{bmatrix} = \begin{bmatrix} -5 & 3 \\ 2 & -1 \end{bmatrix}\begin{bmatrix} 13 \\ 21 \end{bmatrix} = \begin{bmatrix} -2 \\ 5 \end{bmatrix}$. $x=-2$, $y=5$.

7. $\begin{vmatrix} -7 & 2 & 0 \\ 5 & -9 & 0 \\ 3 & 7 & 2 \end{vmatrix} = 0\begin{vmatrix} 5 & -9 \\ 3 & 7 \end{vmatrix} - 0\begin{vmatrix} -7 & 2 \\ 3 & 7 \end{vmatrix} + 2\begin{vmatrix} -7 & 2 \\ 5 & -9 \end{vmatrix} = $

   $2(63-10) = 106$.

8. $a_n=3n^2 - 2n$. 1st four terms: 1, 8, 21, 40.

9. $a_n=3n-1$. $S_n= \frac{n}{2}[a_1+a_n]$. $S_7= \frac{7}{2}[a_1+a_7] = \frac{7}{2}[2+20]=77$.

10. $a_1=2$, $d=-3$. 6th term. $a_n=a_1+(n-1)d$. $a_6=2+(6-1)(-3) = -13$.

11. $S_6=48$, $d=3$. $S_n= \frac{n}{2}[2a_1+(n-1)d]$, $48 = \frac{6}{2}[2a_1+(6-1)(3)]$,
    $48=3[2a_1+15]$, $16=2a_1+15$, $2a_1=1$, $a_1=1/2$.

12. $a_1=3$, $r=2$. 7th term. $a_n=a_1(r)^{n-1}$, $a_7=3(2)^{7-1}=192$.

13. $a_n=3(2)^{n-1}$. Compare with $a_n=a_1(r)^{n-1}$. $a_1=3$, $r=2$. Thus

$$S_n = \frac{a_1(1-r^n)}{(1-r)}, \quad S_5 = \frac{3(1-2^5)}{(1-2)} = 93.$$

14. (a) $6! = 6\cdot5\cdot4\cdot3\cdot2\cdot1 = 720$.

    (b) $\binom{7}{3} = \frac{7!}{4!3!} = \frac{7\cdot6\cdot5\cdot4!}{4!3!} = \frac{7\cdot6\cdot5}{3!} = 7\cdot5 = 35$.

    (c) $\binom{250}{248} = \frac{250\cdot249\cdot248!}{2!248!} = \frac{250\cdot249}{2} = 31125$.

    (d) $_7P_3=7\times6\times5=210$. (e) $_8C_5=8!/(3!5!)=(8\times7\times6)/(1\times2\times3)=56$.

15. $(2x-y)^9$, 4th term. $n=9, r=4$. $\binom{9}{3}(2x)^6(-y)^3 = -5376x^6y^3$.

16. 8 choices for 1st digit. 10 choices for other 6 digits. 8×10^6 possibilities. 8,000,000.

17. 4 senior, 3 junior, 2 sophomore & 2 freshman from 8 seniors,10 juniors, 12 sophomores and 14 freshmen.
    # ways $=_8C_4\cdot_{10}C_3\cdot_{12}C_2\cdot_{14}C_2=50450400$.

18. Die is tossed. (a) $P(6)=1/6$. (b) $P(1 \text{ or } 3)=2/6=1/3$.
    (c) $P(\text{\# greater than } 2)=4/6=2/3$.

19. 57 students want to get in. room for 25. # ways $=_{57}C_{25}$.
    Mary gets in, John does not. Thus 24 places between 55 students. $P=(_{55}C_{24})/(_{57}C_{25})=0.2506$.

20. 45 students, 8 scholarships. 28 female and 17 male apply.
    $P(4 \text{ female } \& 4 \text{ male}) = (_{28}C_4)(_{17}C_4)/(_{45}C_8)=0.2261$.

21. Let the company manufacture x Swirl, y Rapid. $x\geq0, y\geq0$.
    Time constraint: $6x+4y\leq2360$. Cost constraint: $550x+700y\leq308000$.
    Profit $f=55x+50y$. Vertices:$A(0,440),B(210,275),C(393.33,0)$.
    See fig below. $f_A=22000, f_B=25300, f_C=19666.5$. Max $f=25300$ at B.
    Maximum profit is \$25,300 by making 210 Swirl and 275 Rapid.
    See figure below.

22. $D = \begin{bmatrix} 0 & 3 & 2 & 3 & 1 \\ 1 & 0 & 3 & 4 & 2 \\ 2 & 1 & 0 & 1 & 2 \\ 4 & 3 & 2 & 0 & 1 \\ 3 & 2 & 1 & 2 & 0 \end{bmatrix} \begin{matrix} 9 \\ 10 \\ 6 \\ 10 \\ 8 \end{matrix}$ gives $M_3, M_5, M_1, (M_2 \text{ and } M_4)$

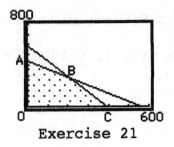

Exercise 21

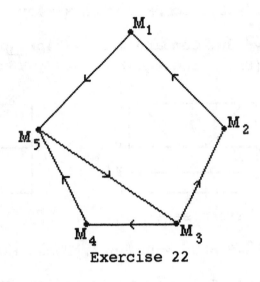

Exercise 22

Appendix A Conic Sections

1. $x^2=12y$. Compare with $x^2=4py$. p=3. Focus F(0,3), directrix y=-3.

3. $y^2=8y$. Compare with $y^2=4px$. p=2. Focus F(2,0), directrix x=-2.
   (Enter the equation as $Y_1=\sqrt{8x}$, $Y_2=-\sqrt{8x}$.)

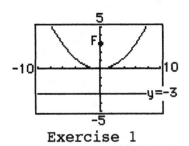
Exercise 1

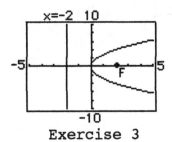
Exercise 3

5. $x^2-8y=0$. $x^2=8y$. p=2. Focus F(0,2), directrix y=-2.

7. $2x^2=-5y$. $x^2=-5y/2$. p=-5/8. Focus F(0,-5/8), directrix y=5/8.

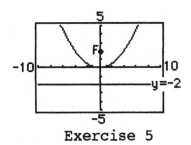
Exercise 5

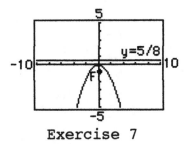
Exercise 7

9. Focus (0,3). Compare with $x^2=4py$ which has vertex at origin
   and focus (0,p). p=3. $x^2=12y$.

11. Directrix y=-1. Compare with $x^2=4py$ which has vertex at origin
    and directrix y=-p. p=1. $x^2=4y$.

13. Focus (6,0). Compare with $y^2=4px$ which has vertex at origin
    and focus (p,0). p=6. $y^2=24x$.

15. Focus (-1,0). p=-1. $y^2=-4x$.

17. Directrix $y=-\frac{5}{3}$. $p=\frac{5}{3}$. $x^2=\frac{20y}{3}$.

19. $\frac{x^2}{4}+\frac{y^2}{1}=1$. Compare with $\frac{x^2}{a^2}+\frac{y^2}{b^2}=1$. a=2, b=1. Vertices
    (-2,0),(2,0),(0,-1),(0,1). $c^2=a^2-b^2=3$, $c=\sqrt{3}$. Foci$(-\sqrt{3},0)$,$(\sqrt{3},0)$.

256

21. $\frac{x^2}{9} + \frac{y^2}{16} = 1$. a=4,b=3. Vertices (-3,0),(3,0),(0,-4),(0,4).
$c^2=a^2-b^2=7$,c=$\sqrt{7}$. Foci(0,-$\sqrt{7}$),(0,$\sqrt{7}$).

23. $\frac{x^2}{1} + \frac{y^2}{25}$ =1. a=5, b=1. Vertices(-1,0),(1,0),(0,-5),(0,5).
$c^2=a^2-b^2=25-1=24$, c=$\sqrt{24}$. Foci(0, -$\sqrt{24}$),(0, $\sqrt{24}$).

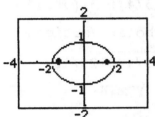
Exercise 19

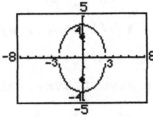
Exercise 21

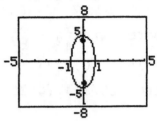
Exercise 23

25. $\frac{x^2}{9} - \frac{y^2}{1} = 1$. Compare with $\frac{x^2}{a^2} - \frac{y^2}{b^2} = 1$. a=3, b=1. Vertices
(-3,0),(0,3). $c^2=a^2+b^2=9+1=10$, c=$\sqrt{10}$. Foci(-$\sqrt{10}$,0),($\sqrt{10}$,0).
Asymptotes, y=$\pm\frac{b}{a}$x, y=$\pm\frac{1}{3}$x.

27. $\frac{x^2}{4} - \frac{y^2}{4} = 1$. Compare with $\frac{x^2}{a^2} - \frac{y^2}{b^2} = 1$. a=2, b=2. Vertices
(-2,0),(2,0). $c^2=a^2+b^2=4+4=8$, c=$\sqrt{8}$. Foci(-$\sqrt{8}$,0),($\sqrt{8}$,0).
Asymptotes, y=$\pm\frac{b}{a}$x, y=$\pm$x.

29. $\frac{y^2}{4} - \frac{x^2}{9} = 1$. Compare with $\frac{y^2}{a^2} - \frac{x^2}{b^2} = 1$. a=2, b=3. Vertices
(0,-2),(0,2). $c^2=a^2+b^2=4+9=13$,c=$\sqrt{13}$. Foci(0,-$\sqrt{13}$,(0,$\sqrt{13}$).
Asymptotes, y=$\pm\frac{a}{b}$x, y=$\pm\frac{2}{3}$x.

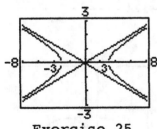
Exercise 25

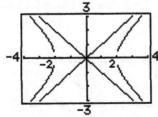
Exercise 27

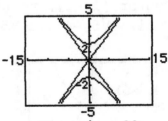
Exercise 29

31. $x^2+4y^2=16$. Divide by 16. $\frac{x^2}{16} + \frac{4y^2}{16} = 1$. $\frac{x^2}{4^2} + \frac{y^2}{2^2} = 1$.

    Compare with $\frac{x^2}{a^2} + \frac{y^2}{b^2} = 1$. a=4,b=2. Ellipse: Vertices (-4,0), (4,0),(0,-2),(0,2). $c^2=a^2-b^2=12$,c=$\sqrt{12}$. Foci(-$\sqrt{12}$,0),($\sqrt{12}$,0).

33. $16y^2-9x^2=25$. Divide by 25. $\frac{16y^2}{25} - \frac{9x^2}{25} = 1$. $\frac{y^2}{(5/4)^2} - \frac{x^2}{(5/3)^2} = 1$.

    Compare with $\frac{y^2}{a^2} - \frac{x^2}{b^2} = 1$. a=5/4,b=5/3. Hyperbola: Vertices (0,-5/4), (0,5/4). $c^2=a^2+b^2=25/16+25/9$,c=$\sqrt{625/144}=25/12$. Foci(0,-25/12),(0,25/12). Asymptotes, $y=\pm\frac{a}{b}x$, $y=\pm\frac{3}{4}x$.

35. $x^2+9y^2=-36$. Divide by 36. $\frac{16x^2}{36} + \frac{y^2}{4} = -1$. $\frac{x^2}{6^2} + \frac{y^2}{2^2} = -1$.

    Cannot be put in form $\frac{x^2}{a^2} + \frac{y^2}{b^2} = 1$, $\frac{x^2}{a^2} - \frac{y^2}{b^2} = 1$ or $\frac{y^2}{a^2} - \frac{x^2}{b^2} = 1$. Thus not an ellipse or hyperbola.

37. $(x+5)^2=4(y-3)$. Parabola $x^2=4y$ has p=1, vertex (0,0), focus (0,1), axis x=0, directrix y=-1. Shift this parabola left 5, up 3. Represents parabola with vertex V(-5,3), focus (-5,4), axis x=-5, directrix y=2.

39. $(y+3)^2=12(x-4)$. Parabola $y^2=12x$ has p=3, vertex (0,0), focus (3,0), axis y=0, directrix x=-3. Shift this parabola right 4, down 3. Represents parabola with vertex V(4,-3), focus (7,-3), axis y=-3, directrix x=1.

41. $\frac{(x-1)^2}{9} + \frac{(y+4)^2}{4} = 1$. Ellipse $\frac{x^2}{9} + \frac{y^2}{4} = 1$, a=3,b=2. Center (0,0), vertices (-3,0),(3,0),(0,-2),(0,2). $c^2=a^2-b^2=5$,c=$\sqrt{5}$. Foci(-$\sqrt{5}$,0),($\sqrt{5}$,0). Shift this ellipse right1, down4. Center (1,-4), vertices (-2,-4),(4,-4),(1,-6),(1,-2). Foci(1-$\sqrt{5}$,-4),(1+$\sqrt{5}$,-4).

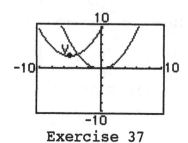
Exercise 37

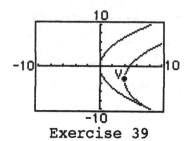
Exercise 39

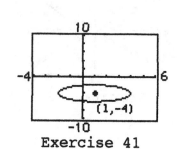
Exercise 41

43. $\frac{(x-3)^2}{1} - \frac{(y-2)^2}{9} = 1$. Hyp $\frac{x^2}{1} + \frac{y^2}{9} = 1$, a=1,b=3. Center (0,0),
   vertices (-1,0),(1,0). $c^2=a^2+b^2=10$,c=$\sqrt{10}$.
   Foci(-$\sqrt{10}$,0),($\sqrt{10}$,0). Asymptotes, y=$\pm\frac{b}{a}$x, y=$\pm$3x.
   Shift this hyperbola right3, up2. Center (3,2),
   vertices (2,2),(4,2). Foci(3-$\sqrt{10}$,2),(3+$\sqrt{10}$,2).
   Asymptotes (y-2)=$\pm$3(x-3), y=3x-7 and y=-3x+11.

45. x^2-4x+16=4y. $(x-2)^2$+12=4y. $(x-2)^2$=4y-12. $(x-2)^2$=4(y-3).
   Parabola x^2=4y has p=1, vertex (0,0), opens up. Shift this
   parabola right2, up3. Vertex becomes (2,3).

47. x^2+4y^2-2x+8y+1=0. $(x^2-2x+1)+4(y^2+2y+1)+1=+1+4$.
   $(x-1)^2+4(y+1)^2=4$. $\frac{(x-1)^2}{4} + \frac{(y+1)^2}{1} = 1$. a=2,b=1.
   Ellipse, center (1,-1). Vertices (-1,-1),(3,-1),(1,-2),(1,0).

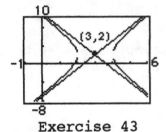
Exercise 43

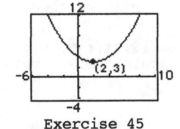
Exercise 45

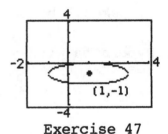
Exercise 47

49. x^2-25y^2+2x+100y-124=0. $(x^2+2x+1)-25(y^2-4y+4)-124=1-100$.
   $(x+1)^2-25(y-2)^2=25$. $\frac{(x+1)^2}{25} - \frac{(y-2)^2}{1} = 1$. a=5,b=1.
   Hyperbola, center (-1,2). Vertices (-6,2),(4,2).

51. $y-x^2$-6x-13=0. $y=x^2$+6x+13. $y=(x+3)^2$+4. Parabola with vertex
   (-3,4), opening up.

53. Parabola, vertex (1,4), axis x=1, y-intercept=6.
   $(x-1)^2$=4p(y-4).When x=0,y=6. 1=4p(2),p=1/8. $(x-1)^2$=(1/2)(y-4).

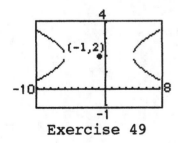
Exercise 49

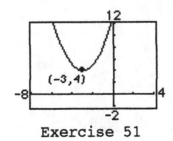
Exercise 51

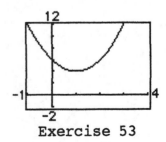
Exercise 53

55. Parabola, vertex (1,2), vert axis, through point (5,3).
   $(x-1)^2$=4p(y-2).When x=5,y=3. 16=4p(1),p=4. $(x-1)^2$=16(y-2).

57. Parabola, focus (1,3) & directrix x=-1. Vertex will be (0,3).
Horizontal axis. $(y-3)^2=4px$. Directrix is x=-p, thus p=1.
$(y-3)^2=4x$.

59. Parabola, vertex (2,-2) and focus (2,-1). Vertical axis.
$(x-2)^2=4p(y+2)$. Directrix will be y=-3, thus p=1.
$(x-2)^2=4(y+2)$.

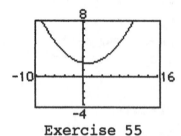
Exercise 55

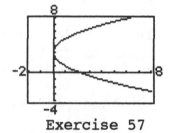
Exercise 57

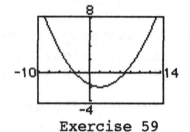
Exercise 59

61. Ellipse with vertices (3,0),(-3,0),(0,7),(0,-7).
See sketch below. Center will be (0,0). a=3,b=7. $\frac{x^2}{9} + \frac{y^2}{49} =1$.

63. Hyperbola with vertices (6,0),(-6, 0), & asymptotes y = ±2x.
See sketch below. Horizontal axis. Center (0,0). a=6.
Asymptotes $y=\pm\frac{b}{a}x =\pm2x$. b=2a=12. $\frac{x^2}{36} - \frac{y^2}{144} =1$.

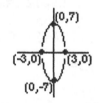
Exercise 61

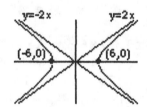
Exercise 63

Introduction

As indicated in the Preface at the beginning of this manual, printed versions of the problems on the *Math in Practice: An Applied Video Companion CD-ROM* comprise the second part of this manual. These problems starting on page A-1 have been formatted to allow you room to complete them and turn your work in to your instructor. On the following pages are two correlation charts that provide a breakdown of the problems. The chart on page 451 gives a correlation based upon mathematical concepts. The chart on page 452 presents a correlation based upon the application topics. Answers to all the problems are given starting on page A-97.

Applications by Concept for *Math in Practice: An Applied Video Companion CD-ROM*

Math Concept	Application	Page	Corresponds to
Basics	Acoustics	A-1	Chapter R
	Aeronautics	A-3	
	Piano Pitch	A-5	
	Beetle Infestation	A-7	
	Air Pollution	A-9	
	Fighting Fires	A-11	
Roots and Radicals	Line of Sight	A-13	Chapter R
	Fighting Fires	A-15	
	Falling Objects	A-17	
	Pendulums	A-19	
	Dinosaurs	A-21	
	Dinosaurs	A-23	
	Piano Pitch	A-25	
Polynomial and Rational Functions	Lighting a Stage	A-27	Chapter R
	Optical Lab	A-29	
	Electronics	A-31	
	Oceanographer	A-33	
	Map Evolution	A-35	
Solving Quadratic Equations	Falling Objects	A-37	Chapter 2
	Atlatl	A-39	
	World War II Bombers	A-41	
	Atlatl	A-43	
	Aeronautics	A-45	
	Heating Ducts	A-47	
	Line of Sight	A-49	
Polynomial and Rational Functions	Optical Lab	A-51	Chapter 2
Exponential / Logarithmic Functions	Beetle Infestation	A-53	Chapter 5
	Population Growth	A-55	
	Air Pollution	A-57	
	Metal Alloys	A-59	
	Population Growth	A-61	
	Earthquakes	A-63	
	Earthquakes	A-65	
Systems of Equations	Investment and Debt	A-67	Chapter 6
	Living Soil Crust	A-69	
	Metal Alloys	A-71	
	Flying to Hawaii	A-73	
	Electronics	A-75	
Matrices	Premature Infant Formula	A-77	Chapter 7
	Premature Infant Formula	A-79	
Series & Sequences	Investment and Debt	A-81	Chapter 8
Conics	Sound Dish	A-83	Appendix A
	Sound Dish	A-85	
	Acoustics	A-87	

The four trigonometry applications on pages A-89 to A-95 appear on the CD-ROM but do not apply to Williams' text. Answers for all problems start on page A-97.

Problems by Application Topic

The four trigonometry applications on pages A-89 to A-95 appear on the CD-ROM but do not apply to Williams' text.

Math Application by Topic: Math on Stage: Acoustics
Math Application by Concept: Basics

As mentioned in the video, echoes from the walls of a concert hall can be a concern when designing acoustics. The problems created by echoes are even worse in a large stadium, since the greater distance traveled can cause the echo to return to a performer's ears more than one full second later. This is the reason many performers choose to pre-record their rendering of the National Anthem at major sporting events.

Suppose the performer stands at the center of the field. If sound travels at 344 m/s, find the time necessary for sound to travel to and echo from a grandstand 200 m away.

Math Application by Topic: Math at Work: Aeronautics
Math Application by Concept: Basics

Sudden changes in wind can down an aircraft in seconds. Pilots need to know how to read the skies for signs of danger since winds can be dramatically different with only a small change in altitude.

One way to detect these changing wind conditions is by observing clouds. If the bottom of the cloud is flat and straight while the top of the cloud maintains a fluffy, rounded appearance, the wind under the cloud is likely much greater than the wind above the cloud. Similarly, if the low clouds in the sky are moving in a different direction than the wind you are experiencing on the ground, a dangerous, sudden shift in current may await you in flight.

Suppose a sudden downward wind pushes on a plane during take-off. If the wind causes the plane to lose 55 feet of altitude per second and the plane is 1,300 feet in the air, how many seconds until the plane is pushed back into the ground if no corrections are made?

Math Application by Topic: Math on Stage: Piano Pitch
Math Application by Concept: Basics

As demonstrated and discussed in the video, the frequency (pitch) of a string instrument is inversely proportional to its length according to the equation $f = \dfrac{k}{l}$, where f is the frequency in Hertz, l is the length of the string in meters, and k is a constant. Once the proper tension has been applied to a guitar string, the pitch can then be altered by pressing the strings against the fingerboard, effectively changing the length of the vibrating string.

a. Suppose a guitar string having length 16 inches is tuned to a frequency of 233 Hz (an A sharp). Find the value of the proportionality constant k.

b. How many inches from the end of the fingerboard should the string be pressed in order to create a frequency of 264 Hz (a C note)?

Date _____ Name _____

Course _____ Section _____ Student Number _____

Math Application by Topic: Math in the Environment: Beetle Infestation
Math Application by Concept: Basics

The pine beetle plays an important role in the lodgepole pine ecosystem since it is a recycler of older trees. However, humans can unknowingly tip the balance of these delicate ecosystems in favor of the beetle.

It is believed that the pine beetle was first introduced to British Columbia in the ballast of ships. Entering an ecosystem without sufficient natural predators, the beetle began to thrive. Also, aggressive fire suppression on the part of forest managers left the beetle an abundance of older, brittle bark under which to bore and colonize.

a. It is estimated that the beetle destroyed 200 million trees from 1972 to 1988. Find the average number of trees lost per year during this period.

b. Realizing measures had to be taken, the Ministry of Forests began a pest control plan in 1989 which decreased the number of trees lost to the beetles by 40%. Find the number of trees saved that year by the pest control plan.

c. If the cost for the pest control in 1989 was $50 million, find the average number of dollars spent for each tree saved that year.

Math Application by Topic: Math in the Environment: Air Pollution
Math Application by Concept: Basics

Despite progress in air quality over the last 50 years, Southern California remains the smoggiest region in the country—no other state even comes close. Many factors contribute to the smog-filled air of this region, and the South Coast Air Quality Management District has been working on all aspects of the problem. Let's examine the algebra of some of these factors:

a. Southern Californians have an on-going love affair with the automobile, demonstrated by the fact that 10.47 million Los Angeles residents in 1990 registered 8 million motor vehicles. If the average family size in this region is 3.21 persons, find the average number of cars per family.

b. Emissions from mobile sources such as vehicles account for 87% of the carbon monoxide in the Los Angeles basin. Tighter emissions standards and reformulated gasolines have decreased the daily pollutants by 3.2 million pounds. If this is the equivalent of taking 3.5 million cars off the road, find the number of pounds of pollutants emitted per car per day.

c. Commuting to and from work, area residents drive 100 million vehicles miles per day in cars containing only 1.1 passengers. If the goal of the SCAQMD is to increase that number to 1.5 passengers per car, find the equivalent decrease in vehicle miles driven per day.

Date _____ Name _____

Course _____ Section _____ Student Number _____

Math Application by Topic: Math at Work: Fighting Fires
Math Application by Concept: Basics

Fire fighting is a physically demanding and dangerous occupation. While battling the flames of a raging fire, firefighters put themselves at risk. The goal of a firefighter is to put out the fire with as little loss of life and property as possible. As shown in the video, to reduce the property damage done by water, firefighters need to be trained to estimate flow rates and water pressures.

The flow rate of water G in gallons per minute is given by the equation $G = 26.8D^2\sqrt{P}$ where D is the diameter of the outlet in inches and P is the water pressure in pounds per square inch (psi).

a. Suppose the pressure from a certain hydrant is 40 psi. If the diameter of the hose is 1 inch, find the number of gallons of water per minute discharged from the hose.

b. If it takes 25 minutes of continuous water flow to put out an apartment fire, find the total gallons of water used.

Date _____ Name _____

Course _____ Section _____ Student Number _____

Math Application by Topic: Math of Matter and Motion: Line of Sight
Math Application by Concept: Roots and Radicals

Due to the curvature of the earth, the view to the horizon is limited and depends on altitude. In fact, the distance d in miles that a person can see from a height of h feet is approximately given by $d = \sqrt{1.5h}$. (Notice that in the equation given in the video, $d = 1.2\sqrt{h}$. The quantity $\sqrt{1.5}$ has been rounded off to 1.2.   For greater accuracy, use the revised equation given in this exercise.)

Suppose a pilot needs to have a range of vision of at least 18 miles. Find the minimum altitude that must be achieved.

Math Application by Topic: Math at Work: Fighting Fires
Math Application by Concept: Roots and Radicals

Algebra is a powerful tool which can be used to determine quantities which are otherwise difficult to measure. In the case of firefighters such as those shown in the video, the rate of water flow from a hose in gallons per minute would be very difficult to measure correctly. Too much water can result in water damage costs rivaling those of the fire and smoke. With the help of a few simple equations, however, the flow rate can be determined quite accurately.

a. The water pressure P in pounds per square inch (psi) is related to the velocity of the water at the point of discharge according to the equation $v = 12.14\sqrt{P}$ where v is measured in feet per second. Using a gauge like the one shown in the video, firefighters determine a velocity of 50 feet per second. Find the water pressure in the hydrant.

b. The flow rate G of water in gallons per minute from a fire hose is given by $G = 26.8D^2\sqrt{P}$ where D is the diameter of the hose and P is the pressure at the nozzle. If the hose measures 1.5 inches in diameter, find the rate of flow from the hose.

c. Firefighters need to make fast decisions and quick calculations in order to reduce property damage and save lives. Using the two equations in parts (a) and (b) above, eliminate the factor $\sqrt{P}$ to get a formula relating water flow G to hose diameter D and velocity v.

Math Application by Topic: Math of Matter and Motion: Falling Objects
Math Application by Concept: Roots and Radicals

Most word problems in text books tend to simplify equations by neglecting the forces due to friction. In reality, however, air resistance plays an important role in the algebra of falling objects such as sky divers.

In fact, the force of friction from air resistance increases with the object's speed until it equals the weight of the falling object. This means that after a certain point, the downward acceleration is zero and thereafter the object maintains a constant speed known as "terminal velocity".

For a sky diver, the equation for terminal velocity v_t is given by $v_t = \sqrt{\dfrac{mg}{0.25}}$, where m is the skydiver's mass in kilograms and $g = 9.8$ m/s^2 is the pull of gravity.

a. If a 75-kg skydiver jumps out of an airplane from a sufficient altitude, find the terminal velocity. (Note: Your answer will be in m/s.)
b. What mass m would yield a terminal velocity of 60 m/s?

Date _____ Name _____

Course _____ Section _____ Student Number _____

Math Application by Topic: Math of Matter and Motion: Pendulums
Math Application by Concept: Roots and Radicals

The metric system was designed for ease of unit conversion. Thus, the different units of measurement are interrelated. For example, a gram is defined to be the mass of one cubic centimeter of distilled water at 0°C and 1 atm of pressure. Using pendulums like those shown in the video, we can interrelate length and time.

Given that the period and length of a pendulum are related by the equation $T = 2\pi\sqrt{\dfrac{l}{g}}$, where T is the period in seconds, l is the length in meters and g is the acceleration due to gravity, what length should a pendulum have that will 'beat seconds'? (Use $g = 9.78$ m/s^2, the acceleration due to gravity at sea level at the equator. Let $T = 2$ seconds.)

Date _____ Name _____
Course _____ Section _____ Student Number _____

Math Application by Topic: Math Over Time: Dinosaurs
Math Application by Concept: Roots and Radicals

Paleontologists attempt to recreate animal life 100 million years ago by examing fossil evidence such as the footprints found in the Texas riverbed. Knowing the distance between these tracks and the leg length of the animal, a scientist can determine the rate at which an animal likely moved. All of these factors become part of the debate as to whether dinosaurs were warm-blooded or cold-blooded.

The dimensionless speed D of an object is given by $D = \dfrac{s}{\sqrt{Lg}}$ where s is the speed of the object in meters per second, L is the length of the leg from hip to foot and $g = 9.8$ m/s^2 is gravitational acceleration. If a dinosaur fossil has a leg length of 2.81 meters and is believed to have had a dimensionless speed of 0.4, find the speed at which the animal moved.

Math Application by Topic: Math Over Time: Dinosaurs
Math Application by Concept: Roots and Radicals

When designing special effects for a movie, the concept of dimensionless speed is used to achieve realistic motion in scale models. If this is not done properly, the effects are not believable and can have the characteristics of B-grade science fiction movies made decades ago.

Suppose a war movie is being filmed and the movement of an aircraft carrier is being recreated with a 10-foot model. If the dimensionless speed for an aircraft carrier is 0.015, find the speed at which the model should move on the movie set. Use $D = \dfrac{s}{\sqrt{Lg}}$ where D is the dimensionless speed, s is the speed on the object on the movie set in feet per second, L is the length of the scale model and $g = 32$ ft/sec^2 is gravitational acceleration.

Math Application by Topic: Math on Stage: Piano Pitch
Math Application by Concept: Roots and Radicals

The piano strings shown are a good illustration of how the frequency (pitch) of a string instrument depends on the length of the string, its mass, and the tension applied to the string. The equation relating all of these variables is given by $f = \dfrac{1}{2l}\sqrt{\dfrac{T}{m}}$, where f is the frequency in Hertz, l is the length of the string in meters, T is the tension in newtons and m is the linear mass density per unit length of the string in kilograms per meter. The pitch of the note played by a string will be altered with any change in these variables.

Suppose a string having a length of 0.60 m and a linear mass density of 0.005 kg/m plays an A note when 348.48 newtons of tension are applied.

a. Use the equation above to find the frequency corresponding to the A note.
b. If we wish to alter only one variable so that a D note ($f = 262$ Hz) is played instead, by what amount should we alter the tension? The linear mass density?

Date _____ Name _____

Course _____ Section _____ Student Number _____

Math Application by Topic: Math on Stage: Lighting a Stage
Math Application by Concept: Polynomial and Rational Functions

Abravanel Hall in Salt Lake City is named for Maurice Abravanel, the first and most famous
director of the Utah Symphony. When designing lighting for a stage such as Abravanel Hall,
there are many factors to consider. Some spotlights provide good illumination but also generate
intense heat if placed too close to the performers. If removed too far from the stage, other
spotlights may provide very poor illumination.

The illumination I of a spotlight in foot-candles is inversely proportional to the square of the
distance d of the spotlight from the stage according to the equation $I = \dfrac{k}{d^2}$ where the constant k
depends on the design of the spotlight.

a. Suppose a certain spotlight has an illumination of 250 foot-candles at a distance of 10 feet.
    Find the value of k for this spotlight.
b. If the same spotlight is placed 130 feet from the stage, find the resulting illumination.
c. At what distance from the stage should the spotlight be placed in order to have an
    illumination of 100 foot-candles?

Math Application by Topic: Math of Matter and Motion: Optical Lab
Math Application by Concept: Polynomial and Rational Functions

Farsightedness is a condition wherein objects come into focus only when they are removed a certain distance from the eye. For example, perhaps in order to see printed words clearly, a book must be held at least 40 cm in front of a person. This distance x is called the near point – the nearest point at which objects come into focus for a farsighted person.

If vision is to be corrected so that objects held at a normal reading distance of 25 cm are clear, we need a lens that will form a "virtual image" of the object at or beyond the near point of 40 cm. In effect, the lens moves the object farther away from the eye to a point where the eye is able to focus. To choose such a lens, we must know its focal point f which is given in centimeters.

The relationship between the focal point f, the near point x and the actual distance s of the object from the person is given by $\dfrac{1}{f} = \dfrac{1}{s} - \dfrac{1}{x}$.

a. Use this equation to find the focal point f of the lens needed to correct the vision described above having near point $x = 40$ cm and actual distance $s = 25$ cm.

b. The "power" p of a lens, measured in diopters, is related to the focal point f by $p = \dfrac{100}{f}$.

Find the power of the lens in part (a).

Math Application by Topic: Math of Matter and Motion: Electronics
Math Application by Concept: Polynomial and Rational Functions

As demonstrated in the video, resistors in a circuit can either be joined in series (end-to-end) or in parallel (side-by-side). The manner in which the resistors are connected determines the overall resistance that the circuit will have. Since the resistance R is related to the voltage V and the current I according to the equation $V = IR$, creating the wrong resistance can quickly overload a circuit.

When resistors R_1 and R_2 are connected in series, the total resistance R_{tot} of the circuit is given by $R_{tot} = R_1 + R_2$. When joined in parallel, the relationship among R_{tot}, R_1, and R_2 is given by $\dfrac{1}{R_{tot}} = \dfrac{1}{R_1} + \dfrac{1}{R_2}$.

Suppose one resistor has a resistance of 10 ohms and the other has resistance 5 ohms. They are to be connected in a circuit supplied with voltage $V = 25$ volts.

a. If these are joined in series, find the total resistance of the circuit. What current I will this circuit draw?
b. If these are joined in parallel, find the total resistance of the circuit. What current I will this circuit draw?
c. Suppose a certain radio circuit needs a total of 4 ohms of resistance to function properly. If two resistors are to be connected in parallel and one has resistance 6 ohms, find the resistance of the other.

Math Application by Topic: Math in the Environment: Oceanographer
Math Application by Concept: Polynomial and Rational Functions

Oceanographers like the one shown in the video use ideas from analytic geometry in order to categorize bodies of water for shoreline development. Each section of shoreline is surveyed and the length L of the shoreline is recorded. Then the area A of the water which the shoreline encloses is calculated. Next, a value D is assigned to the section of shoreline, where D is related to L and A by the equation $D = \dfrac{L}{2(\pi A)^{1/2}}$. This value D has a meaning analogous to that of eccentricity in the study of conic sections.

a. Suppose that the section of shoreline were perfectly circular. Then L would equal the circumference of the circle and A the area of the circle. Find the value of D in this case.
b. An oceanographer tells a developer that the piece of shoreline being considered for a recreational complex has $D = 1.35$. The developer measures the length of the shoreline as 2.6 miles. Determine the area of water enclosed by this development.

Math Application by Topic: Math Over Time: Map Evolution
Math Application by Concept: Polynomial and Rational Functions

The video shows computer images of the changes in a shoreline over time. This graphic example of shoreline erosion illustrates the dynamic nature of these coastal areas. Coastline loss is heightened during storms such as hurricanes in which large waves beat down the shoreline and wash it out to sea.

To combat the loss of beaches, some cities have on-going sand replenishing projects, while other cities choose to erect barriers such as seawalls and jetties. All of these measures are costly, and city planners must evaluate the cost of preserving the shoreline versus the loss of property if nothing is done.

Suppose the cost in millions of dollars incurred by a coastal city can be modeled by the equation $C(x) = -0.05x^3 + 0.5x^2 - 0.9x + 1.2$ for eight years where $x = 0$ is the year 2000.

a. Find the predicted cost to the city in the year 2004.
b. After an initial investment during which the costs increase rapidly, this model predicts that the costs associated with saving beaches will begin decreasing. Graph the function $C(x)$ using a graphing window of $0 < x < 10$ and $0 < y < 4$. Use the FMAX utility to determine the year during which the cost is expected to be a maximum

Math Application by Topic: Math of Matter and Motion: Falling Objects
Math Application by Concept: Solving Quadratic Equations

Neglecting air resistance, the force of gravity causes the speed of a falling object to increase linearly according to the equation $s = v + gt$, where s is the speed of the object in feet/second, v is the initial velocity of the object and $g = 32$ ft/sec^2 is the pull of gravity.

Suppose a water balloon is thrown down from the observation deck of the World Trade Center 110 stories tall (approximately 1,000 feet).

a. If the height h above the ground after t seconds is given by $h = 1,000 - 10t - 16t^2$, find the time needed to hit the ground (when $h = 0$).

b. Using the time found in part (a) and the equation in the introductory paragraph above, find the speed of the balloon on impact if the initial velocity v is 10 ft/sec.

Math Application by Topic: Math Over Time: Atlatl
Math Application by Concept: Solving Quadratic Equations

The atlatl and dart have been found in ancient cultures on several continents and date back thousands of years. Today, there is a resurging interest in the device for both sport and hunting. There is even a World Atlatl Association with national and international competitions. The simple engineering of the atlatl increases both the distance and accuracy of a thrown dart. While the unaided arm can throw a spear a distance of 30 yards, the atlatl increases that distance to 100 yards for the average thrower, and up to 125 yards for the expert. Quoting one atlatl enthusiast, "It's like having a six-foot long arm with an extra elbow."

The kinetic energy of objects in motion such as atlatls depends on mass and velocity according to the equation $k = \dfrac{1}{2}mv^2$, where k is the kinetic energy measured in joules, m is the mass in kilograms, and v is the velocity on impact in meters per second.

a. If a dart and atlatl have a total mass of 1.2 kilograms and kinetic energy of 90 joules on impact, find the velocity with which the dart hits the target.
b. The demonstration in the video showed that using the atlatl increases the velocity fifteen-fold. Find the corresponding increase in kinetic energy. Use arbitrary m and note the effect of substituting $15v$ for v in the equation.

Date _____   Name _____

Course _____   Section _____   Student Number _____

Math Application by Topic: Math Over Time: World War II Bombers
Math Application by Concept: Solving Quadratic Equations

The goal of a World War II bomber pilot was to "skip" a bomb across the surface of a reservoir, allowing it to decrease in velocity, then gently sink to the bottom of the reservoir and explode at the base of the dam. This also allowed time for the bomber to fly a safe distance from the exploding dam.

If the plane is flying at the standard World War II bombing altitude of 60 feet, the equation relating the height h of the bomb t seconds after release is given by: $h = -16t^2 + 60$.

Suppose a pilot miscalculated, resulting in a bomb which exploded on impact when $h = 0$. How many seconds until the bomb explodes?

Math Application by Topic: Math Over Time: Atlatl
Math Application by Concept: Solving Quadratic Equations

The flight path of projectiles—whether they be launched rockets, punted footballs, or arrows shot from a compound bow—can be described by quadratic equations. Generalizing the equation found in the video, suppose the distance s of an object above the ground t seconds after release is given by $s = 6 + v_0 t - 16t^2$ where v_0 is the initial velocity of the object. As you might guess, a change in initial velocity can greatly affect the flight of a projectile.

Initial velocity is due to many factors, such as the strength of a punter's leg or the design of a modern compound bow.

a. Suppose a punter on a football team can kick a ball with initial velocity $v_0 = 50$ feet per second. Set $s = 0$ and use the quadratic formula to find the time until the football hits the ground. In football, this time is referred to as "hang time" and is an important statistic used to evaluate punters.

b. If the punter would like to have a hang time of 5.5 seconds, solve for v_0 to find the initial velocity the punter must produce.

Date _____ Name _____

Course _____ Section _____ Student Number _____

Math Application by Topic: Math at Work: Aeronautics
Math Application by Concept: Solving Quadratic Equations

A popular attraction at the National Museum of Naval Aviation in Pensacola, Florida is a 5-minute ride in a flight simulator which generates missions similar to those flown during Operation Desert Storm. For a small fee, patrons of the museum can experience the pitch and roll of high performance turns in a Navy FA-18 Hornet, including the unique forces caused by catapult launches and arrested landings on a naval carrier.

Suppose a similar attraction at another museum charges $2.50 per person and can accommodate 20 passengers per demonstration. The museum is open Monday through Friday and the attraction runs every 15 minutes beginning at 9:00 a.m. with the last demonstration beginning at 4:45 p.m.

a. If each seat is filled for each performance, find the total revenue from the attraction for one week.
b. From past experience, it is known that for each 50-cent raise in price, an average of 2 fewer seats will be sold per demonstration. Thus, if h is the number of price increases, the new ticket price will be $2.5 + 0.5h$ and the number of passengers per run will be $20 - 2h$. Set up an equation for the total revenue R per showing using the fact that Revenue = (ticket price) × (number of tickets sold).
c. Use the equation in part (b) and the formula for finding the vertex of a parabola to find the number of 50-cent price increases h which will yield a maximum revenue per show.

Date _____   Name _____
Course _____   Section _____   Student Number _____

Math by Application Topic: Math at Work: Designing Heating Ducts
Math by Application Concept: Solving Quadratic Equations

As demonstrated in the video, a heating duct is formed from a sheet of galvanized metal 45 inches wide by bending the two sides up to form 90° angles. The air flow achieved by a duct similar to the one shown depends on the cross-sectional area given by $Area = l \cdot w$. Knowing that the maximum value of a quadratic function occurs at the vertex, find the value of x for which the cross-sectional area will be maximized, thus maximizing air flow.

x 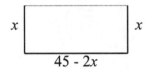 x

45 - 2x

Date _____ Name _____

Course _____ Section _____ Student Number _____

Math Application by Topic: Math of Matter and Motion: Line of Sight
Math Application by Concept: Solving Quadratic Equations

Due to the curvature of the earth, the view to the horizon is limited and depends on the altitude. When the exact curvature of the earth is factored into the equation, the distance d in miles that a person can see from a height of h feet is given by $d = \sqrt{1.5h + 3.587 \times 10^{-8} h^2}$. The other formulas shown thus far have been approximations of this formula.

a. Suppose a pilot is flying at an altitude of 100 feet when he sees a lake appear on the horizon. What is the distance to the lake?

b. What would the pilot's altitude need to be in order to see the same lake from 30 miles away? You will need to square both sides of the equation, set to zero, and then apply the quadratic formula to h. Be sure to keep as many digits of accuracy as your calculator will allow.

Date _____   Name _____

Course _____   Section _____   Student Number _____

Math Application by Topic: Math of Matter and Motion: Optical Lab
Math Application by Concept: Polynomial and Rational Functions

When employees in labs such as an optical lab work together as a team, productivity increases and the time needed to fill customers' orders is reduced.

Suppose two employees, John and Jorge, are working to fill a large order of eyeglasses. Working alone, it would take John 6 hours longer to fill this order than it would take for Jorge if Jorge worked alone. Working together, they can fill the order in 4 hours.

The equation relating their rates of work is given by $\dfrac{1}{x+6}+\dfrac{1}{x}=\dfrac{1}{4}$ where x represents the number of hours needed for Jorge to fill the order working alone. Find the amount of time needed for Jorge to fill this order.

Date _____ Name _____

Course _____ Section _____ Student Number _____

Math Application by Topic: Math in the Environment: Beetle Infestation
Math Application by Concept: Exponential / Logarithmic Functions

If left unchecked, the pine beetle population can cause severe damage to forested areas. Therefore, pesticides are sprayed over large areas in order to control the growth of the pine beetle population.

Suppose two different pesticides are being considered. Pesticide A has been found to be 88% effective in a single application, meaning that, on average, 88 of 100 pine beetles will be destroyed with each application. Pesticide B is only 40% effective, but, because of its emphasis on organic ingredients, it is believed to be less damaging to the environment.

Note that after one spraying of Pesticide B, 60% (0.60) of beetles remain. After two sprayings, $0.60(0.60) = 0.60^2$ or 36% of the beetles remain. Then, after x sprayings, 0.60^x of the beetles remain.

How many applications of Pesticide B would be needed to exceed the effectiveness of a single application of Pesticide A?

Date _____   Name _____

Course _____   Section _____   Student Number _____

Math Application by Topic: Math in the Environment: Population Growth
Math Application by Concept: Exponential / Logarithmic Functions

One factor contributing to the rapid increase in world population is that, due to advances in medical technology, the average life expectancy continues to increase. According to the National Center for Health Statistics, the average life expectancy L at birth for a person born in year x is approximated by $L(x) = \dfrac{79.25}{1 + 9.7135 \times 10^{24} \cdot e^{-0.0304x}}$

a.   What life expectancy does this mathematical model predict for someone born in 1960? In 1970?

b.   In what year of birth does this model predict an average life expectancy of 78 years?

Date _____ Name _____

Course _____ Section _____ Student Number _____

Math Application by Topic: Math in the Environment: Air Pollution
Math Application by Concept: Exponential / Logarithmic Functions

The South Coast Air Quality Management District was established in 1947 to monitor air quality in the Los Angeles Basin. The data collected over the years shows that, while many geographical and meteorological factors contribute to the problem, the majority of air pollutants come from commuters' vehicles. In fact, the measure of pollutants in the air at 6:00 a.m. consistently falls below the maximum level set by the state. As the morning commute progresses, however, pollutants build in the air, the warm Southern California sun "cooks" the mixture, and by 2:00 p.m., smog alerts can be in effect.

Suppose pollutant levels on a certain day can be modeled by the equation

$P(t) = \dfrac{0.004}{0.02 + 0.18e^{-0.6t}}$ where $P(t)$ is the measure of pollutants in parts per million (ppm) t hours after 6:00 a.m. (that is, $t = 0$ is 6:00 a.m.).

a. Find the level of pollutants P at 9:00 a.m. ($t = 3$). For comparison, the state standard is 0.09 ppm.

b. The federal standard for air quality is exceeded when P reaches 0.12 ppm. Find the time of day that this standard is exceeded.

c. A health advisory is issued when pollutant levels exceed 0.15 ppm. Since 1990, health advisories have been issued an average of 76 days per year. According to the equation above, will a health advisory be issued on this day? At what time?

Math Application by Topic: Math of Matter and Motion: Metal Alloys
Math Application by Concept: Exponential / Logarithmic Functions

When working with metal alloys, ingredients must be mixed precisely and forged at proper temperatures in order to ensure strength. Using too low a temperature results in an alloy that is soft and pliable, while excessive temperature causes an alloy to become brittle. The alloy shown in the video was forged at a temperature of 2200° Fahrenheit for three hours and withstood 1800 pounds of pressure before splitting.

If a hot object such as a bar of newly formed alloy is allowed to cool at room temperature, the temperature T of the bar after t hours could be modeled by $T(t) = 72 + 2128e^{-0.228t}$.

a. Find the temperature of the metal at the moment it is removed from the kiln ($t = 0$).
b. Find the temperature of the bar after 6 hours.
c. If the metal can be handled at a temperature of 300° F, how long must the bar be allowed to cool before it can be moved?

Date _____ Name _____

Course _____ Section _____ Student Number _____

Math Application by Topic: Math in the Environment: Population Growth
Math Application by Concept: Exponential / Logarithmic Functions

The video presents an interesting and oft-encountered real-life scenario: Suppose a group of city planners is making a decision on whether or not to expand a wastewater treatment plant. Operating at maximum capacity, the current treatment plant can meet the needs of 90,000 people. The planners know that population growth is modeled by the equation $P(t) = P_0 e^{kt}$ where P_0 is the initial population, k is the rate of growth of the city, and $P(t)$ is the population after t years. To aid their decision, the planners are using the last two census figures for the town, showing a population of 65,000 at the beginning of 1990 and a population of 80,500 at the beginning of the year 2000.

a. Use the given census figures (using 65,000 as P_0) to determine the growth rate k of the city.

b. Calculate the year in which the present wastewater treatment plant will no longer be sufficient for the needs of the growing town. (Remember: $t = 0$ is the year 1990.)

Date _____ Name _____

Course _____ Section _____ Student Number _____

Math Application by Topic: Math in the Environment: Earthquakes
Math Application by Concept: Exponential / Logarithmic Functions

Named for its developer Charles Richter, the Richter scale gives a rating R for an earthquake based on the intensity of the quake relative to the "zero-level quake" i_0. This relationship is given by the equation $R = \log(I / i_0)$. Thus, an earthquake having an intensity 10,000 times stronger than i_0 ($I = 10{,}000\,i_0$), would have a Richter reading of $R = \log(10{,}000 i_0 / i_0) = 4$.

a. Suppose an earthquake has 250,000 times the intensity of i_0 ($I = 250{,}000\,i_0$). Find its reading R on the Richter scale.

b. The largest earthquake ever recorded happened in Concepcion, Chile in 1960. This earthquake had a Richter measure of $R = 9.5$. Find the intensity I of this earthquake in terms of i_0.

Math Application by Topic: Math in the Environment: Earthquakes

Math Application by Concept: Exponential / Logarithmic Functions

The Great North American Earthquake occurred on March 28, 1964 in Anchorage, Alaska. At that time, the quake was assigned a Richter measure of 8.5. However, this number has since been revised to 9.2.

The shaking lasted for 4 minutes, and when it was over, one side of Main Street was 10 feet higher than the other. This incredible earthquake vertically displaced the earth's crust over an area of 200,000 square miles. Amazingly, only nine deaths were reported from the quake itself. The resulting tsunami wave, however, caused another 122 deaths and devastated towns along the Gulf of Alaska with a wall of water over 100 feet high. This tsunami arrived on the shores of Hawaii a mere 8 hours later.

a. Given that the distance from Anchorage to Hawaii is roughly 5,200 kilometers, find the speed of the tsunami wave.

b. The equation $R = \log(I / i_0)$ relates the Richter scale measure R of an earthquake and the intensity I of the quake relative to the "zero-level quake" i_0. How many times stronger is the intensity of an earthquake measuring 9.2 than that of an earthquake measuring 8.5?

Math Application by Topic: Math at Work: Investing and Debt
Math Application by Concept: Systems of Equations / Matrices

As discussed in the video, different investments have different rates of return due to the risk factors involved. Investors must decide what risk they are willing to take in order for a chance at a high return.

Suppose a person has a total of $12,000 to invest and is considering two funds: a conservative government bond with a 4 ½ % annual interest rate and a riskier stock market investment which has averaged 21% annually in the past. The conservative investor would like to invest as little as possible in the risky stock, yet still wants to make at least $1,500 in interest per year.

How should the money be divided between the two investments?

Math Application by Topic: Math in the Environment: Living Soil Crust
Math Application by Concept: Systems of Equations / Matrices

Blue-green algae is one of the oldest forms of life and plays an important role in the formation of soil. In high desert regions like the Colorado Plateau, the sticky fibers of the algae form a "living crust" over the loose sand, providing shelter from eroding winds and water runoff. These living organisms can also store water by swelling up to ten times their normal size. This makes it possible for higher plant forms to become established in otherwise arid conditions. Once these plants are established, the dead plant material adheres to the fibers of the crust and composts into rich, new soils.

When the crust is destroyed by footsteps or vehicles, Park Service personnel attempt to replace it by spraying the area with an inoculum of algae suspended in a nutrient solution.

Suppose that two concentrations of the nutrient solution have been prepared, one a 50% concentration and the other a 12% concentration. After testing, it is found that the 50% mixture is too rich and suffocates the algae, while the 12% is too weak and diminishes the growth potential of the algae. To avoid waste, the two concentrations are to be mixed to form a 35% nutrient concentration. How many gallons of each should be used to make 50 gallons of the 35% mixture?

Date _____ Name _____

Course _____ Section _____ Student Number _____

Math Application by Topic: Math of Matter and Motion: Metal Alloys
Math Application by Concept: Systems of Equations/Matrices

When making an alloy, new recipes are tested through a tightly-controlled process according to a computer design. Each element contributes to the make-up of the new alloy: cobalt increases strength while chromium resists corrosion.

Suppose that during testing, it is found that the amount of cobalt should be 2 ½ times the amount of nickel while the amount of chromium should remain 5/6 the amount of cobalt. If the amount of chromium, nickel and cobalt combined should equal 40 parts, find the amount of each that should be added to make the alloy. Write a system of equations representing the problem and solve using 2 decimal places of accuracy.

Math Application by Topic: Math at Work: Flying to Hawaii
Math Application by Concept: Systems of Equations / Matrices

As the pilot discussed in the video, she never leaves the runway on a flight to Hawaii without knowing her 'point of no return'. This is the moment in the flight after which she will no longer have enough fuel to return to Los Angeles. Because she faces a headwind flying west, but a tailwind returning to the east, this point of no return does not occur at the halfway point.

Suppose an airplane flies at 180 mph and holds enough fuel to fly for 10 hours. The National Weather Service reports a steady 40 mph wind from the west. How many hours can the pilot fly into the wind before reaching the point of no return?

(Set up a system of two equations using t_1 for the time flown into the wind and t_2 for the time needed to return. Remember that when a plane flies at p mph with a headwind of w mph, the actual rate of travel is $p - w$ mph.)

Date _____ Name _____

Course _____ Section _____ Student Number _____

Math Application by Topic: Math of Matter and Motion: Electronics
Math Application by Concept: Systems of Equations / Matrices

Many commonplace circuits cannot be reduced to simple series or parallel combinations such as those discussed in the video. And, as you might expect, more complicated circuits require more intricate mathematics.

For example, the circuit shown below is called a bridge circuit and is used in many types of measuring and control systems. Because the current I in the circuit branches at each junction of the circuit and the differing branches have differing resistance, the current varies from branch to branch.

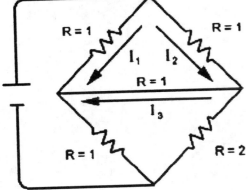

To determine the currents I_1, I_2 and I_3, a system of equations must be set up and solved using a property known as Kirchhoff's Law. (The physics required to set up the system have already been done for you.) The system which describes this circuit is given below.

Now, use either a matrix or the elimination method to solve the system below for the currents I_1, I_2 and I_3. Notice that you will need to rearrange the terms in the equations first.

$$\begin{cases} 13 - I_1 - (I_2 + I_3) = 0 \\ -I_2 - 2(I_2 - I_3) + 13 = 0 \\ -I_1 + I_3 + I_2 = 0 \end{cases}$$

Math Application by Topic: Math Over Time: Premature Infant Formula
Math Application by Concept: Systems of Equations / Matrices

Because the organs of premature babies are not fully developed, the rate at which these infants absorb nutrients can vary from day to day. A blood sample is drawn from each baby every evening to evaluate the existing levels of potassium, protein and sodium in the infant. The samples are processed overnight and, based on these results, each baby receives a customized formula for the next day.

Suppose the nutrients needed for these babies come in three different pre-mixed formulas: Formula A, Formula B and Formula C.   Formula A supplies 2 units of potassium, 1 unit of protein and 3 units of sodium. Formula B contains 1 unit of potassium, 3 units of protein and 2 units of sodium. Formula C contains 1 unit of potassium, 3 units of protein and 1 unit of sodium. It is determined that a baby needs 15 units of potassium, 20 units of protein and 22 units of sodium. How should the three formulas be combined in order to meet this baby's nutritional requirements?

To solve, set up a system of three equations. Let x = units of Formula A, y = units of Formula B and z = units of Formula C. Write one equation for the amount of potassium contained in the mixture and set it equal to 15. Similarly, write equations for protein and sodium, respectively. Then solve this system of three equations using matrices.

Math Application by Topic: Math Over Time: Premature Infant Formula
Math Application by Concept: Systems of Equations / Matrices

Because the organs of premature babies are not fully developed, the rate at which nutrients are absorbed varies greatly from day to day. Therefore, a nutrient formula for each baby must be re-evaluated every morning.

Suppose the nutrients are supplied by combining three different formulas: each ounce of Formula I contains 2 units of potassium, 1 unit of protein and 3 units of sodium. Each ounce of Formula II contains 1 unit of potassium, 3 units of protein and 2 units of sodium. Each ounce of Formula III contains 1 unit of potassium, 3 units of protein and 1 unit of sodium.

Each mixture for each baby each day produces a system of linear equations which has the same coefficient matrix A when written in matrix equation form $AX = B$. The column matrix B will be different for each baby based on that baby's need for the nutrients potassium, protein and sodium on a particular day. The solution matrix is $X = \begin{pmatrix} x \\ y \\ z \end{pmatrix}$ where x is the number of ounces of Formula I, y is the number of ounces of Formula II and z is the number of ounces of Formula III.

a.   From the information given above, the coefficient matrix A is given by $A = \begin{bmatrix} 2 & 1 & 1 \\ 1 & 3 & 3 \\ 3 & 2 & 1 \end{bmatrix}$. Find the inverse of this matrix.

b.   Now that the inverse has been found, the correct mixture for each baby can be quickly calculated. Using the mathematics of matrix equations, the solution $X = \begin{pmatrix} x \\ y \\ z \end{pmatrix}$ for each system

$AX = B$ is given by $X = A^{-1}B$. Suppose it is determined that a certain baby needs 12 units of potassium, 32 units of protein and 20 units of sodium.

Set $B = \begin{pmatrix} 12 \\ 32 \\ 20 \end{pmatrix}$ and use matrix multiplication to solve for X, thereby determining the number of

ounces of each formula that should be used.

Math Application by Topic: Math at Work: Investing and Debt
Math Application by Concept: Series & Sequences / Conics

Saving for retirement is an issue that becomes more important to people as they get older. Ironically, however, retirement needs to be planned early in order for an adequate balance to accrue with realistic paycheck deductions.

Most retirement plans include a fixed amount of money set aside each month. Although each of these payments is equal, the fact that they are made at different times means they will compound to different amounts: Suppose A dollars are invested each month for 4 months, where i is the interest per month, and the compounding is monthly. Then

$A(1+i)^4 =$ value of first deposit after 4 months

$A(1+i)^3 =$ value of second deposit after 3 months

$A(1+i)^2 =$ value of third deposit after 2 months

$A(1+i) =$ value of fourth deposit after 1 month

$A \ =$ value of a fifth deposit

The total investment balance is given by the sum of these values and forms a geometric series $A + A(1+i) + A(1+i)^2 + A(1+i)^3 + A(1+i)^4$, having first term A and common ratio $1+i$. If these deposits continue for n months, then the balance B at the end of n months is the partial sum of a geometric series. Thus $B = \dfrac{A(1-(1+i)^n)}{1-(1+i)} = \dfrac{A((1+i)^n - 1)}{i}$ where B is the balance, n is the total number of months the investment is made, and i is the interest earned *per month*.

a. Solve the given equation for A.

b. Suppose a retirement balance of $B = \$100{,}000$ is desired at age 65. If the interest per month i is a constant 0.75% (9% annually), what amount A should be deducted each month if a person begins investing at age 25? Age 35? Age 55?
    Hint: At age 25, let $n = 12(40) = 480$.

Date _____ Name _____

Course _____ Section _____ Student Number _____

Math Application by Topic: Math on Stage: Sound Dish
Math Application by Concept: Series & Sequences / Conics

The reflective properties of parabolas make it possible to redirect sound waves. You may have seen parabolic microphones even larger than the one shown in the video being used by television sound technicians on the sidelines of a professional football field. By using these large parabolic microphones, sounds on the field can be heard clearly even at great distances.

Suppose the outline of a parabolic dish corresponds to the equation $y^2 = 3x$ where all measurements are given in feet. In order to capture sound most efficiently, the microphone must be placed the proper distance p from the vertex of the parabola at the point known as the focus. Using the equation above and the standard formula for a parabola, the correct distance from the vertex p for placement of the microphone can be found by simply setting $3 = 4p$. Find this distance.

Math Application by Topic: Math on Stage: Sound Dish
Math Application by Concept: Series and Sequences / Conics

As shown in the video, the New Mexico installation known as the Very Large Array is made up of 27 dishes covering 38 miles. This radio telescope detects waves emitted from the stars and solar systems, allowing scientists from around the world to see a radio portrait of the heavenly objects irregardless of darkness or cloud cover. These 27 dishes are controlled by a central processor and move together to form the equivalent of a telescope 20 miles across. Some discoveries made using this powerful instrument include the surprising existence of icy poles on the planet Mercury and a likely location of a black hole.

Each of these dishes, being parabolic, has shape corresponding to a quadratic equation $y = ax^2$.

a. If the dish measures 25 meters in diameter and is 3.5 meters deep, then the corresponding parabola can be thought of as passing through the vertex and the point (12.5, 3.5). Use this information to find the value of the coefficient a in the equation above.
b. For proper reception, the receiver must be placed at the focus located p units above the vertex of the dish where p is related to the value of a found in part (a) by the equation $4a = 1/p$. Find the height p of the receiver above the dish.

Date _____ Name _____

Course _____ Section _____ Student Number _____

Math Application by Topic: Math on Stage: Acoustics
Math Application by Concept: Series & Sequences / Conics

The Tabernacle on Temple Square in Salt Lake City, Utah, is home to one of the world's most famous choirs, the Mormon Tabernacle Choir. This historic building was completed in 1867 and is known for its excellent acoustics. As discussed in the video, the room in which music is performed can be viewed as an extension of the instrument, enhancing the sound of each performance.

The acoustics of the Tabernacle arise due to the fact that the building is elliptical, therefore having the property that any sound emitted from one focus will reflect off the walls and travel through the other focus. On guided tours of the facility, tourists situated at one focus can hear the sound of a pin dropped by a guide located at the other focus. This feature can have drawbacks, also, since every whisper by concert audience members seated at the focus can be heard clearly by the choir conductor who stands at the other focus.

If the Tabernacle is 250 feet long and 150 feet wide, use the equations for ellipses found in your text to find the distance between the two focal points.

Math Application by Topic: Math of Matter and Motion: Pendulums
Math Application by Concept: Trigonometry

Due to differences in the forces exerted on a pendulum in motion, each swing of a Foucault
pendulum such as the one shown in the video has a slightly different orientation than the last.
This phenomenon is called 'precession'. The number of hours T needed for a pendulum to
completely precess around its base depends on the latitude θ of the pendulum according to the
equation $T = \dfrac{24}{\sin\theta}$

a. If a pendulum is placed in a location having latitude 38.6°, how long will it take the
    pendulum to complete one full revolution of its base?

b. If it takes a certain Foucault pendulum 35 hours and 24 minutes to precess completely, what
    is the latitude of the pendulum?

Date _____ Name _____

Course _____ Section _____ Student Number _____

Math Application by Topic: Math at Work: Designing Heating Ducts

Math Application by Concept: Trigonometry

The video shows how heating ductwork is formed from a flat sheet of metal 45 inches wide by bending 12 inches from each side.   Suppose that new ductwork is being designed for a house, but, due to design restrictions, instead of bending the sides to the usual angle of 90°, they will be bent to an angle θ as shown in the diagram.

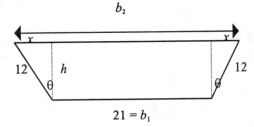

a.   Using the basic definitions of sine and cosine, express h and x in terms of θ.

b.   The airflow in the duct depends on the cross-sectional area of the duct. Since this duct is trapezoidal, its area is given by $A = \dfrac{h}{2}(b_1 + b_2)$ where $b_2 = 21 + 2x$. Using this equation and substituting for x, h, b_1 and b_2, express the cross-sectional area A in terms of θ.

c.   Use your graphing calculator to find the angle θ for which the duct will have a maximum cross-sectional area. Use Degree Mode and set the viewing window to $0 \le x \le 90$ and $0 \le y \le 400$.

Date _____ Name _____
Course _____ Section _____ Student Number _____

Math Application by Topic: Math at Work: Flying to Hawaii
Math Application by Concept: Trigonometry

During flights covering long distances over water, pilots have no landmarks from which to get their bearings. Such flights demand more skill and knowledge on the part of the pilot, and pilots of small craft often use a navigational technique referred to as "dead reckoning". Taking into account the effect of wind, they determine the length of their flight and the necessary course of travel. They then take off, hoping to see their destination at the conclusion of that time. Due to changing winds, this method used alone is not always successful.

Suppose a plane has an air speed of $|\mathbf{p}| = 210$ miles per hour and the wind $\mathbf{w}$ is blowing due east at 51 miles per hour. The actual course of the plane relative to the ground is the resultant vector $\mathbf{c}$, and the speed of the plane relative to the ground can be derived from the Law of Cosines as follows: $|\mathbf{c}|^2 = |\mathbf{p}|^2 + |\mathbf{w}|^2 - 2|\mathbf{p}||\mathbf{w}|\cos 55°$, so $|\mathbf{c}| \approx \sqrt{34{,}415} \approx 185.5$ mph.

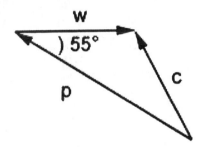

a. Use the equation given above to find the actual ground speed $|\mathbf{c}|$ of the plane.
b. If the distance to Hawaii is 1,194 miles, estimate the total flight time.
    Suppose that the wind had not been taken into account and the pilot had assumed that the speed of 210 mph shown on the controls was the speed relative to the ground. Find the amount of time which would have been predicted for the flight.

Math Application by Topic: Math Over Time: World War II Bombers
Math Application by Concept: Trigonometry

World War II bomber pilots had neither the time nor the instrumentation necessary in order to make precise measurements of ever-changing flight conditions. Thus, many calculations were simplified for them by pre-determining a set of standard flying factors, such as speed and altitude. For instance, the pilot mentioned that a standard bombing altitude of 60 feet was used.

Pilots knew they were flying at the correct altitude when the beams from two lights, one mounted at the nose of the plane and the other at the tail, converged on the ground.

Suppose the two lights are mounted so that each makes the same angle a with the plane, causing the lights to converge directly below the midpoint of the plane. Use the definition of

$\tan \alpha = \dfrac{\text{opp}}{\text{adj}}$ to determine the angle α necessary in order for the lights to converge at 60 feet

altitude if the length of the bomber is 53 feet.

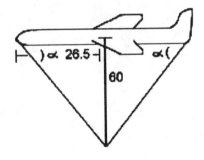